THE ROYAL HORTICULTURAL SOCIETY

PESTS &
DISEASES

THE ROYAL HORTICULTURAL SOCIETY

PESTS&
DISEASES

PIPPA GREENWOOD
ANDREW HALSTEAD

LONDON, NEW YORK, MUNICH,
MELBOURNE, DELHI

PROJECT EDITORS *Laura Langley, Martha Swift*
DESIGNER *Rachael Parfitt*
MANAGING EDITOR *Louise Abbott*
MANAGING ART EDITOR *Lee Griffiths*
DTP DESIGNERS *Chris Clark, Matthew Greenfield*
PRODUCTION CONTROLLER *Martin Croshaw*
ILLUSTRATIONS *Sandra Pond and Will Giles*

2009 EDITION
EDITORS *Helen Fewster, Caroline Reed*
DESIGNER *Pamela Shiels*
PRODUCTION EDITOR *Kavita Varma*
RHS EDITORS *Rae Spencer-Jones, Simon Maughan*

First published in Great Britain in 1997
by Dorling Kindersley Limited,
80 Strand, London WC2R 0RL
A Penguin Company

Revised 2003, 2007
This revised edition published in 2009

A CIP catalogue record for this book is available
from the British Library

ISBN 978-1-40534-177-6

Reproduced in Italy by Scanner Services SRL
Printed and bound in China by Hung Hing

Discover more at
www.dk.com

CONTENTS

INTRODUCTION 7

HOW TO USE THIS BOOK 8

GALLERY OF SYMPTOMS
LEAVES 12 • STEMS & BUDS 33 • FLOWERS 42
FRUITS, BERRIES, NUTS & VEGETABLES 47 • SOIL, ROOTS, TUBERS & BULBS 53
LAWNS 58 • BENEFICIAL GARDEN CREATURES 60

GARDEN HEALTH AND PROBLEMS
ECOLOGY OF THE GARDEN 66 • TYPES OF PESTS AND DISEASES 72
CONTROLLING PESTS AND DISEASES 78

A–Z OF PESTS, DISEASES AND DISORDERS
97–194

INDIVIDUAL PLANT PROBLEMS
GARDEN TREES 196 • SHRUBS & CLIMBERS 198 • HERBACEOUS PERENNIALS 200
LAWNS 201 • BULBOUS PLANTS 202 • ANNUALS & BIENNIALS 203 • GREENHOUSE & HOUSE PLANTS 204
FRUITS & NUTS 206 • VEGETABLES & HERBS 207

REFERENCE SECTION
CONTROLS CURRENTLY APPROVED 210
GLOSSARY 211 • INDEX 216
CONVERSION TABLES 223
ACKNOWLEDGMENTS &
USEFUL ADDRESSES 224

INTRODUCTION

THE IDENTIFICATION AND DIAGNOSIS OF A PROBLEM IS HALFWAY TO
SOLVING IT. THIS BOOK HELPS THE GARDENER TO CONSIDER ALL THE
FACTORS IN THE GARDEN THAT AFFECT GROWING PLANTS, SO THAT
PROMPT ACTION CAN BE TAKEN IF NEEDED.

IT IS IMPOSSIBLE to garden without coming across pests, diseases and
disorders on your plants. They are unwelcome in any garden, large or
small. At the very least they spoil a plant's appearance, but if left to spread
and develop unchecked, the more serious ones may dramatically reduce
the plant's vigour, impairing growth, flowering ability or yield. Some may
create wounds, which are of little significance in their own right, but which
allow the entry of secondary organisms which then wreak havoc.

The key to a flourishing garden lies not only in quick identification and
treatment of pests and diseases, but also in understanding how to keep
plants healthy, thereby preventing problems occurring in the first place.
Many common problems originate in cultural disorders (the conditions in
which a plant is growing), or are only capable of serious damage on a
plant which is already stressed by having to grow in an unsuitable site, or
by poor cultivation. Once a cultural problem has been identified and
understood, it is often quite simple to overcome it and so avoid
a whole host of possible additional problems.

It is vital to identify correctly the pests, diseases and disorders
themselves. If you are uncertain about what is attacking a plant it is
impossible to tell whether or not it is something you need worry about,
and what, if anything, you should do about it. *RHS Pests and Diseases*
makes a wealth of information on all aspects of the problem side of
gardening available to gardeners and professional horticulturists alike.
By looking at the garden as a whole and then identifying the specific
problem, it allows you to stay one step ahead of pests, diseases
and disorders in the garden.

HOW TO USE THIS BOOK

THE EASY TO USE format provides three ways to help you identify a plant pest, disease or disorder: by symptom, by plant and by problem. Symptoms of pests and diseases are shown in colour photographs in the *Gallery of Symptoms*, which is cross-referenced to the *A–Z of Pests, Diseases and Disorders*, a dictionary of most possible garden problems. *Individual Plant Problems* is a plant-by-plant listing, enabling you to look up the problems your particular plant could be suffering from and find their reference in the A–Z and Gallery. A section on *Garden Health and Problems* looks at the garden and plant cultivation as a whole, discussing the background to problems that might occur, while at the back of the book, the comprehensive *Reference Section* contains a glossary and index, and tables that relate to the use and correct application of all the most common commercially available garden chemicals.

IDENTIFYING AND CONTROLLING PROBLEMS:

BY SYMPTOM

See Gallery of Symptoms, *pp.10–63*

The Gallery section contains hundreds of photographs, enabling you to identify the cause of the symptoms on your plant and then referring you to the A–Z section. It is divided into six parts: Leaves; Stems and Buds; Flowers; Fruits, Berries, Nuts and Vegetables; Soil, Tubers, Roots and Bulbs; and Lawns. At the end are four pages enabling the identification of Beneficial Garden Creatures.

Colour photograph of affected plant part with symptoms clearly visible

Section heading, colour-coded for easy reference

Whether problem is a pest, a disease or a disorder

Name of problem, corresponding with detailed A–Z entry

Page reference in A–Z

APPLE CAPSID
PEST • *See p.100*

This pest sucks sap from the fruitlets and causes yellowish-brown, raised, corky swellings. The damage is only skin deep and does not affect the fruits' keeping qualities.

PLANTS AFFECTED Apple trees
SEASON Late spring to early autumn

Plants affected by that particular problem

Brief description of visible and other symptoms

Season when problem is most likely to occur

FRUIT PROBLEMS

FRUITS, BERRIES, NUTS AND VEGETABLES

IF THE EDIBLE PARTS of a plant are attacked by a pest or pathogen, the whole reason for growing the plant may well be lost. Fruits and berries are particularly prone to attack, as they usually have a high sugar and moisture content. Nuts, peas and beans are also valuable food sources for many creatures.

In addition, some ornamental trees and shrubs are very commonly grown mainly for their attractive show of berries, and when these are eaten, removed or spoiled by disease, the plant's appearance is ruined.

In many cases prolonged protection of the plant may be needed, perhaps by using barriers or traps to keep pests away from the flowers or the developing and ripening fruits. Chemicals can be used with great success, but again these may sometimes need to be applied before the first signs of damage are noticed.

APPLE SAWFLY PEST • *See p. 101*
Most infested fruits drop off in early summer. Some do develop as ripe fruit but have distinctive ribbon scars where the young larvae fed under the skin earlier in the summer.
PLANTS AFFECTED Apples
SEASON Mid-spring to early autumn

APPLE SCALD DISORDER • *See p. 101*
Reddish-brown patches of discoloration develop on the skin of the fruit. The flesh beneath is usually undamaged, but it may be discoloured. Exposed surfaces are most badly affected.
PLANTS AFFECTED Apples
SEASON Mainly summer

BIRD DAMAGE PEST • *See p. 107*
Blackbirds, starlings and other birds will peck holes in ripening fruits. Wasps may enlarge this damage, and brown rot frequently develops once the skin has been broken.
PLANTS AFFECTED Apples, pears, plums
SEASON Mid-summer to early autumn

APPLE CAPSID PEST • *See p. 100*
This pest sucks sap from the fruitlets and causes yellowish-brown, raised, corky swellings. The damage is only skin deep and does not affect the fruits' keeping qualities.
PLANTS AFFECTED Apples
SEASON Late spring to early autumn

CODLING MOTH PEST • *See p. 118*
This is the cause of maggoty apples in late summer. The caterpillar feeds at the core and tunnels out through the side or the eye end of the ripening fruit when it is fully developed.
PLANTS AFFECTED Apples, pears
SEASON Mid-summer to early autumn

WASPS PEST • *See p. 188*
These familiar insects enlarge damage started by birds, but can also initiate feeding on softer-skinned fruits (*see p.49*). Their presence can make fruit picking hazardous.
PLANTS AFFECTED Tree fruits, grapes
SEASON Mid-summer to early autumn

BROWN ROT DISEASE • *See p. 111*
Soft, brown rot develops on the fruit (here on apple) and soon becomes studded with raised, creamy-white pustules (*see also p.49*). These are often arranged in concentric rings.
PLANTS AFFECTED Tree fruits
SEASON Late summer and autumn

47

BY PLANT

See Individual Plant Problems, *pp.195–208*

A comprehensive listing of garden plants by type, with listings of the general and specific ailments to which each is liable. The problem can then be identified by referring to the *Gallery of Symptoms* or to the A–Z section.

Pests, diseases or disorders that the plant might suffer from, listed alphabetically

APPLE CANKER **36**, 100
APPLE CAPSID **47**, 100
APPLE LEAF MINER **19**, 101
APPLE SAWFLY **47, 48**, 101
APPLE SCAB **16**, 101
APPLE SOOTY BLOTCH 101
APPLE WATER CORE 102

Page reference for colour photograph in Gallery of Symptoms

Page reference for entry in A–Z of Pests, Diseases and Disorders

---

Example page: Fruit and Nut Problems

FRUITS AND NUTS

MANY GARDENERS are happy to use chemicals on ornamentals, but when it comes to cropping plants, most prefer to avoid them wherever possible. Good cultivation, preventive measures taken at the correct time, and the use of resistant cultivars will all help. Some problems affect many fruits (*see right*). Others are more host-specific; plants so affected are listed below. More resistant species or cultivars may be suggested in the entry for the problem in question in the *A–Z of Pests, Diseases and Disorders*.

COMMON PROBLEMS

The most common problems affecting plants and trees bearing edible fruits and nuts are:

APHIDS 99 / RABBITS 168
BIRD DAMAGE ON FRUITS 107 / SQUIRRELS 180
CORAL SPOT 120 / WASPS 190
GREY MOULD 135
HONEY FUNGUS 138
MAGNESIUM DEFICIENCY 147
POWDERY MILDEWS 167

Page reference in bold type indicates that symptoms are illustrated in the Gallery section

---

BY PROBLEM

See A–Z of Pests, Diseases and Disorders, *pp.97–194*

Comprehensive entries for the pests, diseases and disorders that may be found on garden plants clearly describe the symptoms, cause and control of each problem. Preventive and control measures include organic, chemical and biological methods as appropriate. In addition, suggestions are given for techniques for avoiding and minimizing further damage, and for resistant cultivars, if they are available. It is cross-referenced to the gallery section and to other associated entries.

Example A–Z page excerpt:

APPLE AND PEAR CANKER

SYMPTOMS Areas of bark sink inwards (*see p.36*), usually starting near a wound or bud. The bark becomes discoloured and then shrinks and cracks, forming concentric rings of flaky bark. The branch may become swollen around the cankered area. As the canker enlarges, the shoot may be girdled and so foliage and growth above it starts to die back. Stems and trunks may be attacked and very occasionally semi-mature fruits are infected and show rotting. In summer raised white fungal pustules develop on the canker, and in winter raised, red fruiting bodies develop. Apples and pears are frequently affected; poplars (*Populus*), beeches (*Fagus*), willows (*Salix*), hawthorn (*Crataegus*) and *Sorbus* may also succumb. **CAUSE** The fungus *Nectria galligena*, which is spread mainly during the spring as wind-borne spores enter through wounded areas such as leaf scars, pruning cuts, scab infection, frost crack, attack by woolly aphid. **CONTROL** Prune out whole spurs or branches where possible. On larger branch or trunk infections, carefully pare away all infected and damaged bark and wood, cutting back to completely clean tissue, and treat the wound with a proprietary wound paint. Dispose of prunings and parings carefully. Spray affected apple trees with Bordeaux mixture to help suppress canker. Improve general growing conditions, in particular soil drainage, since unsuitable conditions such as wet soil may aggravate the disease. Avoid growing apples known to be particularly susceptible, such as 'Cox's Orange Pippin', 'James Grieve', 'Worcester Pearmain' and 'Spartan'. The following cultivars seem to show a degree of resistance under most circumstances: 'Laxton's Superb', 'Newton Wonder', 'Bramley's Seedling', and 'Lane's Prince Albert'.

APPLE BITTER PIT

SYMPTOMS The skin of the apple fruit may develop sunken brown spots, each less than 1mm in diameter (*see p.48*). The flesh is spoiled by numerous pale brown spots or freckles. Affected apples may have an unpleasant, slightly bitter taste. Large fruits are especially susceptible, as are fruits produced by heavy-cropping trees. Symptoms may appear while the fruits are on the tree, but often only develop during storage. **CAUSE** Calcium deficiency in the fruit. This results in the collapse of small groups of cells within the flesh, so that

brown flecks develop. Calcium levels in the soil may be perfectly adequate, but in dry conditions, the tree cannot take up the amount it needs. Large fruits and heavy-cropping trees are more prone to this disorder because of their greater demand for calcium. Too high a concentration of magnesium or calcium in fruits may also be a cause of bitter pit. **CONTROL** Keep apple trees well watered and mulched to maintain soil moisture levels. Maintain even growth by feeding with a balanced fertilizer; avoid excessive use of high-nitrogen feeds. Spray developing fruits with calcium nitrate solution from early summer until early autumn. The cultivars 'Bramley's Seedling', 'Discovery' and 'Crispin' may be damaged by this chemical, so care should be taken when treating them.

APPLE BLOSSOM WEEVIL

SYMPTOMS Apple flower buds start to develop but the petals fail to open and turn brown (*see p.46*). If these "capped" blossoms are pulled apart, a creamy-white grub, up to 7mm long, or the pupal stage, will be seen. Damaged blossoms cannot produce fruits. **CAUSE** A weevil, *Anthonomus pomorum*, which lays eggs in the unopened buds during mid- and late spring. The adult weevil is brown with whitish markings on its upper surface and is 4–6mm long. **CONTROL** Light infestations have little impact on the yield of fruits. Spraying with bifenthrin when flowers are at the green cluster stage is only worthwhile if heavy

attacks have occurred in previous years.

APPLE CAPSID

SYMPTOMS Ripe apples have raised yellowish-brown corky bumps or scabby areas on the skin (*see p.47*). Leaves at shoot tips may be distorted, with many small holes. **CAUSE** Apple capsid, *Plesiocoris rugicollis*, which is a bright green sap-feeding insect up to 7mm long. It injects a toxic saliva into the plant as it feeds on the foliage and immature fruits during early summer. This kills some plant cells and, as the developing leaves grow and expand, these dead areas tear into small holes. Fruits respond to the capsid's saliva by developing raised bumps or scabby areas. **CONTROL** The fruit damage is only skin-deep and light attacks can be tolerated. If damage has been extensive in previous years, spray with bifenthrin or pyrethrum at petal fall.

APPLE FROST DAMAGE

SYMPTOMS Russeting of the skin, usually on the most exposed face of the fruit. The more protected faces are usually unharmed. Damaged areas may develop a corky, roughened appearance and the fruit may become deformed and uneven in growth. Flowers and young foliage subjected to frost wither and die off rapidly. **CAUSE** Frost on young, unhardened plant tissues. This is exacerbated if the plant is growing in a position where, following overnight frosts, it is then subjected to rapid warming by early morning sun.

GALLERY OF SYMPTOMS

Even for experienced gardeners, the correct identification of the exact cause of a plant problem, be it a pest, disease or disorder, can be difficult. For ease of comparison with affected plants, the detailed photographs in this section are grouped by the parts of plants attacked, and show similar types of symptom together so that differences can be seen.

LEAVES

DAMAGE TO THE FOLIAGE can spoil a plant's appearance and, if any part of the plant is edible, the potential crop may be reduced. Leaf markings may take the form of spots, either flat (*pp.12–14*) or raised (*pp.14–17*) to the touch, often with the pycnidia (fruiting bodies) of fungi visible; blotches (*pp.17–20*); mottling (*pp.20–22*) or discoloration (*pp.22–25*). Leaves may also be distorted (*pp.25–27*), dying (*pp.28–29*) or damaged (*pp.29–31*). Pests may be visible (*p.32*).

The majority of leaf markings are caused by diseases. Weaker fungal diseases are only likely to cause extensive damage to a plant already lacking in vigour. Plants are particularly vulnerable if they are growing in unsuitable conditions, have not been well cared for, or have suffered some other serious attack. Both cultural and chemical remedies may be necessary to both improve the general health of the plant and to control the specific problem.

Leaf marking can occur at any stage, but older foliage is usually worst affected, especially in the autumn. Variegated leaf areas are especially prone to damage.

MARKINGS ON LEAVES

ORCHID VIRUS
DISEASE • *See p.155*

Dark brown or black, or occasionally yellow, markings appear on the leaf. These are usually seen as streaks or spotting. Flowering is often affected and the plant's overall vigour will decline.

PLANTS AFFECTED Most orchids
SEASON All year

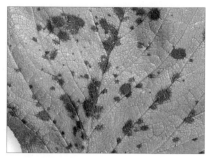

STRAWBERRY LEAF SPOT
DISEASE • *See p.181*

Reddish-purple spots are found on the leaves, which may later enlarge (*see p.17*). Off-white fungal growth may develop over the surface. Flowers can be affected.

PLANTS AFFECTED Strawberries
SEASON Mainly mid-summer to autumn

BEAN CHOCOLATE SPOT
DISEASE • *See p.105*

Chocolate-brown spots are seen on the leaves and there are similar coloured streaks on the stems, pods and flowers. The spots may join together.

PLANTS AFFECTED Broad beans
SEASON Mainly late spring to autumn

DELPHINIUM BLACK BLOTCH
DISEASE • *See p.123*

Black blotches appear on the leaves, which enlarge and often spread to the leaf stalks and stems. The leaves may eventually die off.

PLANTS AFFECTED Delphiniums
SEASON Late spring to autumn

LEAF SPOT (FUNGAL)
DISEASE • *See p.144*

Rounded, grey or brown spots develop on the leaves (here of a castor oil plant) and may join together. Numerous pycnidia are often apparent on the spots.

PLANTS AFFECTED Most plants
SEASON Varies with host plant

CELERY LEAF SPOT
DISEASE • *See p.116*

Dark brown or black, pinprick-sized spots appear on the leaves, which may discolour and die off. The leaf stalks can also be affected and the crop reduced.

PLANTS AFFECTED Celery and celeriac
SEASON All season

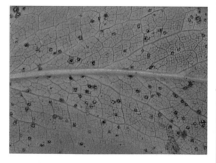

MAHONIA RUST
DISEASE • *See p.148*

Dark brown, spore-filled pustules are found on the lower leaf surface, with bright orange spots on the upper surface (*see p.24*). Leaves may fall early, but overall plant health is rarely affected.

PLANTS AFFECTED Mahonias
SEASON Late spring to early autumn

MARKINGS ON LEAVES *CONTINUED*

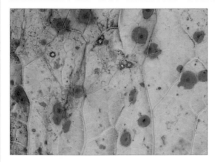

BRASSICA DARK LEAF SPOT
DISEASE • *See p.111*

Grey-brown, round spots are seen on the leaf surface, often with concentric rings of pycnidia on them. Affected areas may drop away, leaving a ragged leaf.

PLANTS AFFECTED Brussels sprouts, cauliflowers
SEASON Mainly mid-summer to late autumn

LEAF SPOT (BACTERIAL)
DISEASE • *See p.144*

Circular or angular spots of dead tissue are found on the leaves (here of hibiscus), usually surrounded by a bright yellow "halo". No fungal bodies present.

PLANTS AFFECTED Many plants
SEASON Varies with host plant

PANSY LEAF SPOT
DISEASE • *See p.155*

Purple-brown or buff-coloured spots appear on leaves and may join together. The affected areas may fall away, leaving holes. Older plants are usually the most badly affected.

PLANTS AFFECTED Pansies
SEASON Mainly mid-summer to late autumn

ROSE BLACK SPOT
DISEASE • *See p.172*

Purple-black spots are seen on leaves. The spots may join together or remain separate. Leaf yellowing and premature fall are common.

PLANTS AFFECTED Roses
SEASON Mainly early summer to autumn

Blotches appear shiny and may be up to 1.5cm in diameter

Main part of leaf tissue remains healthy green colour

Surface of spot is rough and slightly raised

LEAF SPOT (FUNGAL)
DISEASE • *See p.144*

Brown spots are seen on the leaves (here of a primula), often with a grey centre and sometimes accompanied by leaf yellowing. Pycnidia may be visible.

PLANTS AFFECTED Most plants
SEASON Varies with host plant

TAR SPOT OF ACER
DISEASE • *See p.183*

Large, black blotches with bright yellow "halos" develop on upper leaf surfaces. Leaves may fall early and the tree can look disfigured, but overall vigour is not usually affected.

PLANTS AFFECTED Acers (usually, sycamore)
SEASON Late spring to autumn

MARKINGS ON LEAVES CONTINUED

ARBUTUS LEAF SPOT
DISEASE • *See p. 102*

Small spots appear, each with a dry, pale grey centre and a purple edge. These may enlarge but will rarely join together. Leaf fall may be slightly premature.

PLANTS AFFECTED Strawberry tree (*Arbutus*)
SEASON Mainly mid-summer to autumn

RHODODENDRON LEAF SPOT
DISEASE • *See p. 170*

Brownish-purple spots on leaves often have a pink or purple ring around the edge of each one. Concentric ringing is also seen, with pycnidia in the centre.

PLANTS AFFECTED Rhododendrons
SEASON All year (mainly late summer onwards)

PEA LEAF AND POD SPOT
DISEASE • *See p. 156*

Yellow or brown, often sunken, spots are found on the leaves and pods. There is a general yellowing of adjacent tissues. Pycnidia are usually apparent.

PLANTS AFFECTED Peas
SEASON All season, but mainly summer

RAISED MARKINGS ON LEAVES

CHRYSANTHEMUM WHITE RUST
DISEASE • *See p. 118*

Raised, warty, buff-coloured pustules appear beneath the leaf surface, corresponding with yellow, sunken areas above. Leaves may die back.

PLANTS AFFECTED Chrysanthemums
SEASON All season (mainly summer and autumn)

WHITE BLISTER
DISEASE • *See p. 191*

Shiny, white, raised spots are seen on the lower leaf surface of brassicas (here on a cabbage), often arranged in concentric rings. The upper leaf surface is yellowed and usually distorted.

PLANTS AFFECTED Edible brassicas
SEASON Mainly early summer to late autumn

LIME NAIL GALL MITE
PEST • *See p. 147*

Red or yellowish-green, tubular structures containing microscopic mites grow on the upper leaf surfaces. Heavily infested leaves may be distorted.

PLANTS AFFECTED Lime trees (*Tilia*)
SEASON Late spring to autumn

RAISED MARKINGS ON LEAVES CONTINUED

VIBURNUM WHITEFLY
PEST • *See p.187*

Black, oval pupae encrusted with white wax are present on the lower leaf surface for much of the year. White-winged adults are found during the summer.

PLANTS AFFECTED Laurustinus; also *Arbutus*
SEASON All year

CUSHION SCALE
PEST • *See p.122*

Honeydew and sooty mould develop on the upper leaf surfaces. Oval, yellow or brown scales live on the underside (here of a camellia leaf) and in early summer white egg bands are deposited.

PLANTS AFFECTED Mainly camellias and hollies
SEASON All year

PERIWINKLE RUST
DISEASE • *See p.160*

Numerous small, dark brown, spore-filled pustules are found on the underside of the leaf. Leaves become pitted and distorted. The whole plant may become distorted and die back.

PLANTS AFFECTED Periwinkles (*Vinca*)
SEASON All season (mainly summer and autumn)

PELARGONIUM RUST
DISEASE • *See p.159*

Dark brown, fungal pustules are seen on the lower leaf surfaces, usually arranged in concentric rings. There is yellow discoloration on the upper surfaces.

PLANTS AFFECTED Zonal pelargoniums
SEASON All season

IRIS RUST
DISEASE • *See p.141*

Numerous raised, orange pustules are found on leaves. Leaves yellow and may die back prematurely. Older leaves are usually the worst affected. The plant's vigour is rarely affected.

PLANTS AFFECTED Irises
SEASON Mainly late summer to autumn

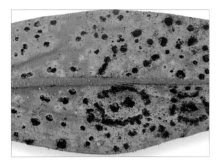

ANTIRRHINUM RUST
DISEASE • *See p.99*

Concentric rings of tiny, dark brown, spore-filled pustules are seen on lower leaf surfaces. A yellow "halo" appears around the edge of each spot.

PLANTS AFFECTED Snapdragons (*Antirrhinum*)
SEASON All season (mainly spring and summer)

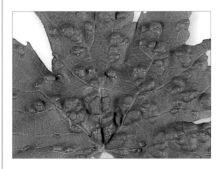

VINE ERINOSE MITE
PEST • *See p.188*

Parts of the leaf bulge upwards; the underside of affected areas is covered in creamy-white hairs which turn brown. Microscopic mites live amongst the hairs.

PLANTS AFFECTED Grape vines
SEASON Late spring to autumn

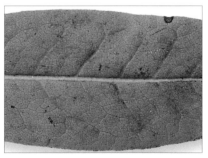

RHODODENDRON POWDERY MILDEW
DISEASE • *See p.171*

Pale brown to buff, felty patches of fungal growth develop under leaves, with yellow blotches above. Leaves fall early and the plant may be weakened.

PLANTS AFFECTED Rhododendrons
SEASON All year on evergreen rhododendrons

HYDRANGEA SCALE
PEST • *See p.140*

Oval, white, waxy egg masses appear on the leaves and stems. Flat, oval, yellowish-brown scale insects are also present, and live on the underside of the leaves next to the leaf veins.

PLANTS AFFECTED Mainly hydrangeas
SEASON All year (eggs late spring to mid-summer)

RAISED MARKINGS ON LEAVES CONTINUED

Leaf tissue between pustules usually remains green

Affected leaves may fall rapidly, thereby weakening the plant

Bright orange or yellow spots are seen on lower leaf surface

ROSE RUST
DISEASE • *See p.173*

Bright orange clusters of spores appear on the lower leaf surfaces, with dark brown, overwintering spores developing from mid-summer. Leaves fall early.

PLANTS AFFECTED Roses
SEASON Mainly summer and autumn

PLUM LEAF GALL MITE
PEST • *See p.162*

Whitish-green structures containing microscopic mites develop on the upper leaf surface, especially around the leaf margins.

PLANTS AFFECTED Wild and cultivated plums
SEASON Late spring to autumn

APPLE SCAB
DISEASE • *See p.101*

Greyish-green or khaki, scabby spots develop on the leaf surface. Slight blistering or puckering may also occur. Leaves yellow and fall prematurely and the tree is weakened.

PLANTS AFFECTED Edible and ornamental *Malus*
SEASON Mainly late spring and summer

PEAR LEAF BLISTER MITE
PEST • *See p.158*

The foliage develops many yellowish-green or pink, slightly raised patches, which later turn black. These patches form a band on each side of the midrib.

PLANTS AFFECTED Pear trees
SEASON Mid-spring to autumn

HEMISPHERICAL SCALE
PEST • *See p.137*

Convex, dark brown, hemispherical scale insects infest leaves (here of a fern) and stems. They are often accompanied by sticky honeydew and sooty mould.

PLANTS AFFECTED Many greenhouse plants
SEASON All year

ELM GALL MITE
PEST • *See p.126*

Hard, pale green swellings containing microscopic mites develop on the upper surface of elm leaves. Almost every leaf may be affected, but the tree's vigour is not usually reduced.

PLANTS AFFECTED Elms (*Ulmus*)
SEASON Mid-spring to autumn

RAISED MARKINGS ON LEAVES CONTINUED

LEEK RUST
DISEASE • *See p.145*

Bright orange, oval pustules are found on the leaves, the outer leaves usually being the worst affected. The leaves may yellow and die off, usually becoming papery and brown.

PLANTS AFFECTED Leeks, chives, onions, shallots
SEASON Mainly mid-summer to late autumn

HOLLYHOCK RUST
DISEASE • *See p.137*

Orange, raised pustules appear on the lower leaf surface (here of a mallow), later turning buff-brown. The upper leaf surface has corresponding yellow or orange spots. Leaves may die back.

PLANTS AFFECTED Hollyhocks and mallows
SEASON All season

ACER GALL MITES
PEST • *See p.98*

Numerous small, round, reddish or yellowish-green structures (galls) form on the upper leaf surface, or felty hairs develop on the undersurface. Heavily infested leaves may become distorted.

PLANTS AFFECTED Maples and sycamores (*Acer*)
SEASON Late spring to autumn

OAK GALL WASP (SPANGLE GALLS)
PEST • *See p.153*

Flat, circular discs or galls containing tiny grubs form on the lower leaf surfaces. They are reddish at first, but yellowish-brown when mature.

PLANTS AFFECTED Deciduous oaks (*Quercus*)
SEASON Late summer to mid-autumn

OAK GALL WASP (SILK BUTTON GALLS)
PEST • *See p.153*

Golden-brown discs with central depressions develop on the underside of oak leaves. Each disc contains a single grub which will become a gall wasp.

PLANTS AFFECTED Deciduous oaks (*Quercus*)
SEASON Late summer to autumn

WILLOW BEAN GALL SAWFLIES
PEST • *See p.192*

Coffee bean-sized swellings, which are red or yellowish-green, form on willow leaves. Each gall contains a small, caterpillar-like larva.

PLANTS AFFECTED Willows (*Salix*)
SEASON Early summer to early autumn

BLOTCHY MARKINGS ON LEAVES

BEET LEAF MINER
PEST • *See p.106*

Pale green blotches later dry up and become brown. White maggots live within the mined areas and later turn into brown pupae, from which adult flies will emerge.

PLANTS AFFECTED Beetroot and spinach beet
SEASON Late spring to late summer

STRAWBERRY LEAF SPOT
DISEASE • *See 181*

Early spotting (*see p.12*) develops into large blotches, each with a grey centre. These may appear slightly raised. Off-white fungal growth may develop.

PLANTS AFFECTED Strawberries
SEASON Mainly mid-summer to autumn

HOLLY LEAF MINER
PEST • *See p.137*

Irregular, yellowish-purple blotches occur on the upper leaf surfaces. There may also be linear tunnels in the leaf coming from the central blotch.

PLANTS AFFECTED Hollies (*Ilex*)
SEASON All year

G A L L E R Y   O F   S Y M P T O M S

BLOTCHY MARKINGS ON LEAVES CONTINUED

WILLOW LEAF BEETLES
PEST • *See p.192*

Whitish or brown dried-up areas form where the adult beetles or their larvae have grazed away part of the leaves. The larvae often feed in groups of up to 20.

PLANTS AFFECTED Willows, poplars and aspens
SEASON Early summer to early autumn

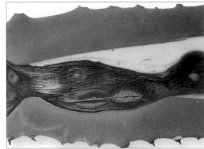

YUCCA LEAF SPOT
DISEASE • *See p.194*

Circular, brown or grey-brown spots appear on leaf surfaces with concentric rings and central pycnidia. The spots may join together.

PLANTS AFFECTED Yuccas
SEASON All year (mainly mid-summer onwards)

SHOT-HOLE
DISEASE • *See p.177*

Spots or blotches of discoloured tissues develop. The damaged areas then fall out, leaving holes (*see p.30*). No insects are present, but fungal growth is sometimes visible.

PLANTS AFFECTED Ornamental and fruiting *Prunus*
SEASON Mainly mid- to late summer

FUCHSIA RUST
DISEASE • *See p.131*

Numerous tiny, pale orange pustules are found on the lower leaf surface, with the upper surface showing yellow-orange or purple discoloration. Leaves may wither and fall prematurely.

PLANTS AFFECTED Fuchsias and willowherbs
SEASON Mainly early summer to autumn

DELPHINIUM LEAF MINER
PEST • *See p.123*

Brown, dried-up areas extend from the leaf tips to the leaf stalk. Small, white grubs live within the mined areas. Adult flies cause pale spots on the leaves.

PLANTS AFFECTED Delphiniums and monkshood
SEASON Early to late summer

LABURNUM LEAF MINERS
PEST • *See p.142*

Roughly spherical, whitish-brown leaf mines are caused by small caterpillars. In heavy infestations the mines join together to form irregular blotches.

PLANTS AFFECTED Laburnum
SEASON Early summer to early autumn

ROSE SLUGWORM
PEST • *See p.174*

Pale green caterpillars with brown heads graze the leaves, causing white or brown dried-up areas. They usually feed on the lower leaf surface, but may occur on the upper surface in shaded areas.

PLANTS AFFECTED Roses
SEASON Early summer to early autumn

RASPBERRY VIRUS
DISEASE • *See p.169*

Yellow-green flecks, mosaics, mottles or ring spots are seen on the leaves. The leaves may become small and distorted and the plant will have a reduced crop. The plant's overall growth is very poor.

PLANTS AFFECTED Cane fruits
SEASON Spring to autumn

RASPBERRY LEAF AND BUD MITE
PEST • *See p.169*

Pale yellow blotches form on the upper leaf surface, with reduced hairiness of the lower leaf surfaces. Leaves at the shoot tips are often distorted.

PLANTS AFFECTED Raspberry
SEASON Late spring to early autumn

BLOTCHY MARKINGS ON LEAVES CONTINUED

Galls may form between the leaf veins or along them

Spots are usually creamy-white, but on purple-tinged leaves they are often red or pink

Felt galls disfigure but cause little real damage to plants

PEAR AND CHERRY SLUGWORM
PEST • *See p.158*

Small caterpillars covered in black slime graze the leaf surface, causing it to dry up and turn brown. The larvae often feed on the upper leaf surface.

PLANTS AFFECTED Cherry, pear, plum, hawthorn
SEASON Early summer to mid-autumn

APPLE LEAF MINER
PEST • *See p.101*

Small caterpillars make white or brown lines where they have bored through the leaves. Pupation occurs in white silk cocoons spun on the outside of the leaf.

PLANTS AFFECTED Apples and cherries
SEASON Early to late summer

FELT GALL MITES
PEST • *See p.127*

Microscopic mites cause the abnormal growth of dense patches of hairs (here on beech), usually on the underside of the foliage. In late summer the hairs dry up and become brown.

PLANTS AFFECTED Some deciduous trees
SEASON Late spring to autumn

CELERY LEAF MINER
PEST • *See p.116*

Brown, dried-up patches develop on the leaves. White maggots live within the mined areas and later become brown pupae.

PLANTS AFFECTED Celery, parsnip, parsley
SEASON Early summer to early autumn

LILAC LEAF MINER
PEST • *See p.147*

Irregular, brown blotch mines develop where whitish-green caterpillars have fed inside the foliage. The caterpillars then roll up the leaf tips with silk threads and complete their feeding there.

PLANTS AFFECTED Lilac, privet and ash
SEASON Early summer to early autumn

OEDEMA
DISORDER • *See p.154*

Raised, pale green and, later, brown spots are found on the lower leaf surface (here of a pelargonium). The upper leaf surface may show corresponding yellow spotting.

PLANTS AFFECTED Many plants
SEASON All year

19

GALLERY OF SYMPTOMS

BLOTCHY MARKINGS ON LEAVES CONTINUED

ROSE POWDERY MILDEW
DISEASE • See p.173

White, powdery fungal growth is found on the leaves. They may become distorted, generally yellowed and fall early. Stems can also be affected.

PLANTS AFFECTED Roses
SEASON All season (mainly summer to autumn)

HELLEBORE BLACK DEATH
DISEASE • See p.136

Black streaking and mottling of the tissues along or between the leaf veins. Leaves and stems become stunted and distorted. Flowers may also be affected.

PLANTS AFFECTED Hellebores
SEASON All year

HELLEBORE LEAF BLOTCH
DISEASE • See p.136

Slate-grey or grey-brown blotches are seen on the leaves. The lesions may eventually join together. Leaf and flower stems and flowers may also be affected.

PLANTS AFFECTED Hellebores
SEASON Mainly early summer to autumn

Pycnidia can develop on lesions

Affected areas become dry

HORSE CHESTNUT LEAF BLOTCH
DISEASE • See p.139

Tan-brown, irregular blotches appear, each with a bright yellow margin, on the leaves. The leaf margins and tips are most severely affected.

PLANTS AFFECTED Horse chestnuts (*Aesculus*)
SEASON Mid-summer to autumn

MOTTLED MARKINGS ON LEAVES

GLASSHOUSE LEAFHOPPER
PEST • See p.133

Coarse, pale spots are found on the upper leaf surface, caused by small, active insects feeding from the lower leaf surface.

PLANTS AFFECTED Many greenhouse plants
SEASON All year

FRUIT TREE RED SPIDER MITE
PEST • See p.130

The foliage of apple and plum trees becomes increasingly dull and yellowish, usually with fine, pale mottling, and may fall prematurely.

PLANTS AFFECTED Apple and plum
SEASON Late spring to early autumn

CARNATION TORTRIX MOTH
PEST • See p.115

Pale green caterpillars with brown heads bind leaves together with silk threads and graze away the inner surfaces. The leaves will then dry up and turn brown.

PLANTS AFFECTED Greenhouse and garden plants
SEASON All year

MOTTLED MARKINGS ON LEAVES CONTINUED

PELARGONIUM VIRUS
DISEASE • See p.159

Yellow flecks, spots, mottles or sometimes ring spots are seen on the leaves. Leaves may be small and distorted, and overall growth poor.

PLANTS AFFECTED Pelargoniums
SEASON All season

EUONYMUS SCALE
PEST • See p.127

Pale yellow mottling and elongate white scale insects are found on the foliage. Stems become encrusted with the dark brown, pear-shaped female scales. Heavy infestations may cause dieback.

PLANTS AFFECTED Evergreen spindles (*Euonymus*)
SEASON All year

ROSE LEAFHOPPER
PEST • See p.173

A coarse, pale mottling develops on the upper leaf surface. Pale yellow insects live on the underside of the leaves. Roses growing in warm, sheltered places can become heavily infested by summer.

PLANTS AFFECTED Roses
SEASON Mid-spring to mid-autumn

RHODODENDRON LACEBUG
PEST • See p.170

A coarse, pale yellow mottling develops on the upper leaf surface. A rusty brown deposit and insects are found on the lower leaf surface.

PLANTS AFFECTED Rhododendrons
SEASON Early spring to late summer

ROSE VIRUS
DISEASE • See p.174

Yellow flecks, spots, mottles, vein-banding or ring spots are found on the leaves. Flowers may show slight colour-breaking symptoms and growth may be affected, but these are both rare.

PLANTS AFFECTED Roses
SEASON All season

CUCUMBER MOSAIC VIRUS
DISEASE • See p.121

Yellow mottles, mosaicing and blotches are seen on leaves (here of a marrow), which may be distorted (*see p.25*). The whole plant is stunted and dies early.

PLANTS AFFECTED Wide host range
SEASON All season

MOTTLED MARKINGS ON LEAVES CONTINUED

BANDED PALM THRIPS
PEST • See p. 104

Black, white-banded, narrow-bodied, elongate insects cause a silvery-brown discoloration of the upper leaf surfaces of the younger foliage.

PLANTS AFFECTED Some house plants
SEASON All year

GLASSHOUSE RED SPIDER MITE
PEST • See p. 133

Early stages of spider mite attack show as a fine, pale mottling of the upper leaf surface. Tiny, yellow-green mites are found on the underside of the leaves.

PLANTS AFFECTED Greenhouse and garden plants
SEASON Mid-spring to late autumn

PLUM RUST
DISEASE • See p. 163

Numerous dark brown, spore-filled pustules are found on the underside of the leaf. Corresponding bright yellow or orange spots are seen on the upper surface. Leaves fall prematurely.

PLANTS AFFECTED Plums and anemones
SEASON Mainly late summer and autumn

CAMELLIA YELLOW MOTTLE
DISEASE • See p. 114

Yellow or white mottled markings and patches of discoloration form on leaves. General growth is not affected. Usually only isolated branches are affected.

PLANTS AFFECTED Camellias
SEASON All year

TULIP FIRE
DISEASE • See p. 185

Buff-coloured or bleached flecks develop on the leaves. Early infection causes extensive leaf distortion and the development of fuzzy, grey fungal growth. Flowers may also be affected.

PLANTS AFFECTED Tulips
SEASON Mid- to late spring

PRIVET THRIPS
PEST • See p. 167

Elongate, narrow-bodied insects cause green leaves (*above right*) to develop a silvery-brown discoloration (*above left*), especially during late summer. The thrips are found on the upper leaf surfaces.

PLANTS AFFECTED Privet and lilac
SEASON Early summer to early autumn

DISCOLOURED LEAVES

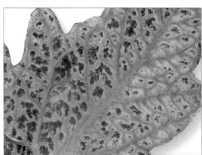

MAGNESIUM DEFICIENCY
DISORDER • See p. 147

Yellow, brown or otherwise discoloured areas form between the leaf veins and around the leaf edge (here on a tomato). Older leaves are the worst affected.

PLANTS AFFECTED Many plants
SEASON Mainly late spring to autumn

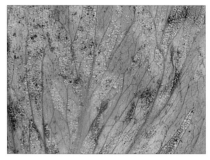

LETTUCE DOWNY MILDEW
DISEASE • See p. 146

Angular, yellow discoloured areas on the leaves rapidly turn pale brown and papery or soft. Off-white fungal growth may develop beneath the leaf.

PLANTS AFFECTED Lettuces
SEASON All season (mainly summer and autumn)

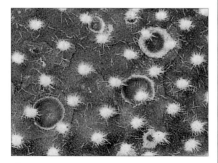

CACTUS CORKY SCAB
DISORDER • See p. 114

Pale, roughened areas appear on the surface and may become brown and sunken. In extreme cases holes develop as the affected tissue dies.

PLANTS AFFECTED Cacti, succulents
SEASON All year

DISCOLOURED LEAVES *CONTINUED*

VIRUS
DISEASE • *See p.189*

Virus symptoms include yellow vein-clearing (here on a sweet pea), mottling, mosaicing, spotting or ring spotting seen on the leaves. Leaves may be distorted (*see p.25*) and general growth is stunted.

PLANTS AFFECTED Wide host range
SEASON Mainly late spring to late summer

HORSE CHESTNUT BLEEDING CANKER
DISEASE • *See p.138*

Symptoms include yellowing foliage, premature leaf drop, and eventually crown death. Leaves may be smaller than those on healthy trees. The elongate discolouration in the centre of the leaflets seen above may also indicate the presence of horse chestnut leaf-mining moth (*see p.139*).

PLANTS AFFECTED Horse chestnut (*Aesculus*)
SEASON Mainly spring to autumn

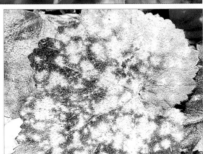

BEGONIA POWDERY MILDEW
DISEASE • *See p.107*

White or pale grey, powdery fungal growth is found mainly on upper leaf surfaces. Affected leaves turn papery and soft. May affect stems and buds.

PLANTS AFFECTED Begonias
SEASON All year

NARCISSUS VIRUS
DISEASE • *See p.152*

Yellow streaking or flecking is seen on leaves. Plants can become stunted and occasionally symptoms are apparent on flower stems or petals. Growth and performance may be unaffected.

PLANTS AFFECTED Daffodils, narcissi
SEASON Mainly early to late spring

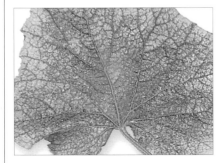

GLASSHOUSE RED SPIDER MITE
PEST • *See p.133*

If heavily infested, leaves lose much of their green colour and dry up. Fine silk webbing may appear between the leaves and the mites swarm over the plant.

PLANTS AFFECTED Greenhouse and garden plants
SEASON Mid-spring to late autumn

WEEDKILLER DAMAGE – CONTACT
DISORDER • *See p.191*

Yellow spots and blotches appear (here on a rose), which may turn brown as the leaf tissue dies off. Plant growth can be stunted and the plant may die.

PLANTS AFFECTED Any plant
SEASON Mainly spring to summer

LEEK MOTH
PEST • *See p.145*

Whitish-brown patches develop on the leaves where the leaf-mining caterpillars have been feeding. Stems and bulbs are also tunnelled. Small, oval, net-like cocoons may be found on the foliage.

PLANTS AFFECTED Leeks and onions
SEASON Early summer to mid-autumn

DISCOLOURED LEAVES CONTINUED

IRON AND MANGANESE DEFICIENCY
DISORDER • See p.141

Yellowing between the veins on leaves (here of a rhododendron) is seen. The youngest leaves are affected most badly. Overall growth may be affected.

PLANTS AFFECTED Many lime-hating plants
SEASON Varies with host plant

OAK POWDERY MILDEW
DISEASE • See p.153

Powdery, white fungal growth appears on the leaf surface. Leaves, particularly young ones, may be distorted and die off prematurely.

PLANTS AFFECTED Oaks (*Quercus*)
SEASON Spring to autumn

CHERRY BLACKFLY
PEST • See p.117

Leaves at the shoot tips become curled, with black aphids on the underside. Affected leaves often turn brown in mid-summer.

PLANTS AFFECTED Cherry trees
SEASON Late spring to late summer

Circular holes eaten in foliage by adult weevils

Foliage becomes ragged

POTASSIUM DEFICIENCY
DISORDER • See p.164

Leaf tips and edges are scorched and may show brownish-purple spotting on the underside (here on a pelargonium). Flowering and fruiting will be poor.

PLANTS AFFECTED Many plants
SEASON Mainly mid-summer to autumn

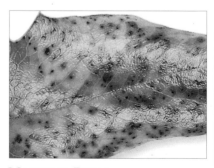

BEECH LEAF-MINING WEEVIL
PEST • See p.106

The leaf tips dry up and turn brown where the leaf-mining larvae have been feeding. The dried-up areas will then break off.

PLANTS AFFECTED Beech (*Fagus*)
SEASON Mid-spring to late summer

MAHONIA RUST
DISEASE • See p.148

Bright, discoloured spots appear on the upper leaf surface, with pustules on the underside (*see p.12*). Leaves may fall prematurely, but the overall health of the plant is rarely affected.

PLANTS AFFECTED Mahonias
SEASON Late spring to early autumn

DISCOLOURED LEAVES *CONTINUED*

GERANIUM DOWNY MILDEW
DISEASE • *See p.132*

Pale green or light brown, angular patches of discoloration are seen on the leaves. Off-white, slightly fuzzy fungal patches develop beneath.

PLANTS AFFECTED *Geranium* species and cultivars
SEASON All season

CUCUMBER MOSAIC VIRUS
DISEASE • *See p.121*

Leaves may be distorted and reduced in size, and also discoloured, with blotches and mosaicing (*see p.21*). On some hosts, flowers show colour-breaking.

PLANTS AFFECTED Wide host range
SEASON All season

CHRYSANTHEMUM LEAF MINER
PEST • *See p.118*

White or brown sinuous lines develop where the leaf-mining grubs have been feeding. In heavy attacks most of the leaf area becomes discoloured.

PLANTS AFFECTED Chrysanthemum, cineraria
SEASON Mid-spring to late autumn

POWDERY MILDEW
DISEASE • *See p.167*

White, powdery fungal growth develops on the upper leaf surface (here of a geranium). The leaf can become distorted and the leaf stalk may also be affected.

PLANTS AFFECTED Wide range of plants
SEASON Varies with host plant

LEAF AND BUD EELWORMS
PEST • *See p.143*

Brown areas enclosed by the larger leaf veins form islands or wedges in the leaves. Eventually the whole leaf will become discoloured.

PLANTS AFFECTED Many plants
SEASON Mainly late summer to late autumn

CARROT MOTLEY DWARF VIRUS
DISEASE • *See p.115*

Foliage turns yellow and pink and may become twisted and stunted. Leaf stalks are also twisted. Foliage dies off prematurely and the crop is reduced.

PLANTS AFFECTED Carrots and parsley
SEASON Mainly summer

DISTORTED LEAVES

AMERICAN GOOSEBERRY MILDEW
DISEASE • *See p.98*

Powdery fungal growth is seen on puckered leaves, usually on the upper surfaces. Young leaves wither and die off. Stems and fruits may be infected.

PLANTS AFFECTED Gooseberries and blackcurrants
SEASON All season

APPLE POWDERY MILDEW
DISEASE • *See p.101*

Powdery, white fungal growth develops on leaves. On young leaves the upper and lower surfaces may be attacked, and the foliage withers and dies.

PLANTS AFFECTED Edible apples and crab apples
SEASON Spring to autumn

VIRUS
DISEASE • *See p.189*

Growth may be distorted and stunted (as on this tobacco plant) and the plant usually dies off early. Other symptoms include yellow mottling, spotting or mosaicing on the leaves.

PLANTS AFFECTED Most plants
SEASON Mainly summer and autumn

DISTORTED LEAVES CONTINUED

Withered remains of leaf still attached to gall

Texture is firm but compressible and becomes softer with age

Gall is rounded, but may also be forked with finger-like protrusions

CAMELLIA GALL
DISEASE • *See p.114*

Large galls, up to 18cm long, develop in place of leaves. The surface becomes covered in spores and so may appear white in places. The plant's overall vigour is rarely affected.

PLANTS AFFECTED Camellias
SEASON Late summer to autumn

PEACH LEAF CURL
DISEASE • *See p.157*

Leaves become puckered; they are pale green at first, but soon turn red and purple. A powdery, white spore layer develops on the surface.

PLANTS AFFECTED Peaches, nectarines, almonds
SEASON Spring and early summer

BAY SUCKER
PEST • *See p.104*

Leaves at the shoot tips become yellow, with thickened, down-curled margins. Later, the damaged areas dry up and turn brown. Fluffy, white sucker nymphs may be seen on the lower leaf surface.

PLANTS AFFECTED Bay (*Laurus nobilis*)
SEASON Late spring to mid-autumn

BLACKCURRANT GALL MIDGE
PEST • *See p.108*

Leaves fail to expand, remaining tightly crumpled and later drying up. Small, white maggots may be found between the folds of infested leaves.

PLANTS AFFECTED Blackcurrant
SEASON Late spring to late summer

BOX SUCKER
PEST • *See p.110*

Shoot extension is severely restricted and the cupped leaves form cabbage-like structures. Flattened, pale green, aphid-like insects may be present and excrete a white, runny liquid.

PLANTS AFFECTED Box (*Buxus*)
SEASON Mid-spring to mid-summer

IRREGULAR WATERING
DISORDER • *See p.141*

Leaves (here of a rhododendron) become puckered and distorted, but retain their normal colour. Not all leaves will be affected.

PLANTS AFFECTED All plants
SEASON All season (mainly spring and summer)

DISTORTED LEAVES CONTINUED

VIRUS
DISEASE • *See p. 189*

The plant (here a *Passiflora*) may be distorted and stunted and will usually die off early. Leaves often also display yellow mottling, spotting, mosaicing or veining.

PLANTS AFFECTED Most plants
SEASON Mainly summer and autumn

WEEDKILLER DAMAGE – BROADLEAVED
DISORDER • *See p. 190*

Leaves (here of a pelargonium) become cup-shaped, often with the veins close together. Stems and petioles may be curled or distorted (*see below right*).

PLANTS AFFECTED Any plant
SEASON Mainly spring and summer

ROSE LEAF-ROLLING SAWFLY
PEST • *See p. 172*

The margins of affected leaflets are rolled inwards to form slender tubes in which small, pale green larvae feed. Rolled leaflets often hang downwards.

PLANTS AFFECTED Cultivated and wild roses
SEASON Late spring to mid-summer

VIOLET GALL MIDGE
PEST • *See p. 188*

Infested leaves have greatly swollen and curled leaf margins. Pale orange maggots or white silk cocoons can be found inside. Galled leaves stay on the plant over the winter.

PLANTS AFFECTED Violets
SEASON All year

PLUM LEAF-CURLING APHID
PEST • *See p. 162*

In the spring leaves become tightly curled and crinkled, due to greenfly feeding on the underside of the leaves. Leaves remain distorted all summer.

PLANTS AFFECTED Plums, damsons and gages
SEASON Early to late spring

CURRANT BLISTER APHID
PEST • *See p. 121*

Leaves at the shoot tips develop raised, red or yellowish-green, puckered areas. Pale yellow aphids may be found on the underside of affected leaves.

PLANTS AFFECTED Red, white and blackcurrants
SEASON Mid-spring to early summer

TARSONEMID MITES
PEST • *See p. 183*

Microscopic mites live and feed at shoot tips and in flower buds. Their feeding causes scarring on the stems and progressive distortion and stunting of the foliage. Growing tips may be killed.

PLANTS AFFECTED Many greenhouse plants
SEASON All year

GLEDITSIA GALL MIDGE
PEST • *See p. 134*

Leaves at the shoot tip fail to expand and resemble small seed pods. Each galled leaflet contains several whitish-orange maggots.

PLANTS AFFECTED Honey locust (*Gleditsia*)
SEASON Early to late summer

WEEDKILLER DAMAGE – BROADLEAVED
DISORDER • *See p. 190*

Stems and petioles (here of a tomato plant) may become curled and distorted. Leaves narrow or curl up (*see top, centre*) and stems may be roughened.

PLANTS AFFECTED Any plant
SEASON Mainly spring and summer

LEAVES DYING OFF

FROST DAMAGE
DISORDER • *See p.130*

The foliage (here of a hydrangea) becomes withered and hangs limply. It may become blackened or turn pale green or brown. The exposed parts of the plant are most severely affected.

PLANTS AFFECTED Many plants
SEASON Mainly winter and spring

KEITHIA THUJINA NEEDLE BLIGHT
DISEASE • *See p.142*

Tiny, black, slightly raised spots are embedded in the foliage, leaving pits or holes as the spores are released. Foliage turns brown and dies back.

PLANTS AFFECTED *Thuja plicata*
SEASON Mainly late summer and early autumn

POTATO BLIGHT
DISEASE • *See p.164*

Brown patches develop on the leaves, mainly around the tips and edges. Fluffy, white fungal growth can develop. Leaves eventually wither and die. Top growth may collapse.

PLANTS AFFECTED Potatoes and tomatoes
SEASON Mid- to late summer

DROUGHT
DISORDER • *See p.125*

Leaves (here of a hosta) wilt and later dry out or shrivel, often turning brown and papery. In the early stages lowered temperatures and watering may revive the plant; later, dieback or death occurs.

PLANTS AFFECTED Many plants
SEASON Mainly summer

PEA WILT
DISEASE • *See p.157*

Leaves turn yellow and die back. Internally the stem's vascular system is stained brown, but this is only visible if the outer stem is stripped away. The whole plant wilts and is killed rapidly.

PLANTS AFFECTED Peas and sweet peas
SEASON Mainly summer

LETTUCE GREY MOULD
DISEASE • *See p.146*

Fluffy, grey fungal growth appears on the leaves. This is often combined with brown or orange slimy rotting of the stem and base of the plant.

PLANTS AFFECTED Lettuces
SEASON Mainly spring to autumn

LEAVES DYING OFF *CONTINUED*

NARCISSUS LEAF SCORCH
DISEASE • *See p.152*

Leaves are red-brown and scorched as they emerge. Brown spots with pycnidia are found on the leaves. Bulbs do not rot or show discoloration.

PLANTS AFFECTED Narcissi, amaryllis, crinums
SEASON Spring

VERTICILLIUM WILT
DISEASE • *See p.187*

Foliage wilts and dies and soft stems wilt. Usually only part of the plant is affected initially. The leaf veins and vascular tissue are discoloured.

PLANTS AFFECTED Many plants
SEASON All year

LEAF SPOT (BACTERIAL)
DISEASE • *See p.144*

Spots develop on the foliage (here of a calathea), often with a water-soaked appearance. A distinct yellow "halo" may develop around each spot (*see p.13*).

PLANTS AFFECTED Depends on bacterium involved
SEASON Varies with host plant

HOLES IN LEAVES

IRIS SAWFLY
PEST • *See p.141*

Brownish-grey, caterpillar-like larvae eat V-shaped notches from the leaf margins. Older larvae cause extensive defoliation, especially of the upper parts of the leaves.

PLANTS AFFECTED Waterside irises only
SEASON Early to mid-summer

SLUGS
PEST • *See p.178*

Slugs feed mainly at night and eat irregular holes in foliage or graze away the surface tissues. A silvery slime trail is sometimes left on plants. Snails (*see right*) cause similar damage.

PLANTS AFFECTED Many plants
SEASON All year

SNAILS
PEST • *See p.178*

Irregular holes are eaten in the foliage and a silvery slime trail may be found nearby. Slugs (*see left*) cause similar damage and a torchlight inspection may be required to identify the culprits.

PLANTS AFFECTED Many plants
SEASON Early spring to late autumn

CAPSID BUG
PEST • *See p.115*

Leaves at the shoot tip (here of a phygelius) develop many small, brown-edged holes. The leaves are often distorted and flower buds may be killed or damaged.

PLANTS AFFECTED Many plants
SEASON Late spring to late summer

FLEA BEETLES
PEST • *See p.128*

Small, rounded holes are scalloped out of the upper leaf surface (here of a wallflower). Small, blue-black or black, yellow-striped beetles, which jump off the foliage, may be seen.

PLANTS AFFECTED Brassicas and related plants
SEASON Mid-spring to late summer

VINE WEEVIL (DAMAGE BY ADULT)
PEST • *See p.188*

Adult vine weevils eat irregular notches from the leaf margins (here of a rhododendron). On shrubs the leaves closest to the ground are attacked.

PLANTS AFFECTED Many plants
SEASON Mid-spring to mid-autumn

HOLES IN LEAVES *CONTINUED*

ELEPHANT HAWK MOTH
PEST • *See p.126*

Large holes are eaten in the leaves by blackish-brown or green caterpillars. The caterpillar's eye markings give them a snake-like appearance.

PLANTS AFFECTED Mainly fuchsias in gardens
SEASON Mid- to late summer

LEAF-CUTTING BEES
PEST • *See p.144*

Semi-circular or elliptical pieces are removed from the leaf margins (here of a rose). The smooth outline of the missing pieces distinguishes leaf-cutting bee damage from other leaf-eating pests.

PLANTS AFFECTED Many plants, especially roses
SEASON Early to late summer

CABBAGE MOTH
PEST • *See p.113*

Infested plants have many holes of various sizes in their foliage. On cabbages the caterpillars bore into the heart leaves and soil the edible parts with their droppings.

PLANTS AFFECTED Brassicas, onions, pelargoniums
SEASON Early summer to early autumn

WATER LILY BEETLE
PEST • *See p.190*

Both the adult beetles and their larvae feed on the upper leaf surface, grazing elongate strips which rot away to form holes. Damaged leaves turn yellow and decay prematurely.

PLANTS AFFECTED Water lilies
SEASON Late spring to late summer

CHINA MARK MOTH
PEST • *See p.117*

Elliptical pieces of various sizes are cut from the leaf margins of water lilies and used to form protective cases for the caterpillars. The larval cases can be found on the underside of the leaves.

PLANTS AFFECTED Water lilies
SEASON Early to late summer

SHOT-HOLE
DISEASE • *See p.176*

Leaves develop spots (*see p.18*) where the leaf tissue dies and falls away, leaving holes, usually with a slight browning around their edges, in the leaves (here of a *Prunus*).

PLANTS AFFECTED Many plants
SEASON Mainly summer and autumn

ARUNCUS SAWFLY
PEST • *See p.102*

Pale green caterpillars initially eat small holes between the leaf veins. However, by the time they are fully grown, they may have eaten everything except the larger veins.

PLANTS AFFECTED Goatsbeard (*Aruncus dioicus*)
SEASON Early to late summer

PIGEONS
PEST • *See p.161*

Pigeons rip pieces off leaves with their beaks, causing ragged tears on the foliage (here of a cabbage). In heavy attacks most of the soft parts of the leaves will be removed.

PLANTS AFFECTED Brassicas and peas
SEASON All year

LACKEY MOTH
PEST • *See p.143*

Hairy caterpillars, with red, blue and white stripes running along their bodies, cause extensive defoliation (here of a *Malus*). The feeding area is covered by silk webbing spun by the caterpillars.

PLANTS AFFECTED Many trees and shrubs
SEASON Mid-spring to early summer

HOLES IN LEAVES CONTINUED

SOLOMON'S SEAL SAWFLY
PEST • *See p.179*

Whitish-grey caterpillars with black heads eat slots in the middle of the leaves. By the time they are fully grown, they will have caused extensive defoliation.

PLANTS AFFECTED Solomon's seal (*Polygonatum*)
SEASON Late spring to mid-summer

VAPOURER MOTH
PEST • *See p.187*

Irregular holes are eaten in the leaf margins by very hairy caterpillars. Plumes of black hairs project from their head and rear ends, with four tufts of buff hairs on the upper surface.

PLANTS AFFECTED Mainly trees and shrubs
SEASON Late spring to mid-summer

MINT BEETLE
PEST • *See p.149*

Irregular holes are eaten in the foliage by iridescent green beetles and their rotund, black larvae. Infestations are rarely heavy enough to cause serious damage to the plant.

PLANTS AFFECTED Wild and cultivated mints
SEASON Mid- to late summer

GOOSEBERRY SAWFLY
PEST • *See p.135*

The soft parts of the foliage are rapidly devoured, leaving just the stalks and larger veins. The larvae are pale green with many black spots or uniformly green in colour.

PLANTS AFFECTED Gooseberries, red/white currants
SEASON Mid-spring to late summer

WINTER MOTH
PEST • *See p.193*

Small holes are eaten in newly-emerged spring foliage; these holes increase in size as the leaves expand. The pale green caterpillars loosely bind leaves together with silk threads.

PLANTS AFFECTED Fruit and other trees
SEASON Early to late spring

PEA AND BEAN WEEVIL
PEST • *See p.156*

Uniform, U-shaped notches are eaten from the leaf margins (here of a broad bean leaf). The greyish-brown beetles do not usually cause extensive damage.

PLANTS AFFECTED Broad beans and peas
SEASON Late spring to late summer

INSECTS ON LEAVES

BEECH WOOLLY APHID

PEST • See p.106

Pale yellow aphids covered with a fluffy, white wax form dense colonies on the underside of beech leaves. Foliage at the shoot tips may be distorted.

PLANTS AFFECTED Beech (*Fagus*)
SEASON Late spring to late summer

CEDAR APHID

PEST • See p.116

Greyish-brown aphids feed at the base of the leaves (needles). Honeydew excreted by the aphids makes foliage sticky and allows the growth of sooty mould. Leaves may turn yellow and fall.

PLANTS AFFECTED Cedars (*Cedrus*)
SEASON Late spring to mid-summer

BLACK BEAN APHID

PEST • See p.107

Dense colonies of black aphids with small, white markings on their upper surface infest the foliage, and sometimes the shoot tips and flowers. Ants often collect the honeydew they excrete.

PLANTS AFFECTED Beans, nasturtiums, dahlias
SEASON Late spring to late summer

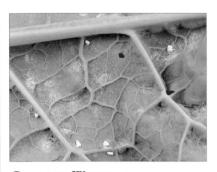

CABBAGE WHITEFLY

PEST • See p.114

Small, white-winged insects live on the underside of leaves and readily fly up when disturbed. Flat, oval, scale-like immature nymphs also occur on the lower leaf surface.

PLANTS AFFECTED Cabbage and other brassicas
SEASON All year

MEALYBUGS

PEST • See p.148

Flattened, oval insects covered in a white, mealy or fluffy wax live on the underside of leaves and in the leaf axils. These sap-feeding insects excrete honeydew which makes the foliage sticky.

PLANTS AFFECTED Many greenhouse plants
SEASON All year

MEALY CABBAGE APHID

PEST • See p.148

Dense colonies of grey-white aphids live on the underside of brassica leaves and on the shoot tips of young plants. Mottled yellow patches develop on leaves.

PLANTS AFFECTED Brassicas and swedes
SEASON Late spring to early autumn

ALDER SUCKER

PEST • See p.98

Flattened, greenish-black insects covered in a fluffy, white wax live on the stems and undersides of alder leaves during the spring. The fluffiness disappears in early summer when the nymphs mature.

PLANTS AFFECTED Alders (*Alnus*)
SEASON Mid-spring to mid-summer

JUNIPER WEBBER MOTH

PEST • See p.142

Dead patches develop in juniper bushes as the foliage turns brown and dries up. The patches are bound together by silk webbing produced by the caterpillars.

PLANTS AFFECTED Junipers (*Juniperus*)
SEASON Late spring to early summer

LARCH ADELGID

PEST • See p.143

Small, black, aphid-like insects concealed under fluffy, white wax feed by sucking sap from the foliage. Heavy attacks can make the foliage sticky with the pests' excrement.

PLANTS AFFECTED Larches (*Larix*)
SEASON Mid-spring to late summer

STEMS AND BUDS

THE STEMS OF A PLANT can be compared to the skeleton of animals, their support giving the plant its characteristic shape and holding the leaves, flowers and buds in the position in which they function best. Since the stems contain the plant's vascular, or transport, system they are also fundamental to its function and when stems are damaged by a pest or by disease, the buds, flowers and leaves carried on them may deteriorate too.

Stems may be attacked in a variety of ways, by pests that gnaw the outer tissues (sometimes ringing the whole stem), feed from them by sucking sap or even tunnel through them. Pest damage may then allow pathogens to enter and so cause further infections, which in turn may spread down the stem and into the crown of the plant; if this occurs, the whole plant may be killed. The tissues of the stem may also be prevented from working by the presence of fungal wilts, which literally clog up the vascular system, so preventing it from functioning.

STEM PROBLEMS

CURRANT CLEARWING MOTH
PEST • See p.122

A white caterpillar with a brown head bores into the stems of currants, making them liable to snap. Emergence holes may be seen on stems in early summer.

PLANTS AFFECTED Currants
SEASON Late spring to late summer

VERTICILLIUM WILT
DISEASE • See p.187

Soft stems and leaves wilt (*see p.29*). Conducting tissue is stained brown, causing longitudinal streaking which is only visible if the bark is removed.

PLANTS AFFECTED Many trees and shrubs
SEASON Symptoms most apparent in summer

SHOT-HOLE BORERS
PEST • See p.177

Small holes in the bark, resembling damage by shotgun pellets, are the emergence holes of small, brownish-black beetles. Attacks mainly occur on trees growing under stressful conditions.

PLANTS AFFECTED Plums, cherries, almonds
SEASON Late spring to mid-summer

PHLOX EELWORM
PEST • See p.160

Stems are stunted and abnormally swollen and there may be splitting at the base. Leaves at shoot tips are narrow and may consist of little more than the midribs (as on the top stem, above).

PLANTS AFFECTED Perennial and annual phlox
SEASON Mid-spring to mid-summer

DIDYMELLA STEM ROT
DISEASE • See p.124

Blackish-brown, sunken blotches appear on the stem base (here of a tomato). Adventitious roots may form above soil level. Older leaves turn yellow.

PLANTS AFFECTED Tomatoes and aubergines
SEASON Mainly summer

LILAC BLIGHT
DISEASE • See p.146

Young shoots wilt and die, with the older stems developing cankers. Angular brown spots appear on leaves which then die off. Grey mould growth is often present, masking the symptoms.

PLANTS AFFECTED Lilacs (*Syringa*)
SEASON Late spring and early summer

BARK BEETLES
PEST • See p.104

The larvae of several species of beetle feed underneath the bark of recently dead or dying trees and shrubs. This network of tunnels was made by elm bark beetle grubs.

PLANTS AFFECTED Many trees and shrubs
SEASON All year

GALLERY OF SYMPTOMS

STEM PROBLEMS CONTINUED

Compost is sodden

Flowers, leaves and stems have lost all turgidity

DAMPING OFF
DISEASE • *See p.123*

Seedlings (here of lettuce) collapse. The stem base is discoloured and may shrink inwards, often appearing water-soaked initially. Fluffy white fungal growth may be present.

PLANTS AFFECTED All seedlings
SEASON Mainly late winter and spring

WATERLOGGING
DISORDER • *See p.190*

Foliage wilts and yellows, as if suffering from drought, but the condition does not improve if more water is applied. The entire top growth may flop over (as with the cyclamen above).

PLANTS AFFECTED Any plant
SEASON All year

WEEDKILLER DAMAGE – BROADLEAVED
DISORDER • *See p.190*

Leaves are narrowed and may be cupped and leaf veins appear parallel. Here conifer foliage is distorted. Petioles and stems may show spiral twisting.

PLANTS AFFECTED Many plants
SEASON Mainly early to mid-summer

GREY MOULD
DISEASE • *See p.135*

Discoloured patches develop on stems (here of a rose) and may girdle them, causing dieback. Fluffy grey fungal growth and black fungal resting bodies can develop on affected areas.

PLANTS AFFECTED Soft and woody-stemmed plants
SEASON All year

HONEY FUNGUS
DISEASE • *See p.138*

A creamy white mycelium (fungal sheet), smelling strongly of mushrooms, develops beneath the bark on infected stem bases, trunks and roots. The woody material beneath discolours.

PLANTS AFFECTED Trees, shrubs, some other plants
SEASON All year

RASPBERRY CANE SPOT
DISEASE • *See p.169*

Purple spots with silvery-white centres develop on the canes. The spots enlarge rapidly and the canes may split and die off. Leaves and flowers may be affected.

PLANTS AFFECTED Raspberries, other cane fruits
SEASON Early summer to autumn

STEM PROBLEMS *CONTINUED*

BACTERIAL CANKER
DISEASE • *See p.104*

Patches of bark flatten and sink inwards. Amber-coloured, resin-like ooze may appear close by. Foliage withers and dies off or buds fail to open. Branches may be killed by girdling cankers.

PLANTS AFFECTED Ornamental and fruiting *Prunus*
SEASON Mainly spring and autumn

BEECH BARK DISEASE
DISEASE • *See p.106*

Small areas of bark die off and sink inwards. Black, tar-like exudations appear and as the damage increases foliage may yellow and die.

PLANTS AFFECTED Beeches (*Fagus*)
SEASON All year

SEIRIDIUM CANKER
DISEASE • *See p.176*

Foliage (here on a cypress) yellows and dies. Bark becomes roughened as cankers develop. Resin oozes from the infected areas and tiny, black pycnidia appear. Cankers may ring stems.

PLANTS AFFECTED Several conifers
SEASON All year, mainly autumn and winter

SMUT
DISEASE • *See p.178*

Raised, blister-like areas develop on stems (here on *Trollius*). These erupt to produce black, powdery spore masses. Affected stems may be killed if the pustules are sufficiently large.

PLANTS AFFECTED Many plants
SEASON Mainly summer

POOR PRUNING
See p.94–95

This pruning cut – here on birch – has removed a branch too close to the trunk and the wound has failed to heal successfully. Infection has entered the wound, causing the wood to deteriorate.

PLANTS AFFECTED Any tree or shrub
SEASON All year

ROSE ROOT APHID – EGGS
PEST • *See p.173*

Large, dark brown aphids come up from the roots in autumn and lay clusters of eggs on the stems. The eggs are brown when freshly laid but turn shiny black.

PLANTS AFFECTED Roses
SEASON Eggs seen autumn to spring

FIREBLIGHT
DISEASE • *See p.128*

Flattened areas develop on the bark (here of a pear tree) and may exude bacterial ooze. The wood beneath is stained foxy-red brown. Flowers wither and die, followed by adjacent leaves.

PLANTS AFFECTED Pome fruits
SEASON Late spring to summer

WOOLLY APHID
PEST • *See p.194*

Fluffy, white patches develop on branches, especially on pruning wounds and in cracks in the bark. Infestations on young shoots cause knobbly swellings where the aphids have been feeding.

PLANTS AFFECTED Apple, pyracantha, cotoneasters
SEASON Mid-spring to mid-autumn

BEECH BARK SCALE
PEST • *See p.106*

An off-white, powdery wax accumulates in crevices in the bark of beech trees. It is secreted from tiny, sap-feeding scale insects. Heavy infestations are linked with beech bark disease (*see above*).

PLANTS AFFECTED Beech (*Fagus*)
SEASON All year

STEM PROBLEMS CONTINUED

WILLOW BARK APHID
PEST • *See p.192*

Dense colonies of greyish-black aphids, up to 5mm long, develop on willow stems. The bark and ground under the tree becomes sticky with honeydew excreted by the aphids.

PLANTS AFFECTED Willows (*Salix*)
SEASON Mid-summer to mid-autumn

BROWN SCALE
PEST • *See p.112*

Reddish-brown shells cover the bodies of sap-feeding insects or their eggs, especially on shrubs growing in sheltered places (here on a cotoneaster). Heavily infested plants lack vigour.

PLANTS AFFECTED Many woody plants
SEASON All year

SLUGS
PEST • *See p.178*

The softer parts of leaves and leaf stalks are eaten (here of celery, causing elongate grooves between the veins). Damaged parts turn brown and may be infected by secondary rots.

PLANTS AFFECTED Soft tissue of many plants
SEASON Late summer to mid-autumn

SCLEROTINIA
DISEASE • *See p.175*

Stems (here of a Jerusalem artichoke) become discoloured and covered in dense, white, fluffy fungal growth. Large black sclerotia (fungal resting bodies) are embedded in the fungal growth.

PLANTS AFFECTED Wide range of plants
SEASON Mainly early autumn, but all year

APPLE AND PEAR CANKER
DISEASE • *See p.100*

Flattened areas of discoloured bark crack and split, forming concentric rings of flaky bark. White or red fungal bodies may develop on the cankered areas.

PLANTS AFFECTED Mainly apple and pear
SEASON All year

SYCAMORE SOOTY BARK DISEASE
DISEASE • *See p.182*

Green or yellow staining develops under bark. Later, black spore masses or blister-like outgrowths develop on the stems. Large areas of bark die and break off.

PLANTS AFFECTED Sycamore (*Acer pseudoplatanus*)
SEASON All year

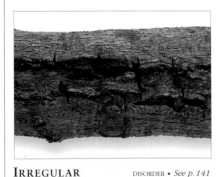

IRREGULAR WATERING
DISORDER • *See p.141*

Longitudinal cracks or splits in bark may penetrate into the woody tissue. Older splits have swollen edges due to the formation of scar tissue.

PLANTS AFFECTED All woody plants
SEASON All year

MUSSEL SCALE
PEST • *See p.151*

Bark may become heavily encrusted with greyish-dark brown scales, which may also spread onto fruits. These sap-feeding insects weaken the plant and cause shoots to die back.

PLANTS AFFECTED Apple trees and many shrubs
SEASON All year

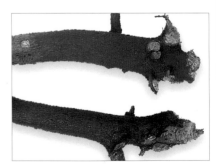

VOLES
PEST • *See p.189*

Branches or whole plants die as a result of these small rodents gnawing away the bark on the stems or roots. Damage often occurs in winter, but the results may not be apparent until summer.

PLANTS AFFECTED Many trees and shrubs
SEASON All year

STEM PROBLEMS CONTINUED

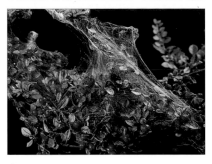

COTONEASTER WEBBER MOTH
PEST • *See p.120*

The foliage turns brown and dries up where small, dark brown caterpillars have been feeding. They cover the infested branches with a silk webbing.

PLANTS AFFECTED Cotoneaster, hawthorn
SEASON Late spring to late summer

JUNIPER SCALE
PEST • *See p.142*

Flat, yellow-white, rounded scales encrust the stems, often causing yellowing of the green parts. Heavy infestations of this sap-feeder can cause stems to die back.

PLANTS AFFECTED Juniper, cypress, thuya
SEASON All year

MEALYBUGS
PEST • *See p.148*

Pinkish-grey or white insects infest the stems and leaf axils. They are often covered with a fluffy, white, waxy substance. Honeydew excreted by the mealybugs allows sooty mould to grow.

PLANTS AFFECTED Greenhouse and house plants
SEASON All year

BRACKET FUNGUS
DISEASE • *See p.110*

Bracket- or shelf-shaped, fungal fruiting bodies of variable size grow out from the tree (here a *Malus*), most commonly at the base of the trunk, but sometimes high in the crown.

PLANTS AFFECTED Most trees and some shrubs
SEASON All year, particularly autumn

FOMES ROOT AND BUTT ROT
DISEASE • *See p.129*

Dark red-brown, woody-textured bracket fungus grows on the trunk or butt of trees. The lower surface is creamy-white and covered in minute pores.

PLANTS AFFECTED Mainly conifers
SEASON All year (mainly autumn)

DAEDALEOPSIS CONFRAGOSA
DISEASE • *See p.122*

Semi-circular, rubbery, annual bracket fungus develops on trunks or branches, reddish-brown and smooth or slightly ridged with a paler undersurface.

PLANTS AFFECTED Deciduous trees
SEASON Mainly autumn

STEM PROBLEMS CONTINUED

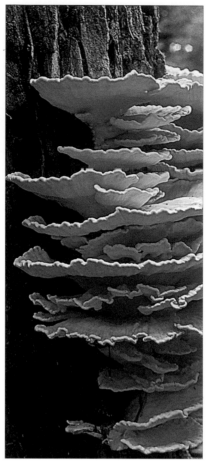

CROWN GALL – AERIAL FORM
DISEASE · See p.120

Roughened, rounded growths break out from within the stem (here of a logan-berry), rupturing it. Galls become woody and if extensive, stems may die back.

PLANTS AFFECTED Many woody plants
SEASON All year, mainly summer

HORSE CHESTNUT BLEEDING CANKER
DISEASE · See p.138

Drops of brown gummy ooze appear in patches on the trunk and main stems. Affected bark oftens cracks, allowing other decay organisms to enter.

PLANTS AFFECTED Horse chestnut (*Aesculus*)
SEASON All year

LAETIPORUS SULPHUREUS
DISEASE · See p.143

Overlapping, annual, bright yellowish-orange fungal brackets are found on the trunk or large limbs of trees. The brackets have wavy edges.

PLANTS AFFECTED Trees
SEASON Mainly autumn

PINEAPPLE GALL ADELGID
PEST · See p.162

Shoot tips are swollen and resemble green pineapples by mid-summer. The galls turn brown after the aphid-like insects emerge in late summer.

PLANTS AFFECTED Spruce (*Picea*)
SEASON All year

BRACKET FUNGUS
DISEASE · See p.110

Single or overlapping, small bracket fungi grow on stems or trunks (here on dead wood). The upper surface may be concentrically zoned in shades of tan, brown, cream and grey.

PLANTS AFFECTED Deciduous trees and shrubs
SEASON All year, particularly autumn

SMALL ERMINE MOTHS
PEST · See p.178

Black-spotted, creamy-yellow caterpillars feed together under white silk webbing that they spin over their feeding area. If numerous they can cause defoliation.

PLANTS AFFECTED A variety of hosts
SEASON Early to late summer

MULBERRY CANKER
DISEASE · See p.150

These are small cankers which may girdle and so kill stems. Tiny red-brown pustules appear around the cankered area. Leaves and shoots may die back.

PLANTS AFFECTED Mulberry
SEASON All year (most obvious in summer)

STEM PROBLEMS CONTINUED

FOMES POMACEOUS
DISEASE • *See p.129*

Hoof-shaped or rounded, fungal fruiting bodies appear on the bark of dead or living stems (here of an apple tree). The upper surface is pale grey-brown.

PLANTS AFFECTED Various deciduous trees
SEASON All year, particularly autumn

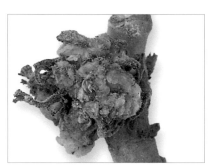

LEAFY GALL
DISEASE • *See p.145*

Numerous small, bunched, distorted, thickened or fasciated leaves grow, usually around the stem base (here of a pelargonium). Normal growth is also present.

PLANTS AFFECTED Annual or herbaceous plants
SEASON Mainly spring and summer

WOOLLY VINE SCALE
PEST • *See p.194*

In early summer this scale insect deposits its eggs under pads of white wax (here on a pyracantha), which can be drawn out in long threads. Heavily infested plants lack vigour.

PLANTS AFFECTED Vines, currants and others
SEASON All year (eggs in early summer)

MERIPILUS GIGANTEUS
DISEASE • *See p.148*

Large, fleshy, golden-brown fungal brackets are found around the base of trees or in the near vicinity of the roots. The undersurface is creamy-white.

PLANTS AFFECTED Oaks, beeches, robinias
SEASON Late summer and autumn

BROOM GALL MITE
PEST • *See p.111*

Buds fail to grow into shoots and become very swollen. These galls gradually dry up and change from green to grey-brown. Heavily infested plants produce few flowers or new growth.

PLANTS AFFECTED Broom (*Cytisus*)
SEASON All year

PINE ADELGID
PEST • *See p.161*

Small, black, aphid-like insects suck sap from the young shoots. The adelgids are hidden under a fluffy, white wax that is secreted from their bodies. This can resemble a mould.

PLANTS AFFECTED Pines (*Pinus*)
SEASON Mid-spring to late summer

WEEDKILLER DAMAGE – BROADLEAVED
DISORDER • *See p.190*

Adventitious roots and other outgrowths form on stems (here of a brassica), causing them to become distorted. The foliage may not be affected.

PLANTS AFFECTED Plants without woody stems
SEASON Mainly late spring and summer

ROBIN'S PIN CUSHION
PEST • *See p.171*

Roughly spherical swellings covered in reddish or yellowish-green, mossy leaves develop around the stems of certain roses and occasionally on foliage.

PLANTS AFFECTED Wild and species roses
SEASON Mid-summer to early autumn

CUCKOO SPIT
PEST • *See p.121*

Globules of frothy white liquid appear on stems and sometimes on foliage. Inside the froth is a yellowish-green, sap-feeding insect, which is the nymphal stage of a froghopper.

PLANTS AFFECTED Many plants
SEASON Late spring to early summer

STEM PROBLEMS CONTINUED

PHYSICAL INJURY
See pp. 70–71

Winds, heavy snow, browsing animals or vandals may cause tearing of bark and injury to the wood tissues beneath. Jagged wounds are slow to heal and harmful pathogens may enter them.

PLANTS AFFECTED Woody plants
SEASON All year

RASPBERRY RUST
DISEASE • See p.169

Bright orange spore masses develop on the stems, often appearing from ruptures in stems. Large fissures remain on the stems (here of a blackberry) when all the spores have been dispersed.

PLANTS AFFECTED Raspberries, blackberries
SEASON Late spring to autumn

HORSE CHESTNUT SCALE
PEST • See p.139

Brownish-black scales secrete a white, waxy substance that covers their eggs. Later the scales die and fall away, leaving the bark covered in white patches.

PLANTS AFFECTED Many trees and shrubs
SEASON All year

ASPARAGUS BEETLE
PEST • See p.103

Both grubs (*above left*) and adults (*above right*) eat the foliage and also gnaw bark from the stems, causing them to dry up and turn a yellowish-brown colour.

PLANTS AFFECTED Asparagus
SEASON Late spring to early autumn

FRUIT TREE RED SPIDER MITE EGGS
PEST • See p.130

The spherical winter eggs are less than 1mm in diameter, but there may be so many that they give the bark a reddish colour, espcially on branch undersides.

PLANTS AFFECTED Apples, plums
SEASON Eggs early autumn to late spring

FORSYTHIA GALL
DISEASE • See p.130

Rough-surfaced, near-spherical woody galls, usually 10–15mm in diameter, develop on stems. Several galls may be fused together. The growth and development of the stem is not affected.

PLANTS AFFECTED Forsythia
SEASON All year

ROSE CANKER AND DIEBACK
DISEASE • See p.172

Stems die back and discoloration occurs, either in patches or all over the affected stem. Tissues may dry out. Growth above the affected areas dies off.

PLANTS AFFECTED Roses
SEASON All year

BLIND SHOOTS
DISORDER • See p.108

Shoots which should bear flower buds form, but no flower buds develop (here on a rose). Leaf and stem growth otherwise appear perfectly healthy, with no sign of dieback.

PLANTS AFFECTED Flowering shrubs
SEASON Mainly summer

PAPERY BARK
DISORDER • See p.156

Bark peels off as a thin, papery brown sheet. Small- to medium-sized shoots are most commonly affected, but any may show symptoms. Shoots can die back. Apple trees are particularly affected.

PLANTS AFFECTED Woody plants
SEASON All year, mainly spring and winter

STEM PROBLEMS CONTINUED

WITCHES' BROOMS
DISEASE • *See p.193*

Closely-packed, densely-branched clusters of twigs grow from normal stems. The stems do not die back, but bear numerous small leaves. Adjacent growth is perfectly normal.

PLANTS AFFECTED Trees
SEASON All year

WILLOW SCAB
DISEASE • *See p.192*

Irregular, dark spots develop on the young stems and may cause girdling. Olive-brown pustules are found on the undersurface of leaves and on affected stems. Stems may die back.

PLANTS AFFECTED Willows (*Salix*)
SEASON All year, particularly spring

CORAL SPOT
DISEASE • *See p.120*

Raised red, orange or coral-coloured, hard pustules develop on dead or dying stems. Affected stems die back further and if the infection spreads to the crown, the whole plant may be killed.

PLANTS AFFECTED Woody plants
SEASON All year

LUPIN APHID
PEST • *See p.147*

Large, whitish-grey aphids cluster on the flower stems and underside of leaves. Heavy infestations cause the plant to wilt. Honeydew and sooty mould soil the foliage.

PLANTS AFFECTED Lupins
SEASON Late spring to late summer

DEER
PEST • *See p.123*

Various species of deer come into gardens and eat foliage, flowers and stems. Nibbled stems invariably have ragged cuts caused by a deer's lack of incisor teeth in its upper jaw.

PLANTS AFFECTED Most plants
SEASON All year

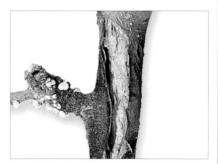

BLOSSOM WILT AND SPUR BLIGHT
DISEASE • *See p.109*

Raised, rounded, buff-coloured fungal pustules appear on twigs and stems (here of an apple tree), often associated with blossom wilt on adjacent flowers.

PLANTS AFFECTED Apple, pear, plum, cherry
SEASON Mainly late spring and summer

BUD PROBLEMS

BLINDNESS OF BULBS
DISEASE/DISORDER • *See p.108*

Flower buds fail to form, or do not develop fully (here on a daffodil) and open partially or not at all. Leaves are unaffected and appear healthy.

PLANTS AFFECTED Many bulbs
SEASON Spring

BIG BUD MITES
PEST • *See p.107*

Infested buds (here on a blackcurrant stem) are abnormally swollen and rounded compared to normal buds. They usually dry up and fail to open in the spring.

PLANTS AFFECTED Blackcurrant, hazel, yew
SEASON Mid-winter to early summer

RHODODENDRON BUD BLAST
DISEASE • *See p.170*

Buds form but turn dry and brown. Numerous tiny, black, bristle-like fungal growths develop on the surface. Affected buds remain on the plant.

PLANTS AFFECTED Rhododendrons
SEASON All year, evident late spring and summer

FLOWERS

MOST GARDENS ARE packed full of plants grown largely because of their flowers, be they annuals, biennials, perennials, shrubs, climbers or trees. In fruit and vegetable gardens, flowers are often equally important, as without them the edible fruits would never develop. Keeping flowers free from pests and diseases is therefore vital.

Flowers are usually short-lived, yet during their brief lifespan there may well be several pests or pathogens which could attack and therefore spoil them. They may be discoloured (*p.42*) or disfigured by infestations of pests (*p.43*), deformed (*pp.44–45*) or damaged by pests (*p.46*). In many cases, however, even if a few blooms are spoilt, later flowers may be unscathed, having missed the attentions of the pest, or the infection period of the pathogen.

Flowering in general is also greatly influenced by cultural and weather conditions, in particular soil moisture and fertility and the level of sunshine. To a large extent, growing a plant in a suitable spot ensures a good display of flowers.

DISCOLOURED FLOWERS

GREY MOULD
DISEASE • *See p.135*

Spots of discoloration appear on the petals (here of a rose). Spots may be edged with a dark ring. No fungal growth is apparent. In extreme cases the whole flower deteriorates.

PLANTS AFFECTED Many flowers
SEASON Mainly spring and summer

RHODODENDRON PETAL BLIGHT
DISEASE • *See p.170*

Spots of discoloration develop on the petals. These enlarge and appear water-soaked. The flower collapses and rapidly discolours all over.

PLANTS AFFECTED Rhododendrons and azaleas
SEASON Spring

VIRUS
DISEASE • *See p.189*

Petals show "colour breaking", with streaks of discoloration (here on a tulip). The petals do not deteriorate. Flowers may be reduced in number and size and show distortion.

PLANTS AFFECTED Most plants
SEASON Mainly spring and summer

WESTERN FLOWER THRIPS
PEST • *See p.191*

A pale flecking develops on petals (here of a gloxinia), and yellowish-brown insects are seen on flowers and foliage. Infested blooms have a shortened life.

PLANTS AFFECTED Mainly greenhouse plants
SEASON All year

PESTS VISIBLE ON FLOWERS

ROSE APHIDS
PEST • *See p.172*

Green, yellowish-green or pink aphids cluster on flower buds and stems. Heavy infestations reduce the quality of the blooms. White cast aphid skins litter the buds and foliage.

PLANTS AFFECTED Roses
SEASON Mid-spring to late summer

POLLEN BEETLES
PEST • *See p.163*

Black or bronzy-green beetles, 2mm long, crawl around the centre of blooms, where they feed on pollen. No direct damage is caused, but they can be a nuisance on cut flowers in the house.

PLANTS AFFECTED Many flowers
SEASON Mid-spring to late summer

MELON COTTON APHID
PEST • *See p.148*

Small black or dark green aphids infest the flowers (here chrysanthemum) and the undersides of leaves. Sooty mould grows on the aphids' excrement.

PLANTS AFFECTED Chrysanthemums and others
SEASON Mainly summer to autumn

SLUGS AND SNAILS
PESTS • *See p.178*

Damage caused by slugs and snails is usually indistinguishable. On daffodils, above, foliage is ignored but petals can be reduced to brown-edged fragments. A silvery slime trail is usually left behind.

PLANTS AFFECTED Wide range of plants
SEASON Early to late spring

APPLE SUCKER
PEST • *See p.102*

Flattened, pale green insects up to 2mm long suck sap from the blossom trusses. Heavy attacks make the petals turn brown and can be confused with frost damage.

PLANTS AFFECTED Apple trees
SEASON Mid- to late spring

CUCKOO SPIT
PEST • *See p.121*

A frothy, white liquid covers yellowish-green insects (here on *Potentilla*), which are the immature nymphs of froghoppers. They also suck sap from foliage and stems (*see p.39*).

PLANTS AFFECTED Many plants
SEASON Late spring to mid-summer

PEONY WILT
DISEASE • *See p.160*

Stem tissue is discoloured and shrinks inwards, commonly at the base or, as above, below flower buds. Fuzzy grey fungal growth develops and black sclerotia are found on or within stems.

PLANTS AFFECTED Peonies
SEASON Spring to autumn

GLASSHOUSE RED SPIDER MITE
PEST • *See p.133*

Although mainly a foliage pest (*see pp.22, 23*), red spider mites also infest flowers, especially chrysanthemums. Yellowish-black mites crawl amongst the petals.

PLANTS AFFECTED Many greenhouse plants
SEASON Mid-spring to mid-autumn

FASCIATION
DISORDER • *See p.127*

Stems become fused together, producing broad, ribbon-like stems (here on delphinium). The stems usually bear leaves and flowers, but become curled and distorted.

PLANTS AFFECTED Many plants
SEASON All year on perennial stems

GALLERY OF SYMPTOMS

DEFORMED AND DISTORTED FLOWERS

SMUTS
DISEASE • *See p.178*

Anthers become swollen and distorted and packed full of masses of dark brown or black spores. Anthers burst to release spore masses. Flower stems may be distorted.

PLANTS AFFECTED Several plants
SEASON Mainly summer

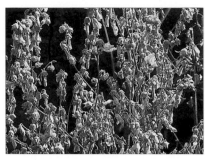

DROUGHT
DISORDER • *See p.125*

Leaves and flowers wilt and soft stems may do so also. If conditions do not improve, wilting may be permanent (as here on a lavatera) and flowers become shrivelled and brown.

PLANTS AFFECTED Most plants
SEASON Mainly summer

FROST DAMAGE (BUDS AND FLOWERS)
DISORDER • *See p.130*

Petals become discoloured (here of a camellia), usually brown and either soft or, less frequently, dry. Exposed flowers are most badly affected.

PLANTS AFFECTED Many flowers
SEASON Mainly spring and early winter

ROSE BALLING
DISORDER • *See p.172*

Outer petals on an unopened or only partially opened bud are pale brown and dry. Initially, inner petals are unaffected, but they may die off when secondary organisms invade.

PLANTS AFFECTED Roses
SEASON Mainly summer

BOLTING
DISORDER • *See p.109*

Plants grown for their leaf crop become elongated and produce flowers unusually early (here on a brassica). This limits the cropping period. Early cultivars are particularly susceptible.

PLANTS AFFECTED Beetroot, brassicas, lettuces
SEASON Mainly summer

PROLIFERATION
DISORDER • *See p.167*

Flower buds form within the centre of the flower (here a rose). These remain as buds or develop into stems bearing flowers and occasionally leaves. Only the flowers or buds are affected.

PLANTS AFFECTED Roses, also pears and apples
SEASON Late spring and summer

DEFORMED AND DISTORTED FLOWERS CONTINUED

MICHAELMAS DAISY MITE
PEST • See p.149

The flowers are converted by the feeding activities of this tiny pest into rosettes of green leaves. Infested plants are often stunted, with scarring on the stems.

PLANTS AFFECTED *Aster novi-belgii*
SEASON Early summer to mid-autumn

CHRYSANTHEMUM PETAL BLIGHT
DISEASE • See p.118

Small brown spots or larger water-soaked lesions develop on the petals. Flowers rot and die off rapidly. Leaf growth is not affected.

PLANTS AFFECTED Mainly chrysanthemums
SEASON Mainly late summer

WIND DAMAGE
DISORDER • See p.193

Petals (here of a camellia) become scorched and discoloured, usually brown. Damaged areas are usually dry, not soggy. Flowers on exposed areas of the plant are most susceptible.

PLANTS AFFECTED Many plants
SEASON Mainly spring

Fuzzy fungal growth begins to cover whole flowerhead

Flowerhead loses shape and dies off

HEMEROCALLIS GALL MIDGE
PEST • See p.136

Infested buds are squat and abnormally swollen; they fail to open, go brown and dry up. Many small white maggots live in the buds at the base of the petals.

PLANTS AFFECTED Daylilies (*Hemerocallis*)
SEASON Late spring to mid-summer

GLADIOLUS THRIPS
PEST • See p.133

Creamy-yellow or black, narrow-bodied insects, up to 2mm long, feed in the flower buds and on the foliage. Infested blooms have pale mottled petals, but buds may turn brown and fail to open.

PLANTS AFFECTED Gladioli
SEASON Mid- to late summer

GREY MOULD
DISEASE • See p.135

At this advanced stage of grey mould, the flowers rapidly become covered in dense, grey fungal growth (here on a pelargonium). Dieback may spread. For earlier symptoms see p.42.

PLANTS AFFECTED Many flowering plants
SEASON All year

GALLERY OF SYMPTOMS

DAMAGED FLOWERS

NECTAR ROBBING
PEST • See p.152

Some short-tongued bumblebees cannot reach the nectar from the front of some flowers and so bite a hole at the base of the flower. Bees taking nectar in this way do not assist with pollination.

PLANTS AFFECTED Often runner and broad beans
SEASON Summer

SPARROWS
PEST • See p.179

House sparrows peck the petals of some flowers (here a crocus) to shreds, leaving fragments scattered nearby on the soil. They sometimes select certain colours, but are not consistent.

PLANTS AFFECTED Crocus, primroses and others
SEASON Early spring to summer

BULLFINCHES
PEST • See p.113

The dormant flower buds (here of a plum) are pecked off and eaten during the winter. The outer bud scales are discarded and dropped to the ground. Bullfinches also eat seeds.

PLANTS AFFECTED Fruits and some ornamentals
SEASON Late autumn to mid-spring

APPLE BLOSSOM WEEVIL
PEST • See p.100

Infested flowers fail to open fully and the petals turn brown. A white grub or the pupa will be found inside these "capped" blossoms.

PLANTS AFFECTED Apples
SEASON Mid- to late spring

Male earwigs (top) have curved pincers, while females (bottom) have straight ones

Ragged holes are eaten in petals

CAPSID BUGS
PEST • See p.115

Flower buds (here of fuchsias) abort so that infested plants (*above left*) are largely flowerless. Foliage is distorted and has many small holes, due to capsid bugs damaging leaves at the shoot tips.

PLANTS AFFECTED Many plants
SEASON Late spring to late summer

EARWIGS
PEST • See p.126

Earwigs emerge after dark, usually in mild weather, to feed on petals (here of a cosmos) and on foliage at shoot tips. This results in holes developing in the blooms and the expanded leaves.

PLANTS AFFECTED Clematis, dahlias and others
SEASON Late spring to early autumn

FRUITS, BERRIES, NUTS AND VEGETABLES

IF THE EDIBLE PARTS of a plant are attacked by a pest or pathogen, the whole reason for growing the plant may well be lost. Fruits and berries are particularly prone to attack, as they usually have a high sugar and moisture content. Nuts, peas and beans are also valuable food sources for many creatures.

In addition, some ornamental trees and shrubs are very commonly grown mainly for their attractive show of berries, and when these are eaten, removed or spoiled by disease, the plant's appearance is ruined.

In many cases prolonged protection of the plant may be needed, perhaps by using barriers or traps to keep pests away from the flowers or the developing and ripening fruits. Chemicals can be used with great success, but again these may sometimes need to be applied before the first signs of damage are noticed.

FRUIT PROBLEMS

APPLE SAWFLY
PEST • See p.101

Most infested fruits drop off in early summer (*see p.48*). Some do develop as ripe fruit but have distinctive ribbon scars where the young larvae fed under the skin earlier in the summer.

PLANTS AFFECTED Apples
SEASON Mid-spring to early autumn

APPLE SCALD
DISORDER • See p.101

Reddish-brown patches of discoloration develop on the skin of the fruit. The flesh beneath is usually undamaged, but it may be discoloured. Exposed surfaces are most badly affected.

PLANTS AFFECTED Apples
SEASON Mainly summer

BIRD DAMAGE
PEST • See p.107

Blackbirds, starlings and other birds will peck holes in ripening fruits. Wasps may enlarge this damage, and brown rot frequently develops once the skin has been broken.

PLANTS AFFECTED Apples, pears, plums
SEASON Mid-summer to early autumn

APPLE CAPSID
PEST • See p.100

This pest sucks sap from the fruitlets and causes yellowish-brown, raised, corky swellings. The damage is only skin deep and does not affect the fruits' eating or keeping qualities.

PLANTS AFFECTED Apples
SEASON Late spring to early autumn

CODLING MOTH
PEST • See p.119

This is the cause of maggoty apples in late summer. The caterpillar feeds at the core and tunnels out through the side or the eye end of the ripening fruit when it is fully developed.

PLANTS AFFECTED Apples, pears
SEASON Mid-summer to early autumn

WASPS
PEST • See p.190

These familiar insects enlarge damage started by birds, but can also initiate feeding on softer-skinned fruits (*see p.49*). Their presence can make fruit picking hazardous.

PLANTS AFFECTED Tree fruits, grapes
SEASON Mid-summer to early autumn

BROWN ROT
DISEASE • See p.112

Soft, brown rot develops on fruit (here on apple) and soon becomes studded with raised, creamy-white pustules (*see also p.49*). These are often arranged in concentric rings.

PLANTS AFFECTED Tree fruits
SEASON Late summer and autumn

FRUIT PROBLEMS *CONTINUED*

ROSY APPLE APHID
PEST • *See p.174*

Small greyish-pink aphids infest the foliage and fruitlets in spring. They cause leaf curling and yellowing, and affected fruits remain small and distorted with a pinched appearance at the eye end.

PLANTS AFFECTED Apples
SEASON Early spring to late summer

PEAR MIDGE
PEST • *See p.158*

Pear fruitlets become abnormally swollen, turn black and drop off the tree in early summer. These fruitlets contain many small, whitish-orange maggots. The whole crop may be lost.

PLANTS AFFECTED Pears
SEASON Late spring to early summer

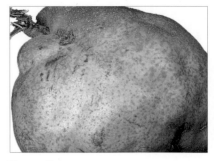

PEAR STONY PIT VIRUS
DISEASE • *See p.159*

Fruits become knobbly and pitted, with hard patches throughout the flesh. Affected fruits usually only appear on isolated branches.

PLANTS AFFECTED Pears
SEASON Late summer to early autumn

PEAR SCAB
DISEASE • *See p.159*

Blackish-brown, scabby patches appear on the skin of the fruits. The fruits are often small and misshapen and may be spoiled by splits, since secondary organisms rapidly cause rotting.

PLANTS AFFECTED Pears
SEASON Late summer to early autumn

QUINCE LEAF BLIGHT
DISEASE • *See p.168*

Numerous small, dark red or blackish-brown spots are found on the skin. Fruits may be distorted and under-sized, and occasionally split.

PLANTS AFFECTED Quinces
SEASON Late summer or early autumn

APPLE BITTER PIT
DISEASE • *See p.100*

Pale brown spots of discoloration appear within the flesh, or sometimes on the fruit's skin. Those on the skin are usually slightly sunken. The fruit is otherwise perfectly normal in shape and size.

PLANTS AFFECTED Apples
SEASON Mainly late summer and in store

APPLE SAWFLY
PEST • *See p.101*

After initially tunnelling under the skin, apple sawfly larvae bore into the core of fruitlets. Wet, blackish-brown excrement comes out of the entry hole. For apple sawfly damage to mature fruit *see p.47*.

PLANTS AFFECTED Apples
SEASON Mid-spring to mid-summer

APPLE FRUIT SPLIT
DISORDER • *See p.101*

Straight or branched splits develop in the skin. The splits are usually single, but several may occasionally appear on one fruit. Secondary organisms such as brown rot may invade and cause rotting.

PLANTS AFFECTED Most fruits, especially apples
SEASON Mid- to late summer

PLUM MOTH
PEST • *See p.162*

Infested plum fruits tend to ripen prematurely. They contain a pinkish caterpillar, which eats the flesh around the stone, leaving behind a trail of brown excrement pellets.

PLANTS AFFECTED Plums, damsons, gages
SEASON Late summer to early autumn

FRUIT AND BERRY PROBLEMS

WASPS
PEST • *See p.190*

These familiar yellow and black stinging insects are attracted to ripe fruits. They can eat out extensive cavities in some softer-skinned fruits. For wasp damage to apples and pears *see p.47*.

PLANTS AFFECTED Tree fruits, grapes

SEASON Mid-summer to mid-autumn

RED BERRY MITE
PEST • *See p.169*

Blackberries fail to ripen completely, with parts or all of the fruit remaining red. Ripening becomes progressively unsatisfactory as the season develops. The cause is microscopic mites.

PLANTS AFFECTED Blackberries

SEASON Late summer to mid-autumn

BROWN ROT
DISEASE • *See p.112*

Fruits (here a cherry) turn brown and rot and rapidly develop numerous raised, creamy-white pustules (*see also p.47*). The fruits then dry out, but remain hanging on the tree.

PLANTS AFFECTED Tree fruits

SEASON Late summer to autumn

POWDERY MILDEW
DISEASE • *See p.167*

Fruits (here grapes) develop a white, off-white or greyish-beige fungal coating in patches or all over their surface. They fail to swell fully, may be small or distorted and may crack or split.

PLANTS AFFECTED Many fruits

SEASON Summer

AMERICAN GOOSEBERRY MILDEW
DISEASE • *See p.98*

Powdery or felty white fungal growth develops on the skin, soon turning buff or pale brown. The fungus may be rubbed off. Fruits may be under-sized.

PLANTS AFFECTED Gooseberries, blackcurrants

SEASON Summer

PYRACANTHA SCAB
DISEASE • *See p.167*

Greenish-black, scabby fungal growth develops on the berries, which may cause them to split and die off. In extreme cases the whole surface of the berry may be obscured.

PLANTS AFFECTED Pyracanthas

SEASON Summer to autumn

FRUIT AND VEGETABLE PROBLEMS

RASPBERRY BEETLE
PEST • See p.168

The stalk end of the fruit dries up and becomes greyish-brown. An elongate, whitish-brown beetle grub, up to 8mm long, may be found feeding in the core or plug of the fruit.

PLANTS AFFECTED Cane fruits
SEASON Mid-summer to early autumn

STRAWBERRY GREY MOULD
DISEASE • See p.181

Soft, pale brown areas develop on the fruit and rapidly increase in size. Fuzzy grey fungal growth develops on the surface and the fruit rapidly rots.

PLANTS AFFECTED Strawberries
SEASON Late spring to early autumn

TOMATO GREENBACK
DISEASE • See p.184

A partial or complete ring of unripened tissue appears around the stalk end of a ripening fruit. The flesh and skin remains either green or yellow.

PLANTS AFFECTED Tomatoes
SEASON Summer

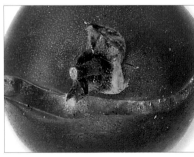

BLOSSOM END ROT
DISEASE • See p.108

A sunken, leathery, dark brown to black patch develops at the bottom end of the developing fruit (here tomatoes). The remainder of the fruit ripens normally, but remains distorted.

PLANTS AFFECTED Tomatoes, peppers
SEASON Summer

TOMATO FRUIT SPLITTING
DISORDER • See p.184

Fruits develop normally and then split shortly before they are ready to pick. The split may dry out or may become infected with secondary fungi.

PLANTS AFFECTED Tomatoes
SEASON Summer or early autumn

GREY MOULD
DISEASE • See p.135

Fruits (here a tomato) turn soft and discolour and become covered in fuzzy grey fungal growth. On tomatoes, ghost spotting symptoms (yellow circles) on otherwise normal fruits may appear.

PLANTS AFFECTED Many fruits
SEASON Summer and early autumn

VEGETABLE PROBLEMS

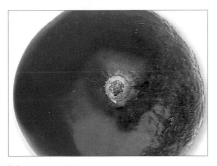

TOMATO BLOTCHY RIPENING
DISORDER • See p.183

Randomly distributed, hard, yellow or green patches of flesh remain unripened. The rest of the fruit develops and ripens normally.

PLANTS AFFECTED Tomatoes
SEASON Summer and early autumn

TOMATO BLIGHT
DISEASE • See p.183

Brown, discoloured areas develop within the fruits and are visible on the surface of both ripe and unripe fruits. Discoloured areas may shrink inwards and suffer secondary rotting.

PLANTS AFFECTED Tomatoes and potatoes
SEASON Summer and early autumn

BEAN SEED BEETLES
PEST • See p.105

Beans, especially broad beans, kept for seed may have circular holes where adult seed beetles have emerged. The grubs feed inside on the cotyledons, but damaged seeds will usually germinate.

PLANTS AFFECTED Beans and peas
SEASON Mid-summer to mid-autumn

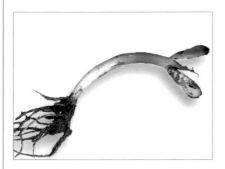

BEAN SEED FLY
PEST • See p.105

White fly maggots feed on the germinating seeds, making holes in the cotyledons and stems. Seedlings may be killed or emerge with the growing point killed.

PLANTS AFFECTED Runner and French beans
SEASON Mid-spring to early summer

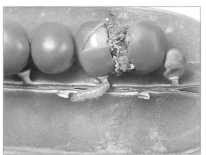

PEA MOTH
PEST • See p.157

Small cream-coloured caterpillars with dark spots feed inside pea pods on the developing seeds. Piles of excrement pellets indicate where the caterpillars have been feeding.

PLANTS AFFECTED Peas
SEASON Early to late summer

PEA LEAF AND POD SPOT
DISEASE • See p.156

Yellow or brown, often sunken, patches develop on the pods. Dark brown or black pycnidia (fungal fruiting bodies) may be visible on affected areas.

PLANTS AFFECTED Peas
SEASON Summer

SWEETCORN SMUT
DISEASE • See p.182

Individual kernels on sweetcorn cobs become enlarged and distorted. Initially pale grey in colour, they rupture to release masses of powdery black spores. The rest of the plant develops normally.

PLANTS AFFECTED Sweetcorn
SEASON Summer

MICE
PEST • See p.149

Rows of germinating peas, beans and other seeds are disturbed by burrowing field mice. Only seeds and corms are eaten; the shoots are left lying on the soil surface.

PLANTS AFFECTED Peas, beans, sweetcorn
SEASON Early spring to mid-summer

CARROT FLY
PEST • See p.115

Elongate, yellow-white fly maggots tunnel in certain tap roots, causing orange-brown discoloured lines on the outside of the roots. Secondary rots may extend the damage.

PLANTS AFFECTED Carrots, parsnips, parsley
SEASON Early summer to late autumn

GALLERY OF SYMPTOMS

VEGETABLE PROBLEMS CONTINUED

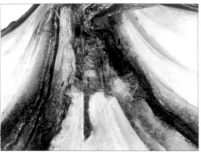

ONION NECK ROT
DISEASE • *See p.155*

The neck area of the onion turns pale brown and soft and appears semi-transparent. Fuzzy grey fungal growth and black sclerotia may develop. The onion can become mummified.

PLANTS AFFECTED Onions, shallots
SEASON Late summer and while in store

ONION EELWORM
PEST • *See p.154*

Microscopic, worm-like animals live within the bulb and leaves. Infested plants are swollen and distorted, with softened tissues that are susceptible to secondary rots.

PLANTS AFFECTED Onions, leeks
SEASON Late spring to early autumn

ONION WHITE ROT
DISEASE • *See p.155*

Dense, fluffy white fungal growth appears around the basal plate. Small black sclerotia may be visible in amongst this and the bulb starts to rot.

PLANTS AFFECTED Onions, leeks, shallots, garlic
SEASON Summer

NUT PROBLEMS

NUT WEEVIL
PEST • *See p.153*

The kernel is eaten by a white grub with a brown head. The fully-fed grub bores a circular hole in the ripe nut shell (here of a hazelnut) when it leaves to pupate.

PLANTS AFFECTED Hazel and cob nuts
SEASON Mid-summer to early autumn

SQUIRRELS
PEST • *See p.180*

Grey squirrels often strip trees before the nuts are fully ripe and the ground becomes littered with broken nut shells. Nuts are buried, sometimes in lawns, for consumption during the winter.

PLANTS AFFECTED Hazel and cob nuts, walnuts
SEASON Late summer to mid-autumn

ACORN GALL WASP
PEST • *See p.98*

Oak acorns are converted into ridged, yellow-brown structures, which are sometimes called knopper galls. The wasp grub feeds in a seed-like capsule inside the gall.

PLANTS AFFECTED Oaks (*Quercus*)
SEASON Late summer to mid-autumn

SOIL, ROOTS, TUBERS AND BULBS

PESTS AND PATHOGENS that attack below ground level are common, but because the symptoms they cause are usually first seen above ground, they may well be misidentified. Often the only way to identify a soil-borne problem conclusively is to dig up all or part of the plant, a process which in itself may obviously cause considerable and long-term damage.

Most soil-borne problems, or pests or pathogens which attack roots, tubers, bulbs or corms directly, are potentially very damaging, simply because the root system or the base of the plant is so fundamental to the survival of the plant as a whole. Early investigation, identification and, where possible, control is therefore essential. Generally, control is difficult, however, since locating and targeting an underground pest or pathogen with a pesticide is not easy.

SOIL PROBLEMS

VINE WEEVIL (DAMAGE BY LARVAE)
PEST • See p.188

Plump, white, legless larvae with pale brown heads feed on roots, corms and tubers, particularly on plants grown in containers. The plant lacks vigour.

PLANTS AFFECTED Many plants
SEASON Late summer to mid-spring

LEATHERJACKETS
PEST • See p.145

Leatherjackets are the larvae of daddy-long-legs. They are greyish-brown, legless maggots with no obvious head. They eat roots and seedling plants and also cause problems in lawns (see p.58).

PLANTS AFFECTED Many garden plants
SEASON Mid-spring to late summer

WOODLICE
PEST • See p.194

Woodlice may be grey or pinkish-brown and up to 15mm long. They live mainly in the surface layers of the soil, where they feed on decaying vegetation and occasionally seedlings.

PLANTS AFFECTED Seedlings
SEASON All year

FUNGUS GNAT (LARVAE)
PEST • See p.131

The larvae of fungus gnats, or sciarid flies, are white, legless maggots with black heads. They feed on decaying vegetation and seedlings.

PLANTS AFFECTED Greenhouse and house plants
SEASON All year

CHAFER GRUBS
PEST • See p.116

Chafer grubs have plump, white bodies, curved like the letter C, with brown heads and three pairs of legs. They eat roots and make cavities in root vegetables and potato tubers.

PLANTS AFFECTED Lawns and garden plants
SEASON Mostly mid-spring to mid-autumn

CUTWORMS
PEST • See p.122

Cutworms are the earth-coloured caterpillars of several species of moths. They kill small plants by severing the stems or by gnawing away the outer bark at soil level.

PLANTS AFFECTED Mostly annuals and vegetables
SEASON Mainly mid-spring to mid-autumn

SWIFT MOTHS (LARVAE)
PEST • See p.182

These are slender, white-bodied caterpillars with brown heads which feed on roots, corms and rhizomes, especially in herbaceous borders.

PLANTS AFFECTED Many garden plants
SEASON Mainly mid-spring to mid-autumn

SOIL PROBLEMS CONTINUED

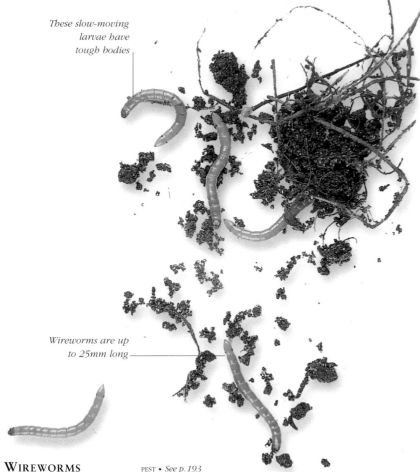

These slow-moving larvae have tough bodies

Wireworms are up to 25mm long

MILLIPEDES
PEST • *See p.149*

These spotted snake millipedes are creamy-white spotted with red, with legs on every body segment. They feed on rotting vegetation and enlarge damage initiated by other pests and diseases.

PLANTS AFFECTED Seedlings, potatoes
SEASON Spring to autumn

WIREWORMS
PEST • *See p.193*

Yellowish-brown grubs with three pairs of short legs at their front eat roots and bore into root vegetables. They cause most damage where grass has been dug to create flower or vegetable gardens.

PLANTS AFFECTED Seedlings, potatoes and others
SEASON All year

SPRINGTAILS
PEST • *See p.180*

Small, white insects come to the surface of pot plants or are washed out of the pot when the plant is watered. They cause no harm to the plant as they feed on dead plant material.

PLANTS AFFECTED Pot plants
SEASON All year

BULB PROBLEMS

BLUE MOULD ON BULBS
DISEASE • *See p.109*

Reddish-brown lesions develop on the flesh of the bulb and may be sunken. Fluffy, white or blue-green fungal growth may appear on top of lesions.

PLANTS AFFECTED Most bulbs in store
SEASON Mainly late summer to autumn

NARCISSUS EELWORM
PEST • *See p.151*

Plants are stunted and distorted. If the bulb is cut in half transversely, brown concentric rings can be seen where the microscopic eelworms are feeding. Infested bulbs rot and die.

PLANTS AFFECTED Daffodils
SEASON All year

NARCISSUS BULB FLY
PEST • *See p.151*

Infested bulbs rot in the soil or produce a few distorted leaves in the spring. One large maggot feeds in the centre of the bulb and fills it with its muddy-brown excrement.

PLANTS AFFECTED Daffodils, amaryllis and others
SEASON Late summer to early spring

BULB PROBLEMS CONTINUED

TULIP FIRE
DISEASE • See p. 185

Tiny, irregularly shaped, black sclerotia (fungal resting bodies) develop on the bulb, clustered around the neck. The bulb may rot or appear firm. Foliage that appears above ground may be withered.

PLANTS AFFECTED Tulips
SEASON Late summer

GLADIOLUS CORM ROT
DISEASE • See p. 133

Brown, ridged areas of discoloration appear on the corm. The corm dries out and becomes mummified. No fungal growths or sclerotia are visible.

PLANTS AFFECTED Gladioli, crocus, bulbous iris
SEASON In storage

GLADIOLUS CORM ROT – FOLIAGE SYMPTOMS
DISEASE • See p. 133

Yellow flecking and, later, striping develops towards the tips of the outer-most leaves. The discoloration spreads and the leaves turn brown and die.

PLANTS AFFECTED Gladioli, crocus, bulbous iris
SEASON Late spring and summer

ROOT AND TUBER PROBLEMS

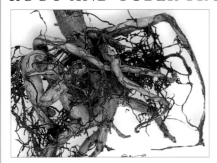

FOOT AND ROOT ROTS
DISEASE • See p. 129

Roots discolour (here on daphne), often turning black or brown and shrinking inwards. Deterioration usually starts at the root tips and roots may disintegrate.

PLANTS AFFECTED Most plants
SEASON All year

IRIS RHIZOME ROT
DISEASE • See p. 140

Roots and rhizomes deteriorate rapidly, turning soft and slimy. The infection usually starts just beneath the leaf, but spreads quickly. Foliage may fail to develop, or yellows and dies back.

PLANTS AFFECTED Rhizomatous irises
SEASON Mainly spring and summer

CLUBROOT
DISEASE • See p. 119

Numerous irregularly shaped swellings develop on roots of cabbage and related plants. Swellings do not contain any insect larvae or grubs. The entire root system may be distorted and swollen.

PLANTS AFFECTED Any in Cruciferae family
SEASON All year

ROOT APHIDS
PEST • See p. 171

Infested plants grow slowly and tend to wilt in sunny weather. Aphids, usually globular and creamy-brown, feed on the roots. They often secrete a white, waxy powder from their bodies.

PLANTS AFFECTED Lettuce, beans and others
SEASON Mid- to late summer

POTATO CYST EELWORMS
PEST • See p. 165

Pinhead-sized, spherical cysts, which may be white, pale yellow or brown, form on the roots. These are the swollen bodies of the female eelworms.

PLANTS AFFECTED Potatoes, tomatoes
SEASON Mid- to late summer

ROOT MEALYBUGS
PEST • See p. 171

These sap-feeding insects have white, elongate bodies up to 2mm long. They secrete a white, waxy powder from their bodies, which coats roots and soil particles.

PLANTS AFFECTED Pot plants
SEASON All year

ROOT AND TUBER PROBLEMS *CONTINUED*

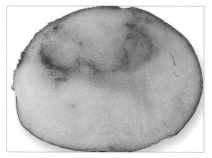

POTATO SPRAING
DISEASE • *See p. 166*

Tan-coloured arcs of discoloration appear in the flesh. Corky tissue may be present around the discoloured areas. The tubers are occasionally distorted. Foliage and stems may be mottled yellow.

PLANTS AFFECTED Potatoes
SEASON Late spring onwards

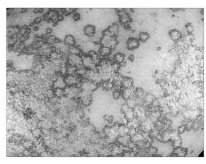

POTATO COMMON SCAB
DISEASE • *See p. 165*

Raised, roughened, scabby patches develop on the skin of the tuber. Scabs have ragged edges. The damage is usually superficial.

PLANTS AFFECTED Potatoes
SEASON Late spring onwards

POTATO SILVER SCURF
DISEASE • *See p. 166*

Pale and inconspicuous silvery-grey markings develop on the tuber skin. The size, shape, flesh and eating quality of the potato are not affected.

PLANTS AFFECTED Potatoes
SEASON Mainly in storage

POTATO BLIGHT
DISEASE • *See p. 164*

Slightly sunken, darkened areas develop on the skin. Internally, the flesh is discoloured by a dry reddish-brown rot. Secondary organisms cause soft rotting and an unpleasant smell.

PLANTS AFFECTED Potatoes and tomatoes
SEASON Late summer onwards

POTATO POWDERY SCAB
DISEASE • *See p. 166*

Small, near-circular scabby areas appear on the tuber skin. The scabs have raised margins and burst to release masses of brown spores.

PLANTS AFFECTED Potatoes
SEASON Summer onwards

WIREWORMS
PEST • *See p. 193*

Elongate, yellowish-brown grubs with three pairs of legs at the head end bore into the tubers. They are mainly a problem where grass has recently been dug over to make a vegetable garden.

PLANTS AFFECTED Root vegetables, potatoes
SEASON Early summer to mid-autumn

ROOT AND TUBER PROBLEMS CONTINUED

VIOLET ROOT ROT
DISEASE • *See p. 188*

The surface of the root (here of a carrot) or tuber is covered in a network of dark purple fungal strands. Sclerotia may be present. Roots and tubers may rot.

PLANTS AFFECTED Various plants
SEASON Mainly spring to autumn

SLUGS
PEST • *See p. 178*

Slugs make small, round holes in the skin of potato tubers but hollow out extensive cavities once they are inside. Some potato cultivars are particularly susceptible to slug damage.

PLANTS AFFECTED Potatoes, root vegetables
SEASON Mid-summer to mid-autumn

POTATO GANGRENE
DISEASE • *See p. 165*

Slightly sunken, round-edged areas develop on the tuber shortly after lifting. Lesions may remain defined or enlarge. The tuber rots, turning pale pink.

PLANTS AFFECTED Potatoes
SEASON Summer and autumn

POTATO DRY ROT
DISEASE • *See p. 165*

One end of the tuber shows wrinkled skin and then shrinks inwards rapidly, forming concentric rings of wrinkles and discoloration. Pink, white or blue-green fungal pustules may develop.

PLANTS AFFECTED Potatoes in store
SEASON Any time of year

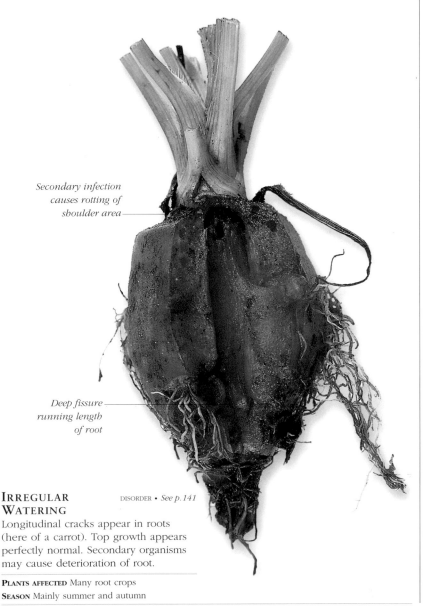

Secondary infection causes rotting of shoulder area

Deep fissure running length of root

PARSNIP CANKER
DISEASE • *See p. 156*

Roughened, reddish-brown, orange or black cankered areas develop on the root, most commonly around the shoulders. Affected tissues may develop secondary rotting.

PLANTS AFFECTED Parsnips
SEASON Mainly autumn and winter

IRREGULAR WATERING
DISORDER • *See p. 141*

Longitudinal cracks appear in roots (here of a carrot). Top growth appears perfectly normal. Secondary organisms may cause deterioration of root.

PLANTS AFFECTED Many root crops
SEASON Mainly summer and autumn

LAWNS

LAWNS ARE VERY POPULAR and one of the most widely used areas of a garden. Even if you do not want a perfect lawn, but prefer an area which is practical, serviceable and a good shade of green, there will still be times when problems need to be dealt with.

There are several fungal infections and pests which are common. The majority of problems, whether caused by a pest, a pathogen or poor cultivation, result in areas of grass turning yellow or brown. At first sight, differentiating between the various causes may not be easy. Most fungal problems do involve the development of visible (if small) fungal growths and a few, such as fairy rings or other toadstools, are very distinct. Lawn pest problems are generally easier to identify and culprits more readily visible.

In all cases, prompt action is advisable, as once turf grasses weaken or their growth becomes thinner, coarse grasses, mosses and weeds colonize rapidly. Among these, dog lichens (*see facing page*) are commonly mistaken for a disease, but in fact are half fungus and half algae.

LAWN PROBLEMS

LEATHERJACKETS PEST • *See p.145*
These daddy-long-legs larvae generally cause damage when they are becoming fully grown. Patches of turf turn yellow-brown and holes may be seen where birds have been searching for the grubs.

PLANTS AFFECTED Turf, roots of many plants
SEASON Mid-summer

SLIME MOULD DISEASE • *See p.177*
Numerous grey, yellow or off-white spheres are clustered together on individual blades of grass. These may appear to "dissolve" when wet. Turf growth and health are not affected.

PLANTS AFFECTED Many plants, including turf
SEASON Mainly summer and autumn

BURROWING BEES PEST • *See p.113*
Small, conical heaps of powdery soil form on the lawn. Each has an entrance hole at the top where a solitary bee leaves and enters. No significant damage is caused to the lawn.

PLANTS AFFECTED Turf
SEASON Mid-spring to mid-summer

CHAFER GRUBS PEST • *See p.116*
Plump, creamy-white grubs with brown heads sever roots, making it easy for foxes, badgers and birds to rip up turf to feed on them. The grubs are curved, with three pairs of legs (*see p.53*).

PLANTS AFFECTED Turf, other plant roots
SEASON Early autumn to early spring

TURF THATCH FUNGAL MYCELIUM DISEASE • *See p.186*
Bleached yellow or reddish patches of grass develop. Dense white or off-white fungal growth appears around the grass roots.

PLANTS AFFECTED Turf
SEASON All year

TURF DOLLAR SPOT DISEASE • *See p.185*
Patches of grass die off and turn straw-yellow. The individual patches are usually 10cm or less in diameter. The fungus causing this disease is most prevalent during warm, damp weather.

PLANTS AFFECTED Turf
SEASON Mainly early autumn

EARTHWORMS PEST • *See p.125*
Some earthworms deposit heaps of muddy excrement on the surface of lawns. These worm casts become smeared by feet or mowers and spoil the lawn's appearance.

PLANTS AFFECTED Turf
SEASON Mainly autumn to spring

LAWN PROBLEMS CONTINUED

ANTS
PEST • *See p. 99*

Heaps of fine soil appear on the lawn's surface above ant nests; if they are disturbed, black or reddish-yellow ants will be seen. Ant heaps get in the way of mowers and make the lawn uneven.

PLANTS AFFECTED Turf

SEASON Late spring to early autumn

TURF RED THREAD
DISEASE • *See p. 186*

Patches of grass begin to deteriorate. Numerous pale red or pink, gelatinous, horn-like fungal growths are attached to individual grass blades. Later, fluffy, pale pink fungal growth develops.

PLANTS AFFECTED Turf

SEASON Mainly late spring to early autumn

DOG LICHENS
See facing page

Small, unevenly shaped, grey, leaf-like growths, off-white beneath, develop amongst grass, particularly when turf is sparse and weak. Lichen is not a pest or disease; it should be treated as a weed.

PLANTS AFFECTED Turf

SEASON All year

TURF SNOW MOULD
DISEASE • *See p. 186*

Patches of dying grass develop and then enlarge and turn brown. Deteriorating grass may be covered in pale pink, fluffy fungal growth.

PLANTS AFFECTED Turf

SEASON Mainly autumn and winter

MOLES
PEST • *See p. 150*

Moles create a network of tunnels in the soil in which they live, and leave the excavated soil in heaps on the surface. Collapsed mole tunnels give the lawn an uneven surface.

PLANTS AFFECTED Turf

SEASON All year

TURF FAIRY RINGS
DISEASE • *See p. 186*

Rings of dead or dying grass, or sometimes other patterns of discoloration, develop on a lawn. One or more rings of lush grass grow adjacent to these. Brown toadstools grow out of the ring.

PLANTS AFFECTED Turf

SEASON Mainly late summer and autumn

BENEFICIAL GARDEN CREATURES

GARDENS ARE INHABITED BY a wide range of insects and other animals, but only a small proportion are plant pests. Most are neither particularly helpful nor harmful to the gardener. There are, however, some which are very beneficial, as they feed on pests, pollinate flowers or improve soil.

Predatory and parasitic insects may not be sufficient to prevent pests causing damage, but they do stop pests from having things all their own way. Pollinating insects are essential for good crops of most tree, bush and cane fruits, and some vegetables. Any insects that move from one flower to another can effect the transfer of pollen.

Most insecticides do not discriminate between pests and beneficial insects. It is therefore important to avoid the unnecessary use of chemicals, saving them for plants which would be seriously damaged by pests or diseases if control measures were not taken.

LADYBIRDS
Most ladybirds (*see p.143*) prey on aphids, but some feed on scale insects, spider mites or mealybugs, and others on powdery mildew fungal spores. Not all are red with black spots; some are yellow with black spots, others are orange or brown with white spots.

FLOWER BUG
Flower bugs, also known as anthocorid bugs, are 3–4mm long and are general predators of aphids, thrips, other small insects, mites and their eggs. They have needle-like mouthparts with which they pierce their prey and suck out the contents.

PARASITIZED APHIDS
Several species of tiny parasitic wasp lay eggs in young aphid nymphs. The wasp grub develops as an internal parasite that kills the aphid, causing it to become very swollen and straw brown in colour. The adult wasp emerges through a hole it makes in the dead aphid's abdomen.

LADYBIRD LARVA
Although less familiar than the adult beetles, ladybird larvae are equally voracious predators of aphids and other pests. The larvae, up to 11mm long, are black with orange or white markings. A larva will eat 200–400 aphids before it is fully fed and ready to pupate.

HOVERFLY LARVA
Only the larval stages (*see p.139*) of hoverflies feed on aphids. The larvae have flattened, semi-transparent bodies, up to 16mm long. The adult flies have yellow and black striped abdomens, giving them a wasp-like appearance, although their bodies usually look flatter.

TACHINID FLIES
There are many species of tachinid flies, between 4 and 18mm long. Most have larvae that are parasites of insects, especially caterpillars. The maggots feed inside the body cavity of their host insect, causing its death by the time the tachinid larvae are ready to pupate.

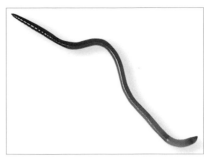

EARTHWORMS
Earthworms (*see p.125*), up to 20cm long, burrow tunnels in the soil through which water can drain and air can reach plant roots. Worms feed on dead plant material and their presence in compost heaps speeds up the conversion of debris to compost.

BROWN CENTIPEDE
This centipede, up to 30mm long, lives in the surface layers of the soil and also shelters under stones, logs and pots. It is active at night, preying on a wide range of small soil animals and their eggs. The front pair of legs curve forward and are modified to act as jaws.

Distinctive yellow and black markings

Wasp grasps victim and uses sting to paralyze victim

SNAKE CENTIPEDES
Snake centipedes live below soil level and are often found when soil is dug. They have slender, yellowish-brown bodies, up to 60mm long, adapted for moving through the soil where they hunt for small insects. If disturbed they writhe in a snake-like manner.

SOLITARY WASPS
Unlike social wasps, these tunnel out individual nests in rotten wood and sandy soil, or make use of beetle tunnels in dead wood. They provision their nests with aphids, flies, weevils or other insects to feed their larvae. There are many species of various sizes.

BUMBLEBEES

Bumblebees make their nests in the ground, often down abandoned mouse holes. They are valuable pollinating insects as they will visit flowers in poor weather when honeybees stay in their hives. At full strength in mid-summer a bumblebee nest houses 200–500 bees.

HONEYBEES

Many plants, especially most fruits, require insects to carry pollen from one flower to another to fertilize the flowers and allow seeds and fruits to set. The hairy bodies of honeybees, 12mm long, become heavily dusted with pollen when they visit flowers to collect nectar.

WOLF SPIDERS

Most spiders spin sticky webs to capture flying or crawling insects, but wolf spiders, 5–8mm long, run rapidly over the soil or vegetation to hunt down their prey. The females are frequently seen in early summer carrying their silken egg sacs under their abdomens.

SHREW

These small, mouse-like predators, which have elongate snouts, are rarely seen but may be numerous in hedgerows and rough grass. They are active during the day and at night, when they feed on a wide range of insects, slugs, woodlice and earthworms.

PARASITIC WASPS

Many species of parasitic wasps lay their eggs in the larvae, pupae or eggs of other insects. The larvae develop inside the host insect, which is killed when the parasite is ready to pupate. Some wasps have long egg-laying organs to enable them to reach concealed insects.

DEVIL'S COACH HORSE

This beetle, 25mm long, is often found sheltering under stones, logs or pots. It is one of the staphylinid or rove beetles, which are characterized by short wing cases that leave most of the abdomen uncovered. It eats slugs and soil pests such as cutworms and leatherjackets.

SONG THRUSH

The song thrush specializes in feeding on snails. It breaks the snail's shell by smashing it against a stone. They often make regular use of certain stones and these "anvils" become littered with pieces of snail shell. Their diet also includes worms, insects and berries.

GREEN LACEWING

Lacewings, 12–15mm long, derive their name from the many veins that give the wings a lace-like appearance. Other types of adult lacewings (*see p.142*) have black or brown bodies. The larval stages are voracious predators of aphids and sometimes other small insects.

CARABID BEETLES

Also known as ground beetles (*see p.135*), these insects live mainly on the soil surface, but some climb up plants in search of insects to eat. They are from 2 to 35mm long and mostly black, brown or metallic green. The larvae prey on a range of insects, slugs and pest eggs.

COMMON TOAD

Toads, like frogs, need ponds in which to breed in the spring, but for the rest of the summer they may be found well away from water. They are mainly active after dark, when they feed on a variety of insects, woodlice, worms and spiders.

Beetle's jaws are strong for killing and eating insects

Long legs to help chase prey

HEDGEHOGS

Although sometimes active in daylight, hedgehogs are largely nocturnal. Their diet includes slugs, worms and insects, particularly beetles. Hedgehogs need somewhere sheltered where they can hibernate undisturbed during the winter months.

TIGER BEETLE

This colourful predator, 11–13mm long, occurs mainly in sandy areas. It is a type of ground, or carabid, beetle. Its larvae live in pits in the soil from which they emerge to seize passing insects. The adults run over the soil surface and readily fly up in sunny weather.

GARDEN HEALTH AND PROBLEMS

GARDEN PLANTS CO-EXIST with many other living things – some beneficial, many benign, but some harmful. Favourable growing conditions and good gardening practice make plants strong and more able to tolerate potentially harmful organisms, but when serious attack by pests or diseases occurs, the gardener must choose an appropriate method with which to bring the problem under control.

THE BALANCE OF LIFE

EVERY GARDEN FORMS a miniature ecological system that is unique, with its own combination of plants, soil conditions, topography and climate. These factors, together with the interaction of the various creatures – including predators, parasites, pathogens, saprophytes and herbivores – that live within the garden, determine what thrives, what survives and what deteriorates or dies. The aim of the gardener is to encourage desirable growth – usually, that of ornamental and cropping plants – but this must be achieved within a healthy ecological balance that involves other life forms; a garden purged of insects or fungi, for example, would soon become unproductive. The activities and life processes of other forms of plant and animal life, together with their death and decay, are crucial to the growth and health of garden plants. Since all the organisms in a garden will influence the overall "working" and functioning of the plants and the garden as a whole, any action taken by the gardener may have an effect on plants or animals other than that being treated. When considering pest or disease control it is therefore important to consider any such effect.

THE ENERGY FOR GROWTH

Plants manufacture energy by the process of photosynthesis, in which the action of sunlight on green leaf pigments enables plants to combine

INTERACTION IN THE GARDEN

Even in the smallest corner of the garden, plants, creatures and environmental factors interact to form a rich and complex ecological system. Any action taken by the gardener may affect numerous links in the varied and interlocking cycles and webs of life.

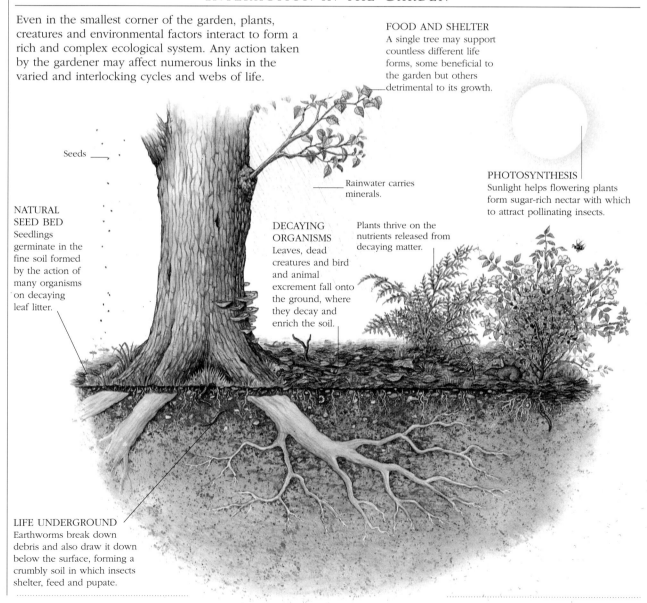

FOOD AND SHELTER
A single tree may support countless different life forms, some beneficial to the garden but others detrimental to its growth.

Seeds

Rainwater carries minerals.

PHOTOSYNTHESIS
Sunlight helps flowering plants form sugar-rich nectar with which to attract pollinating insects.

NATURAL SEED BED
Seedlings germinate in the fine soil formed by the action of many organisms on decaying leaf litter.

DECAYING ORGANISMS
Leaves, dead creatures and bird and animal excrement fall onto the ground, where they decay and enrich the soil.

Plants thrive on the nutrients released from decaying matter.

LIFE UNDERGROUND
Earthworms break down debris and also draw it down below the surface, forming a crumbly soil in which insects shelter, feed and pupate.

carbon dioxide from the air with water to make sugars and starch. Sugars may be used for other purposes, such as aiding pollination and seed dispersal. Many plants are insect-pollinated, and the secretion of sugar-rich nectar by flowers is an effective means of encouraging bees and other insects to visit blooms. While doing so insects pick up pollen when they come into contact with the anthers, and this is transferred to the stigmas of flowers that are subsequently visited.

Sugars also make fruits and berries sweet and attractive to birds and mammals. The seeds are dispersed away from the parent plant when they are spat out or have passed through the animal's gut.

ESSENTIAL NUTRIENTS

The starches and sugars used by plants for growth must be augmented by essential nutrients (*see pp.68–69*), drawn from the soil or, in the case of parasitic plants, from their hosts. For the cycle of life to continue, these nutrients must be returned to the soil. Dead or waste organic material, whether decomposing plant or animal matter, is broken down by the actions of fungi, bacteria, earthworms and others that feed on decaying plant material.

Certain larger fungi and many micro-fungi, described as saprophytic, live on dead organic material and, by breaking it down, may release components which are then used by other organisms in the vicinity. Their primary role, however, is one of clearing up debris which would otherwise accumulate. In a similar way numerous garden creatures live on debris and are to be encouraged because of their recycling function.

RECYCLING ACTIVITY
Woodlice occasionally cause plant problems but have a valuable role in processing garden debris.

As any organic matter deteriorates it releases and uses certain nutrients and so alters the range and quantity of materials available to plants and other organisms in the garden. Although deteriorating plant material may harbour plant pathogens, it may also provide a useful overwintering or breeding site for beneficial insects, so it is important to reach a compromise between extremes of garden tidiness.

FOOD AND FEEDERS

Plants form the basis of all food webs. There is a huge range of invertebrate animals, including nematodes, millipedes, insects, mites, woodlice, slugs and snails, that feed directly on plants. Many

PLANT ASSOCIATIONS
Gardeners' time frames often differ from those of nature. While mistletoe on tree branches (usually, of apple) will weaken the host over many years, most gardeners prefer the short-term attraction and novelty value of its presence.

of them specialize in feeding on particular parts of a plant; some may feed on a variety of plants, while others may be specific in their choice of food plant.

Most pests and disease-causing organisms have a relatively restricted host range, attacking only one type of plant, or perhaps a few closely related ones. An outbreak of one pest or disease could therefore remove the main food source of another potential pest or pathogen, so indirectly minimizing the damage it causes. Plant pests in turn are under attack from predators and parasites which help to prevent pests becoming so abundant that plants are killed.

OTHER BENEFICIAL ASSOCIATIONS

Fungi, bacteria and viruses are equally active in attacking plants, but there are also some which have a beneficial effect. Many plant roots have developed a mutually beneficial relationship with certain fungi or bacteria that help host plants to absorb nutrients from the soil (*see overleaf*).

PEST PREDATOR
A hedgehog is a valuable garden ally, naturally culling slugs and other plant pests, although it also feeds on earthworms.

SOIL AND NUTRIENTS

THE STRUCTURE AND texture of soil, and the range of nutrients it contains, has a great impact on plant growth and development. Texture – basically whether a soil is light and sandy, heavy and clayey, or something in between – influences plants' ability to develop strong root systems through which to take up water and nourishment, and affects the soil's ability to retain moisture and, to a certain extent, nutrients. These nutrients comprise a range of elements and minerals. Although some are present in only small ("trace") quantities, deficiencies, either due to natural factors or caused by heavy demands by plants that go unreplenished, will cause symptoms in plants that are often taken to be the result of attack by pests and diseases (*see table, facing page*). Sometimes, pest and disease problems follow on from deficiency, as the plant is weakened and more vulnerable to attack.

Soil pH or acidity also influences the range of nutrients that are in a form available to plants. For example, soils of a high pH (alkaline, limey or chalky) may "lock up" iron and manganese so that plants become chlorotic. In fact, in an alkaline soil most trace elements are less readily available than in an acid or neutral soil. Similarly a low pH (acid) soil is more likely to produce plants which suffer a deficiency of magnesium. An unusually high level of one nutrient may also influence the levels or

CROSS-SECTION, GARDEN SOIL
Garden soil has a distinct structure, with three layers – the topsoil, the subsoil and the bedrock (not shown here). The topsoil is the uppermost layer and is usually a dark brown colour. It contains a high level of organic matter and soil organisms and is the layer in which you dig and plant. The subsoil has a much lower level of organic matter, nutrients and organisms.

availability of another. Very commonly, tomatoes fed with a high potash feed to encourage a heavy crop of fruits may also show yellowing between the veins and around the edges of older leaves due to magnesium deficiency.

BALANCING NUTRIENT LEVELS

Checking the nutrient status and soil pH is a worthwhile job, especially when taking on a new garden with unfamiliar soil. Various kits are available (*see facing page*) and need not be expensive. These enable you to determine the soil pH and give an outline of the levels of major nutrients. Serious deficiencies are quite rare in gardens and feeding with a general fertilizer or regular use of plenty of organic material and good husbandry is usually enough to keep most plants adequately fed. Occasional use of specific materials to, for example, encourage flowering and fruiting or lush growth, may also be beneficial. It is worth bearing in mind, however, that overfeeding can be damaging both to plants and other soil organisms.

OTHER SOIL LIFE

Many garden creatures, including a number of pests, spend their lives, or a stage in their lives, within the soil. However, in general gardeners are far better occupied in improving soil texture and composition than in attempting to control its inhabitants. Although it is easy to become preoccupied with potentially

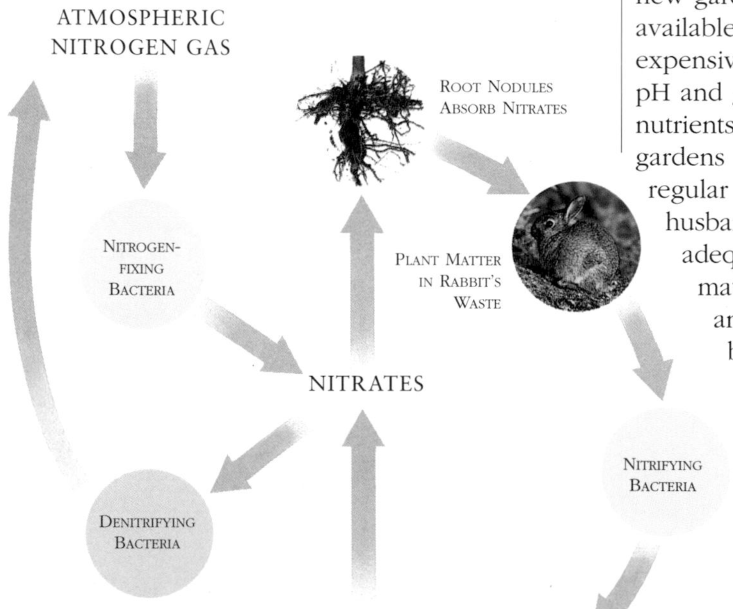

ATMOSPHERIC NITROGEN GAS

ROOT NODULES ABSORB NITRATES

NITROGEN-FIXING BACTERIA

PLANT MATTER IN RABBIT'S WASTE

NITRATES

DENITRIFYING BACTERIA

NITRIFYING BACTERIA

THE NITROGEN CYCLE
Nitrogen is one of the most important elements in the growth and development of all living organisms.

NITROBACTER BACTERIA

NITRITES

USING A KIT TO TEST SOIL NUTRIENTS

1 Preparing. This test for phosphorus is part of a kit to test the pH and major nutrient levels of soil. Mix 1 part soil and 5 parts water and allow to stand until the liquid is fairly clear. Use the pipette to transfer liquid to the testing chambers.

2 Testing. Make sure the water extract reaches the level marks on the colour chart. Open the correct capsule (blue in this test) and pour the powder into the chamber. Fit the cap onto the tester and shake until the powder has dissolved.

3 Results. Allow the colour to develop in the test chamber for a few minutes. Then compare this with the colour chart against a white background and in natural daylight. Read off the results and take further action if required.

damaging garden organisms, the vast majority of creatures in soil are harmless and indeed some are actually beneficial (*see pp.60–63*). The activity of each type of organism affects the population levels of others living in the vicinity and so any action taken by the gardener may well have a knock-on effect on something else.

Some animals and fungi can enhance the productivity or the performance of plants; examples include pest predators and parasites, and mycorrhizae – fungal associations formed on the root of a plant which allow the plant to become more efficient at nutrient uptake. Mycorrhizae may sometimes be visible in the soil as tiny fungal strands attached to the roots – these are often presumed to be harmful. Some also produce fungal fruiting bodies or toadstools around the base of the plant; again it may be presumed, incorrectly, that all toadstools around plants are harmful.

NITROGEN FIXERS

Certain bacteria that "fix" nitrogen, making it available to the plant, inhabit nodules on plant roots. If these roots are left in the soil once the crop has been harvested, they will also improve the nutrient status of the soil for any plants grown there subsequently. The nitrogen-fixing bacteria are invisible to the naked eye but the nodules are seen as distinct swellings on the roots (*see facing page*). The presence of these root nodules on peas and beans can be used to great benefit when following a rotation system for growing vegetables (*see p.96*).

NUTRIENT REQUIREMENTS AND ASSOCIATED PROBLEMS

Element	Specifically aids	Symptoms of deficiency	Possible causes	Remedies
NITROGEN (N)	Leaf and shoot growth.	Pale foliage, poor growth and may develop reddish or yellow discoloration. Older leaves affected first, as the nitrogen is moved from these into the newer growth when in short supply.	Most common on soils with low organic matter or where un-decomposed organic matter is added to soil in large quantities.	See p.152
PHOSPHORUS (P)	Root growth.	The plant's overall growth, including flowering and fruiting and the growth and development of seeds, is affected.	Most common on acid soils or following heavy watering or rain.	See p.161
POTASSIUM (K)	Flowering and fruiting; wood ripening; tissue strengthening.	Leaves may show scorching of the tips. Fruiting and flowering may be reduced and on tomatoes it causes blotchy ripening.	Most common on sandy soils and on chalk.	See p.164
MAGNESIUM (MG)	Chlorophyll formation (the green pigment in plants).	Leaves, particularly the older ones, show yellowing between the veins and around the edges. When it is in short supply, magnesium is moved from the older tissues into the new ones, so young foliage rarely shows any symptoms.	Most common on sandy soils and following heavy watering or rain, or heavy applications of potassium.	See p.147
CALCIUM (CA)	Formation of plant cell walls.	The most commonly seen symptoms are bitter pit on apples and blossom end rot of tomatoes. Foliage may also be reduced in size and distorted.	Most common on acid soils, and on plants growing in any soil which is not adequately watered to allow the plant to take up calcium.	See p.114
IRON (FE)	Chlorophyll formation.	Leaf yellowing, mainly between the veins and most apparent on the young leaf growth.	Most common on alkaline soils.	See p.141
MANGANESE (MN)	Chlorophyll formation.	Yellowing between the veins on young leaves is the most common symptom.	Most common on alkaline soils and waterlogged or poorly draining soils.	See p.148

NATURAL UPSETS

IN ANY NATURAL environment there are factors that are not ideal for plant growth. These naturally occurring problems are often the result of adverse weather conditions that either cause physical wounds through which infections may enter, or, more commonly, simply weaken the entire plant so that it is more likely to succumb to pests and diseases.

Often the gardener can compound problems, either through "contributory negligence" – leaving drought-stricken plants unwatered, for example – or by unintentionally inflicting damage such as that caused by poor pruning or by overtight stem or tree ties. By far the most common contributory factor, however, is the poor choice of plants for a given site. If a plant is not suited to the prevailing conditions and climate, or microclimate, it will rarely thrive and hence will be more prone to

problems. A less than fully hardy plant may be so set back in a cold, exposed site each year that it never makes a good specimen. A moisture-loving plant will fare badly on dry soil; a Mediterranean plant may rot in waterlogged ground. Always check before buying that the plant stands a good chance of survival in the conditions your garden offers.

WATER AND DROUGHT

Water is a fundamental plant requirement, used in photosynthesis (*see pp.66–67*), to maintain turgidity (the ability of stems to stand upright) and to transport nutrients into and around the plant tissues.

Plants vary in their requirement for water and some are better than others at surviving dry conditions. Wilting is the most common symptom of drought; if soil moisture levels are restored fairly

DAMAGE CAUSED BY ENVIRONMENTAL FACTORS

FROST DAMAGE

A scorched appearance is the most common symptom of frost damage. Soft new growth is most vulnerable to damage, with blackening and withering of shoot tips; pruning should not be left so late in the season that regrowth cannot ripen.

Frost-damaged *Pieris japonica* leaves

LIGHTNING DAMAGE

The larger and more ragged wounds are, the more slowly they will heal. Until a wound has completely callused over with scar tissue, it is vulnerable to invasion by harmful pathogens.

ANIMAL VANDALISM

Few plants can withstand repeated injury caused by browsing deer and rabbits, which may cause considerable problems in gardens in rural areas or close to parkland.

Withered maple leaves

DROUGHT

Wilting can be reversed if caught early. Soft, floppy leaves may plump out again if water is added, but leaves that have dried and become crispy will not recover. There are also several wilt diseases with similar symptoms (see p.193).

Poor growth

rapidly it can be reversed. Prolonged drought can, however, cause parts or all of the plant to die. If water levels are consistently low, but never in drastically short supply, the plant may survive, but be small and fail to perform well. Poor flowering, buds that fail to develop, and small fruits with tough skins are common symptoms. Dry soil is most obviously associated with dry weather, but light soil texture and drying winds may also be factors.

WATERLOGGING AND IRREGULAR WATERING

Excess water can be just as harmful as drought. If soil is waterlogged, plant roots may deteriorate rapidly, largely due to the lack of oxygen. Water-logged soil is also more likely to contain harmful pathogens, such as damping-off or foot- and root-rotting fungi. In the early stages, however, chlorosis or yellowing, often associated with the veins of leaves, is a common symptom.

POOR PRUNING
Stubs left after pruning are readily colonized, as here, by rotting fungi that may spread.

Plants also find it difficult to adapt to an erratic water supply. Plants that are growing in containers are very susceptible, as they tend to dry out readily and may then be overwatered by an over-enthusiastic gardener. Symptoms include poor or slow growth, bud drop and cracked or split fruits.

EXTREMES OF TEMPERATURE

In addition to drought symptoms, high temperatures often cause scorching of leaves, particularly on plants growing under glass. Low temperature injury is most commonly seen as frost damage (*see facing page*), causing withering of shoots, foliage or flowers or even death of the whole plant. Plants growing in very cold areas or in containers may also suffer root injury due to freezing of the root ball. This may then be followed by poor growth, dieback or even death. Fruit set is frequently affected by low temperatures as the activity of pollinating insects may be restricted, and flowers themselves damaged.

Snow and hail too can injure leaves or, in the case of snow, weigh down and break branches. This problem is particularly common when a heavy fall of snow has partially melted and re-frozen. In some cases, however, snow may act as an insulating "blanket" and prevent cold wind or frost injury of the plants which it covers.

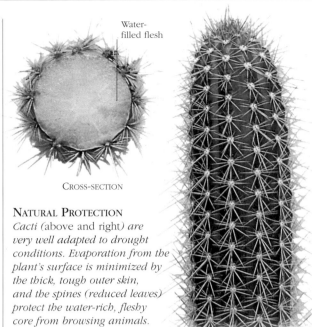

Water-filled flesh

CROSS-SECTION

NATURAL PROTECTION
Cacti (above and right) are very well adapted to drought conditions. Evaporation from the plant's surface is minimized by the thick, tough outer skin, and the spines (reduced leaves) protect the water-rich, fleshy core from browsing animals.

STRONG WINDS

Wind may cause physical injury and scorching of foliage and a prevailing strong wind may also prevent a plant growing to its full potential size, or even alter its shape. Wind blows away moisture shells around a transpiring plant leaf and so will, in turn, increase the plant's water requirements and make it more likely to suffer from drought if a plentiful supply of water is not available. In very windy seasons or sites, pollination may also be poor, again due to pollinating insects being restricted in their activities.

Healthy *Primula obconica* flowers

Flowers dying off

Leaves turn yellow and flop over

WATERLOGGING
The water in saturated soil displaces air between soil particles, creating an anaerobic environment in which plants cannot take up water. The symptoms therefore resemble those of drought, and plants may mistakenly be given yet more water.

TYPES OF PLANT DISEASE

THE DISEASE-CAUSING organisms or pathogens most prevalent in gardens are fungi and viruses. In addition, several bacterial infections and a few diseases caused by mycoplasmas occur.

FUNGAL INFECTIONS

This major group of infections may be caused by microfungi that are extremely small but usually still visible to the naked eye. Examples include downy mildews, powdery mildews, fungal leaf spots and rusts. A few fungal infections are systemic (found throughout the plant), but the majority are restricted to certain plant parts. The larger fungi – those which produce easily visible fungal structures such as fungal brackets or toadstools – are less common but may prove extremely damaging.

VIRUS INFECTIONS

Virus particles are sub-microscopic, that is to say invisible even with the aid of a normal microscope, but can be seen with an electron microscope. The symptoms they cause are, however, generally easily visible, most commonly as yellow leaf markings. Virus particles are systemic so are found throughout an infected plant, but their symptoms may only be apparent on one area, or their appearance may vary from one part of the plant to another. Virus infections may occur alone or in combination with other virus infections. It is impossible to identify precisely which are involved from the symptoms alone. Mycoplasmas are sub-microscopic and similar to viruses but much less common.

BACTERIAL INFECTIONS

Problems caused by bacterial infections are usually localized, but as many cause rapid deterioration of plant tissues, their effect may be quite dramatic. Sometimes the bacteria are apparent as bacterial ooze, for example on a tree infected with bacterial canker or fireblight. Many bacterial infections which result in softening of plant tissues are accompanied by a distinct, and often unpleasant, smell.

HOW DISEASES ARE SPREAD

Most fungal infections are spread from plant to plant or within a plant by spores. These are occasionally large enough to be visible with the naked eye. Their type, size and quantity depends largely on the specific fungus. Most spores are spread either on air currents or by rain splash. Bacteria are spread by similar means. Some infections may also be carried by animals, including insects as they move from plant to plant. Soil-borne infections are usually spread either in association

MICROSCOPIC AND SUB-MICROSCOPIC ORGANISMS

FUNGUS
The majority of fungi produce visible symptoms and in most cases the fungal mycelium and sometimes the spores are apparent too.
The spores of the majority of rust fungi, here hollyhock rust, are released from within raised pustules. Their detailed structure is only visible (inset) with the aid of a microscope.

VIRUS
Infection by virus particles (inset) usually results in visible symptoms, such as leaf distortion (here from cucumber mosaic virus) and yellow patterning, often accompanied by general poor growth and stunting. Virus infected plants may also be symptomless, or the extent of the symptoms may vary greatly.

SPREADING DISEASE
Using cutting tools such as secateurs on an infected plant and then on a healthy one without disinfecting them can spread disease.

COLOUR BREAKING
A characteristic symptom of virus attack is slashing or striping of flower colour. Usually unwanted, it has in some cases, particularly in tulips, been developed through breeding to produce novel petal patterns in otherwise healthy plants.

with soil particles or in soil moisture. Some, such as *Phytophthora*, have spores that can propel themselves, called zoospores, and these spread in water. Weather conditions may affect disease spread and development, and in most cases moisture on the plant surface encourages infection and spread. Humidity and the resultant film of moisture on the plants may also cause a marked increase in disease spread and development.

Viruses are most commonly spread when sap from an infected plant is transferred to another potential host. Pruning tools, physical injury, grafting and the taking of cuttings are all common routes. Even handling – for example, dividing – an infected plant may be sufficient to transfer virus particles to a healthy host. Some viruses are also transmitted by garden creatures which, because of this role, are known as vectors. The most common are sap-feeding pests such as aphids, leafhoppers and nematodes. Viruses may also be seed-borne.

INCREASING VULNERABILITY

A plant which has suffered a severe attack of, say, powdery mildew, with consequent loss of leaves, may be severely stressed, particularly if infection occurs in several consecutive years. The plant may well then be more prone to attack by another pathogen, such as a leaf spot or honey fungus, both of which more readily attack stressed plants. In a similar way an apple which has been attacked by codling moth is very prone to attack by brown rot fungus, which gains entry via the wounded skin of the fruit.

It cannot be reiterated too often that stresses brought about by unsuitable growing conditions or extremes of weather (*see pp. 70–71*) and problems associated with nutrient deficiencies and phytotoxicity (damage to the plant caused by excessive nutrient levels or use of unsuitable chemicals) are often responsible for lowering a plant's ability to fight off infection and disease development.

MOST COMMON PLANT DISEASES

Disease	Type	Method of spread	Conditions making plant more susceptible	Season
POWDERY MILDEWS	Fungi	Water/rain splash; air currents.	Dry soil/root conditions; moist or humid environment around leaves.	Spring, if dry, and summer to autumn.
DOWNY MILDEWS	Fungi	Water/rain splash; air currents.	Humid or wet air around foliage; damp spring and autumn weather; stagnant conditions. Plant debris in vicinity; dead/damaged plant parts.	All year.
RUSTS	Fungi	Water/rain splash; animals; wind.	Damp, warm, stagnant conditions; moisture film on leaves.	Mainly damp summers and autumn.
FUNGAL CANKERS	Fungi	Water/rain splash; air currents.	Damage to bark from pruning wounds; pest attack; frost crack.	All year.
FUNGAL WILTS	Fungi	Pruning tools; infected root material; infested soil.	Damage creating open wounds when spores in vicinity.	All year.
BACTERIAL SOFT ROTS	Bacteria	Water splash, water films; pruning tools; insects.	Damage to the plant's surface allowing entry of bacteria; poor storage conditions.	All year.
FUNGAL LEAF SPOTS	Fungi	Mainly water/rain splash; some air currents.	Poor growing conditions; weather extremes; other diseases.	Mainly spring and summer.
HONEY FUNGUS	Fungus	Underground rhizomorphs (fungal strands); root contact.	Any other predisposing stresses such as other diseases, poor growing conditions, weather extremes or excessive pruning.	All year.
CUCUMBER MOSAIC VIRUS	Virus	Sap-feeding pests; handling; pruning and taking cuttings; occasionally seed-borne.	Presence of any factors which encourage spread (*see left*).	Mainly late spring and summer.
PHYTOPHTHORA	Fungus	Soil- and water-borne.	Injury to plant; over-wet soil conditions.	All year.

TYPES OF GARDEN PEST

ANIMALS BECOME PESTS when their activities begin to have an adverse effect on plants. They may do this by reducing plant vigour so that yields of fruits and vegetables, or the decorative qualities of ornamentals, are impaired. In addition to the direct damage caused by their feeding activities, pests can have indirect effects on plants. Those feeding on the edible parts of fruits and vegetables may not be numerous enough to reduce yields, but nevertheless make them unpalatable due to the presence of insects and their excrement. Many sap-feeding insects soil their host plants with a sugary excrement called honeydew that encourages the growth of sooty moulds.

Pests can also transmit some virus and fungal diseases of plants on their mouthparts or bodies. Some plant diseases are dependent on the presence of pest-damaged tissues and cannot become established without this initial help.

While some garden pest problems are local, others are the result of migrations from elsewhere. Pests may be imported in purchased plants or gifts. Pollen beetles and flea beetles can fly in from agricultural crops, such as oil seed rape, while cabbage white butterflies, which are capable of longer flights, may come to Britain from elsewhere in Europe.

Snail uses its rough "tongue" or radula to break up plant tissues

SOLITARY FEEDER
While snails can occur in great numbers in gardens, they are essentially lone operators when browsing on plants. Each individual is, however, capable of causing extensive damage.

Pests affect plants in a variety of ways. Some attack specific plant parts, while others feed on or damage several different types of plant tissue.

SAP-FEEDING PESTS
Sap-feeding pests such as aphids and scale insects have needle-like mouthparts that are inserted into plants in order to suck sap. No visible holes are made in the leaves, but many sap-feeders have a toxic saliva that results in leaf distortion and/or discoloration. Some sap-feeding pests – including aphids, leafhoppers and thrips – can transmit virus diseases. Most sap-feeders infest the foliage, stems and flowers, but there are some species of aphids and mealybugs that also attack roots.

FLOWER PESTS
Relatively few pests specialize in feeding on flowers or flower buds, but these are attacked by many general plant pests such as earwigs and caterpillars. Pests specifically associated with flowers are pollen beetles, hemerocallis gall midge, apple blossom weevil and Michaelmas daisy mite. Some birds, such as bullfinches and house sparrows, also destroy flowers and flower buds.

GALL-FORMING PESTS
Some pests secrete chemicals into their host plants when they deposit eggs or as they feed. This induces a variety of abnormal growths known as galls that enclose the causal animals. The gall-forming habit is found among many types of insects, mites and nematodes. Plant galls can also be caused by fungi and bacteria.

ROOT-FEEDING PESTS
The larvae of a wide range of insects bore into or destroy roots, corms, bulbs and tubers. Their activities are often slow to be discovered, and can be difficult to eliminate. Some controls involve

PESTS EN MASSE
A single aphid or even a small colony presents no threat to a healthy plant. The speed at which they multiply, however, quickly results in heavy infestations that soon sap plants' vigour.

FEEDING DAMAGE
The depredation caused by feeding is a major factor in pest damage. However, they also create open wounds on plants that may act as entry points for secondary infections.

disposing not only of the affected plant but of the soil in the vicinity of the rootball. Many root pests arrive as eggs laid by winged insects; others may be imported into gardens via container-grown plants. Plants purchased or received as gifts should always be tipped out of their pots and the rootball examined for signs of pest activity before planting.

FRUIT PESTS
Fruits, berries and seeds are attacked by the larvae of many insects, including a variety of moths, sawflies and beetles. Wasps, birds and squirrels can also be a problem. Some pests, such as capsid bugs, reduce the quality of fruits by causing skin blemishes; others render them inedible. Damage to ripening fruits renders them liable to infection by fungal rots, so they cannot be stored.

STEM PESTS
Most pests attacking plant stems feed from the outside and include most of the sap-feeding pests. Bark beetles feed as larvae under the bark of woody plants while the larvae of leopard moth and clearwing moths bore into the centre of stems. Mammals, such as rabbits, deer, voles and squirrels, can damage bark, especially during the winter and spring.

LEAF PESTS
Plant foliage can bear a greater variety of pests than any other part of a plant. In addition to the sap-feeding pests listed above, there are pests that feed on the leaf tissue itself, such as slugs, snails, earwigs, caterpillars of sawflies, moths and butterflies, and adult vine weevils. Some insects, such as the larvae of certain flies, moths, sawflies and weevils, feed within the leaves as leaf miners.

HOW PESTS ATTACK

WHITEFLIES

SAP-FEEDERS
Adelgids, aphids, capsid bugs, froghoppers, leafhoppers, mealybugs, red spider mites, scale insects, suckers, thrips, whiteflies.

EARWIG

FLOWER PESTS
Aphids, apple blossom weevil, bullfinches, caterpillars, earwigs, hemerocallis gall midge, house sparrows, Michaelmas daisy mite, pollen beetles, slugs, snails, thrips.

VINE WEEVIL GRUBS

ROOT-FEEDERS
Cabbage root fly larvae, carrot fly larvae, chafer grubs, cutworms, eelworms, leatherjackets, onion fly larvae, root aphids, root mealybugs, slugs, swift moth caterpillars, vine weevil grubs, wireworms.

GALL WASP LARVAE

GALL-FORMERS
Adelgids, aphids, eelworms, gall midges, gall mites, gall wasp larvae, sawflies, suckers.

WASPS

FRUIT AND SEED PESTS
Apple sawfly, bean seed beetles, birds, codling moth larvae, nut weevil, pea moth larvae, plum moth larvae, plum sawfly, raspberry beetle, squirrels, strawberry seed beetle, wasps.

SCALE INSECTS

STEM PESTS
Many sap-feeding pests (see above), plus clearwing moth larvae, deer, leopard moth larvae, bark beetles, rabbits, squirrels, voles.

CATERPILLAR

LEAF PESTS
All sap-feeding pests (see above), plus adult vine weevils, earwigs, eelworms, flea beetles, leaf-cutter bees, moth and butterfly caterpillars, sawfly larvae, slugs, snails.

PEST LIFE CYCLES

THE LIFE CYCLES of pests are as diverse as the pests themselves, but they can be divided into two broad categories: complete and incomplete metamorphosis. The latter involves little more than a gradual increase in size as the pest reaches maturity, with the immature stages usually resembling the adult animal. This form of growth is shown by eelworms, slugs, snails, millipedes, woodlice, mites and some insects, including earwigs, aphids, leafhoppers, froghoppers, whiteflies, mealybugs, scale insects, capsid bugs and thrips. The feeding habits of the immature stages are often the same as the adult and so all active stages in the life cycle are capable of causing damage.

COMPLETE METAMORPHOSIS

More advanced insects, such as butterflies, moths, beetles, flies, ants, bees, sawflies and wasps, undergo complete metamorphosis. Here, the larval stages are very different in appearance from the adult and there is a pupal stage in which the larva, usually enclosed within a protective casing, changes into the adult insect. With these insects the feeding habits of the larvae are often different to those of the adult and it is usually the larvae that cause

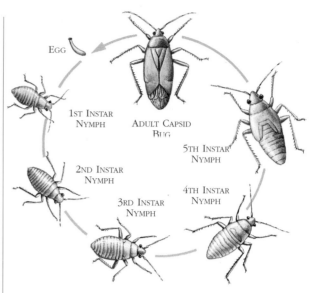

INCOMPLETE METAMORPHOSIS, CAPSID BUG
Pests that undergo incomplete metamorphosis look broadly similar at each stage of their lives and gradually acquire the adult features.

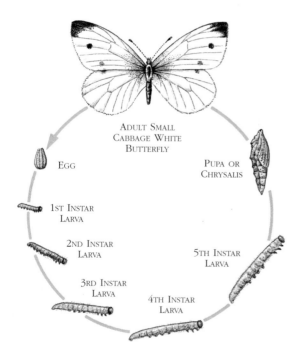

COMPLETE METAMORPHOSIS, CABBAGE WHITE BUTTERFLY
The more advanced types of insect undergo a complete transformation in appearance and often in feeding habits during their lives.

damage to plants. For example, caterpillars of butterflies and moths feed on leaves while the adults feed on nectar. Some beetle pests, however, cause damage at both the larval and adult stages.

HOW PESTS DEVELOP

Pests generally start life as eggs, although aphids give birth to live young for most of the year, only resorting to producing overwintering eggs in the autumn. Invertebrate animals, such as insects, mites, woodlice and millipedes, do not have any bones or internal skeletons. Instead they have a hard outer covering to their bodies. Like a suit of armour, this restricts growth, so must be shed from time to time to allow the animal to increase in size. This process is known as ecdysis. After the old outer skin has been cast off, the soft new skin underneath is stretched by the animal before it hardens, thus creating more space for growth. The number of stages, or instars, between the newly-hatched nymph or larva and the adult animal is usually between four and twelve, depending on the species.

HOW PESTS REPRODUCE

Most adult pests are either male or female and mate in a conventional manner. There are, however, exceptions. Slugs and snails are hermaphrodites with both male and female genitalia. They mate with others of their species and both partners will subsequently lay eggs. Some insects are capable of

APHID GIVING BIRTH
Aphids give birth to live young in spring and summer, but as day length shortens and temperatures fall in autumn, they lay eggs instead, which will overwinter.

breeding without mating (parthenogenesis). Some, like vine weevil and aruncus sawfly, only occur as females when adult. Aphids occur as females only during spring and summer; males are produced in late summer when mating is necessary to produce overwintering eggs. Some gall wasps have alternating asexual all-female generations and sexual generations with males and females.

PEST LIFESPANS

The length of time required to complete the life cycle depends on factors such as temperature, the availability of food and the nature of the pest. Most pests are able to complete their development within a year, but some require much less, while others take longer. Wireworms, for example, feed on roots in the soil for up to five years before they pupate and become adult click beetles. Aphids, by cutting out the egg stage, are able to go from young nymph to adult insect in as little as seven days.

Pests that have more than one generation in a year complete their development more rapidly in warmer weather. Glasshouse red spider mite takes 55 days to complete its life cycle at 10°C (55°F), but only 12 days at 21°C (70°F). Most outdoor pests have one to three generations a year. Glasshouse pests like whitefly and red spider mite can go through as many as 12 generations a year due to the warm, sheltered conditions in which they live.

SEASONAL CHANGES

Pests have evolved alongside the plants they feed on and so they begin to attack their hosts at the time of year when the appropriate plant tissues will be available. This is strongly related to temperature and day length, which not only control plant growth processes such as bud burst, flowering, fruiting and aging, but also bring pests out of winter dormancy and allow egg hatching. Similarly, the life cycles of predatory insects, such as ladybirds, are related to that of their prey, aphids.

SURVIVING THE WINTER

The means by which pests overwinter varies, although each species will usually have one particular stage in its life cycle in which it survives the colder months of the year. This may be the egg, one of the nymphal or larval stages, the pupa or adult. The overwintering stage generally occurs in some sheltered place, such as in the soil, in the crevices of bark or inside dense shrubs such as conifers. Many insects and mites go into a form of hibernation called diapause from which they will not emerge until they have experienced a sufficient period of cold. This prevents them emerging prematurely whenever the winter weather turns mild. Some glasshouse pests, such as whitefly, continue feeding and breeding throughout the year, albeit more slowly in winter than in summer.

ANNUAL POPULATIONS

While some pests are a problem every year, others are only occasional problems. Serious outbreaks may be due to a combination of favourable weather conditions and a relative absence of predators, parasites and diseases of the pest. Minor pests often need a run of several good reproductive cycles before they are numerous enough to become a significant problem. Hot, dry summers favour certain pests such as spider mites and thrips, while slugs and snails thrive under damp conditions. Cold winters do not kill pests as all those native to northern Europe have a cold-tolerant overwintering stage. Mild winters can allow some pests, such as aphids, that would normally go dormant, to continue feeding and breeding, resulting in higher populations in spring.

LADYBIRD LARVA EATING APHID

ADULT LADYBIRD EATING APHID

THE BENEFICIAL LADYBIRD
Both the larval and adult stages of ladybirds feed avidly on aphids, making them doubly valuable to gardeners. Some are also predators of scale insects, thrips, mealybugs or mites.

APPROACHES TO CONTROL

WHEN SOMETHING goes wrong in a garden and pests or diseases are to blame the first consideration is whether control measures are necessary, and if so what action should best be taken. Not all infections or infestations merit control. A powdery mildew infection at the end of the season on a herbaceous plant may look unsightly, but if the leaves are going to die back shortly anyway, the infection has little if any detrimental effect on the plant and may not be worth treating. Similarly, it may be easier and more sensible to pick off one or two caterpillars on a cabbage than attempt to control them with an insecticide.

Some problems cause rapid and severe damage, while others may cause some damage but are probably more worrying to the gardener than they are to the plant. It is therefore important to balance the need to spray with the damage actually caused. Farmers and growers have an "economic threshold" – the point at which a pest or disease is causing or is likely to cause sufficient loss to merit the time and money spent spraying. This is something the gardener should consider, modifying the equation to include the extent to which a problem spoils a plant's appearance, or ability to crop before reaching for the sprays.

USING GARDEN CHEMICALS

There is no doubt that chemical controls (*see pp.82–83*) may sometimes offer rapid, safe and usually complete solutions to pest or disease problems. If correctly used, garden products today have been so carefully formulated and tested (*see pp.80–81*) that adverse side-effects are minimal. Most gardeners

INTEGRATED CONTROL IN THE GARDEN

The integrated approach to controlling problems combines appropriate plant choice and good gardening practice, preventive measures to avoid problems, and sensible use of chemicals only when warranted by serious pest or disease attack.

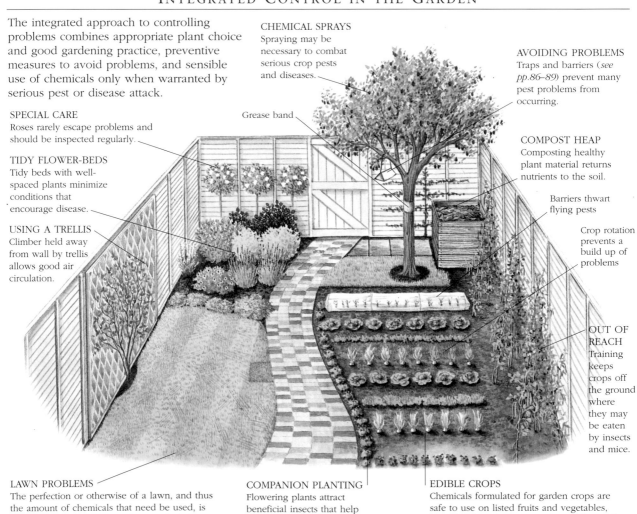

CHEMICAL SPRAYS
Spraying may be necessary to combat serious crop pests and diseases.

AVOIDING PROBLEMS
Traps and barriers (*see pp.86–89*) prevent many pest problems from occurring.

SPECIAL CARE
Roses rarely escape problems and should be inspected regularly.

Grease band

COMPOST HEAP
Composting healthy plant material returns nutrients to the soil.

TIDY FLOWER-BEDS
Tidy beds with well-spaced plants minimize conditions that encourage disease.

Barriers thwart flying pests

USING A TRELLIS
Climber held away from wall by trellis allows good air circulation.

Crop rotation prevents a build up of problems

OUT OF REACH
Training keeps crops off the ground where they may be eaten by insects and mice.

LAWN PROBLEMS
The perfection or otherwise of a lawn, and thus the amount of chemicals that need be used, is generally entirely a matter of personal choice.

COMPANION PLANTING
Flowering plants attract beneficial insects that help to control vegetable pests.

EDIBLE CROPS
Chemicals formulated for garden crops are safe to use on listed fruits and vegetables, provided that harvest intervals are observed.

who do not follow organic gardening methods are also happy to use inorganic or chemical fertilizers. The inorganic approach to gardening is still more widespread than the organic, but increasing numbers of people who would not call themselves organic gardeners use many organic methods.

"GREEN" GARDENING

In recent years organic or so-called "green" gardening has become increasingly popular. Many gardeners have started to question the use of too many garden chemicals and some use purely organic methods. Organic gardening avoids the use of manufactured chemicals, but some organic gardeners will happily use products such as pyrethrum and Bordeaux mixture as these are made from "natural" ingredients.

The organic approach also advocates the use of only naturally occurring materials, such as manure, for improving soil fertility, and materials such as leaf mould for mulching or improving soil texture.

It embraces other environmental concerns, too, such as the use of coir or bark-based growing media in preference to peat-based products, and of recycled materials wherever possible.

Organic gardening also makes full use of what are, essentially, good gardening techniques, such as the pruning out of infected material and good garden hygiene (see pp.92–96). Resistant or tolerant plant cultivars and trap and barrier methods (see pp.86–89) are also important to the organic gardener.

To succeed at organic gardening it is also important to take all possible steps to encourage populations of beneficial creatures such as naturally occurring pest predators and parasites (see pp.90–91). The avoidance of many chemicals will make this easier, but it should be remembered that some of the products accepted by many organic gardeners (for example, pyrethrum) are not selective in their activity and so may damage the beneficial organisms which organic gardeners try so hard to encourage.

INTEGRATED CONTROL IN THE GREENHOUSE

The enclosed environment of a greenhouse can increase problems, but may also assist control. While pests flourish in the warm, sheltered atmosphere, they may also be targeted much more successfully by traps, sprays or the release of the predators or parasites known as biological controls.

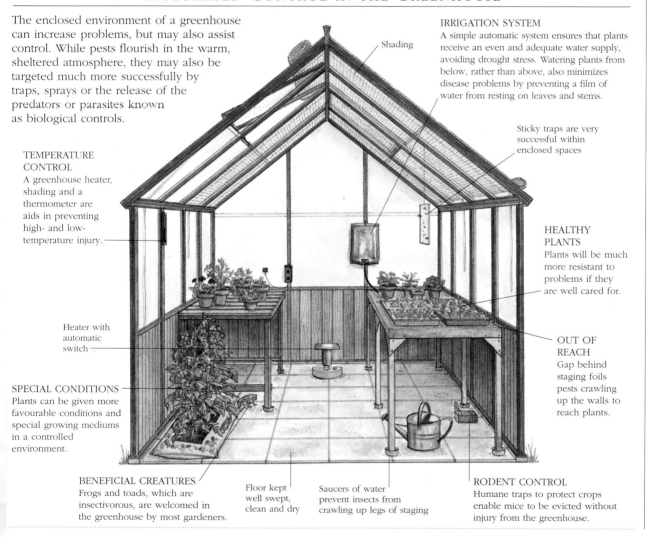

Shading

IRRIGATION SYSTEM
A simple automatic system ensures that plants receive an even and adequate water supply, avoiding drought stress. Watering plants from below, rather than above, also minimizes disease problems by preventing a film of water from resting on leaves and stems.

Sticky traps are very successful within enclosed spaces

TEMPERATURE CONTROL
A greenhouse heater, shading and a thermometer are aids in preventing high- and low-temperature injury.

Heater with automatic switch

SPECIAL CONDITIONS
Plants can be given more favourable conditions and special growing mediums in a controlled environment.

HEALTHY PLANTS
Plants will be much more resistant to problems if they are well cared for.

OUT OF REACH
Gap behind staging foils pests crawling up the walls to reach plants.

BENEFICIAL CREATURES
Frogs and toads, which are insectivorous, are welcomed in the greenhouse by most gardeners.

Floor kept well swept, clean and dry

Saucers of water prevent insects from crawling up legs of staging

RODENT CONTROL
Humane traps to protect crops enable mice to be evicted without injury from the greenhouse.

CHEMICAL CONTROLS

Plant pests and diseases have always been a problem for gardeners and farmers alike. It is, however, only fairly recently that an industry has existed to develop and market products for keeping plants in good health. In the nineteenth century these included some chemicals based on arsenic, cyanide, mercury and other chemicals far too dangerous to be used today. A few insecticides and fungicides, such as the plant-derived pyrethrum, the copper-based Bordeaux mixture and sulphur, survive from that time and are still in use in gardens today.

SAFETY STANDARDS

Most of the chemicals currently in use have, however, been developed since 1945. Some developed at that time, such as the insecticides DDT and aldrin, have subsequently been withdrawn, when it was discovered that they persist in the environment and accumulate in the bodies of birds of prey and other predators. The manufacturers of pesticides and the government departments that regulate them have learned from mistakes such as this and environmental safety is now a high priority.

Before any new product can be marketed it has to be thoroughly researched to test factors such as its effectiveness, its persistence, any effects on

PATENT FUMIGATOR
Chemicals used in the past by gardeners could be toxic; some methods of application were equally intimidating.

non-target plants and animals, and its long- and short-term effects on humans. These tests start in the laboratory and build up to field trials. The data required to support an application to market a new product takes about ten to twelve years to compile and will cost tens of millions of pounds. No other chemicals, apart from pharmaceuticals, are as thoroughly tested before they can be sold to the public.

Insecticides and fungicides are developed for the agricultural market and only a small proportion of the chemicals available to professional growers are marketed for garden use. These are chemicals at the safer end of the toxicity spectrum and are supplied as formulations which can be applied without the need for special protective clothing. The manufacturers and the government regulators have carried out their parts in producing safe pesticides; the final responsibility for safe use lies with the gardener, who must read and follow the instructions before applying a chemical.

PESTICIDE RESISTANCE

It is possible for some pests and fungal diseases to build up an immunity or tolerance to chemicals, so that a pesticide that was once an effective treatment

APPLYING CHEMICALS ON FARMLAND
Garden chemicals are largely developed from those used in agriculture and in the professional horticultural trades. However, the formulations in which they are supplied for use in the garden are much safer, eliminating the need for protective clothing and other special precautions, although sensible safety rules should be observed and the manufacturer's instructions strictly followed.

no longer works. Continual use of the same chemical can exert a powerful selection pressure by killing the more susceptible individuals, but leaving those with some tolerance to pass this characteristic on to the next generation. Pests can develop full resistance in about 40 generations and, since some pests have five or more generations a year, a chemical can fail within a comparatively short time.

Resistance problems are less frequent in gardens than in agriculture or commercial horticulture. This is because most gardeners use chemicals over small areas, and susceptible strains of pests and diseases continually come in from other gardens where different or no chemicals are used. Resistant pests are mainly found in greenhouses, where the enclosed environment and a pest's rapid reproduction rate can make chemicals ineffective against whitefly, red spider mite and aphids. Some diseases, such as grey mould, rose black spot, powdery mildews and apple scab, can become tolerant of fungicides, especially the systemic types.

REDUCING RESISTANCE

Gardeners can reduce the likelihood of resistance occurring by using chemicals sensibly. Where possible, non-chemical alternatives, such as biological controls (*see pp.90–91*) or the removal of badly diseased plants, should be considered. It should be noted that some different active ingredients are chemically related and it is likely that a pest or disease resistant to one will also tolerate other chemicals in the same family. When pesticides are used, the continual use of the same product should be avoided.

PHYTOTOXICITY

Plants are sometimes damaged by insecticides or fungicides; this is known as phytotoxicity. The manufacturer's instructions should be read and

GLASSHOUSE TRIALS
Before potential pesticides are used outdoors, extensive glasshouse trials are conducted. These are designed to test the chemical's efficacy and lack of harmful effects on plants.

followed carefully to ensure that the correct dilution is being applied. Plants known to be harmed by a chemical will be listed by the manufacturer, but this information is not known for all potential plant and chemical interactions.

The risk of such damage occurring can be minimized by not treating plants already under stress as a result of exposure to bright sunlight, extremes of temperature or dryness at the roots. Seedlings are more susceptible than mature plants, and petals are more likely to be scorched than leaves. Where there is doubt and there are a number of plants to be sprayed, it is sensible to treat just one plant and watch for adverse reactions over the next week. Plants that have been damaged may suffer a significant check in growth but will usually make a recovery.

WITHIN A CONTROL DOME

SYSTEM OF CONTROL DOMES

USING CONTROL DOMES TO TEST CHEMICALS
Small-scale tests in enclosed environments, such as the control domes illustrated here, are used to detect possible environmental problems before field trials are held.

CHEMICALS AND THEIR ACTIONS

M OST INSECTICIDES and fungicides are contact in action. This means that insecticides must cover the plant surface thoroughly so that the pest is either hit directly or picks up the pesticide from the treated surface. Similarly, contact fungicides kill fungal spores when they germinate if coverage of the plant surface has been adequate.

Systemic insecticides and fungicides are absorbed into the plant and may be transferred to parts which have not been directly treated. This movement may be within the leaves or up to the shoot tips. Systemic insecticides and fungicides will not move from the foliage down to the roots. Systemic fungicides generally only move within a single leaf. They kill fungal growth within plant tissues and have a curative action, rather than a preventive role. Systematic insecticides are useful in controlling sap-feeding pests (*see pp. 74–75*), including those concealed in distorted leaves that are difficult to reach with contact sprays.

THE WAYS CHEMICALS WORK

Different problems require chemicals to be applied in different ways. Sometimes, a combination of methods of application is needed to achieve complete elimination.

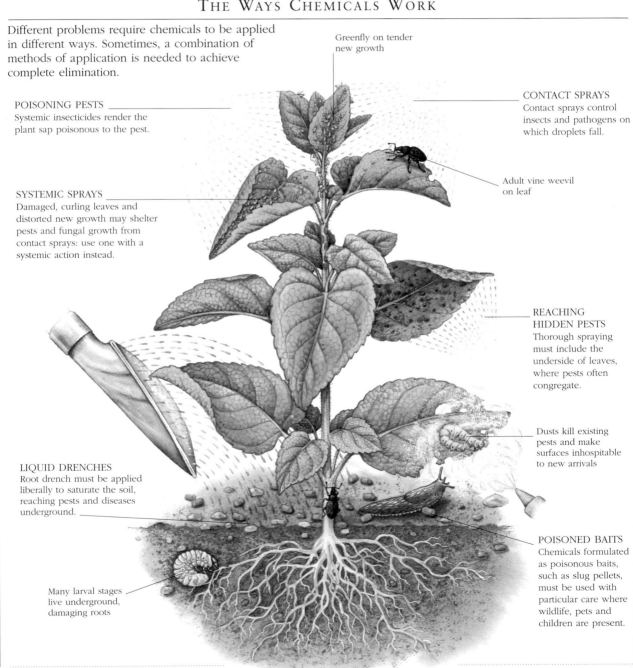

Greenfly on tender new growth

POISONING PESTS
Systemic insecticides render the plant sap poisonous to the pest.

CONTACT SPRAYS
Contact sprays control insects and pathogens on which droplets fall.

Adult vine weevil on leaf

SYSTEMIC SPRAYS
Damaged, curling leaves and distorted new growth may shelter pests and fungal growth from contact sprays: use one with a systemic action instead.

REACHING HIDDEN PESTS
Thorough spraying must include the underside of leaves, where pests often congregate.

Dusts kill existing pests and make surfaces inhospitable to new arrivals

LIQUID DRENCHES
Root drench must be applied liberally to saturate the soil, reaching pests and diseases underground.

POISONED BAITS
Chemicals formulated as poisonous baits, such as slug pellets, must be used with particular care where wildlife, pets and children are present.

Many larval stages live underground, damaging roots

APPLICATION METHODS

Chemical type	Advantages	Disadvantages
LIQUID SPRAY	Contact and systemic types available. Most have a wide range of applications. Inexpensive.	Handling and accurate measuring of concentrate needed. Requires special equipment for application.
LIQUID DRENCH	Contact and systemic types available. Most have a wide range of applications. Inexpensive.	Handling and accurate measuring of concentrate needed. Difficult to work out precise amounts.
DUST	Used direct from pack, so little is wasted. It is easy to apply direct to the target organism.	Only contact types available. May leave unsightly deposit on plant.
READY-TO-USE SPRAYS AND PUMP-GUNS	No need to handle concentrate as contents already correctly diluted. Quick and simple to use.	Expensive and really only suitable for treatment of relatively small areas of infection/infestation.
PELLETS	In measured doses, so easy to apply correct quantity. Repellents in them to deter animals.	May be eaten by pets, children or wildlife. Can become less active after rain.
AEROSOL	Quick, simple and easy to use. Required dose applied, so no wastage. No equipment needed.	Propellant may be potentially flammable and may prove phytotoxic to plants. Expensive.

A few pests can be controlled by chemicals that have been mixed with something palatable to form a poisoned bait. These include rat poisons, slug pellets and ant baits. All such chemicals should be used with care around pets and children.

The pesticides available to amateur gardeners generally control a wide range of insects, including beneficial types if they are on the plants when they are sprayed. Such chemicals are said to have a broad spectrum of activity. There are others, available to professional growers, which are selective, such as pirimicarb, which control aphids but leave most other pests and beneficial insects unharmed.

Other chemicals may be relatively selective by having a short persistence. Organic insecticides, such as pyrethrum, fatty acids and plant oils, are

PROTECTING PONDS
Fish are extremely susceptible to harm from garden chemicals, which should not be used on plants in or near ponds. Avoid chemical fertilizers also where they may leach into pond water.

CHEMICALS IN THE FOOD CHAIN
There is a danger that birds may pick up insects and other pests that have been treated with pesticides, either to eat themselves or to feed to their young. It is difficult to prevent this happening, but it is rare for them to eat sufficient amounts for harmful effects to be observed.

broad-spectrum treatments, but remain effective on the plant for less than 24 hours, so their impact on non-target insects is limited.

PERSISTENCE OF CHEMICALS
The rate at which insecticides and fungicides lose their effectiveness and break down into other less active compounds varies. This is largely due to the degree of stability of the chemical's molecular structure, but is also affected by factors such as temperature and exposure to sunlight. Some organochlorine insecticides, such as DDT and aldrin, are virtually indestructible and this is why they have been withdrawn from use.

Most chemicals currently available to gardeners remain active against pests or diseases for less than two or three weeks after application, and the period of protection is often much less. It is important, therefore, to treat plants thoroughly, and to apply the chemical at the correct time or times in order to get the best results. Insecticides and fungicides used on fruits, vegetables and culinary herbs may require an interval between treatment and harvesting. For the chemicals currently available to amateur gardeners, this harvest interval varies from nothing to three weeks.

HARM TO OTHER CREATURES
Many chemicals, in particular some insecticides, have the potential to be harmful to creatures other than garden pests and other insects. If used with care, and precisely according to the manufacturer's instructions, there should be no danger. Product labels will state clearly where risks exist and it is important to take all possible efforts to keep any creatures which could be harmed out of the area where chemicals are being used. In most cases, once a spray has dried on the plant which has been treated, it no longer poses a threat to pets, children or wildlife.

APPLYING CHEMICALS

GARDEN CHEMICALS are a great help in controlling pest and disease problems. If used with care and only when needed they pose little or no threat to the environment. Changes in legislation and the increasing concerns about safety, both to the environment and the user, mean that the range of chemicals available is constantly being reviewed. All chemicals are also rigorously tested before being marketed.

When considering chemical control, always ascertain whether the situation really necessitates their use. Slight pest infestations or those which occur towards the end of the season may not pose a sufficient threat to the plant.

Wherever possible select a chemical which controls just the pest organism. Formulations which contain both a fungicide and an insecticide are only appropriate when both the specified pest and disease are present at a level that warrants control.

READING THE LABEL

Always consult the product label to determine precisely what the chemical can be used for and whether it is formulated for use against the problem and on the plant you have in mind.

Products which carry a label recommendation for use on edible crops may also specify a minimum interval which must elapse between the application of the chemical, and the consumption of the fruit or vegetable. Provided that this interval is observed, it should be perfectly safe to eat the crop.

The timing, application rate and frequency of application are also very important and again information about all of these is available on the label. Applying a chemical at the incorrect rate,

HEALTHY VEGETABLES
To make sure that vegetables and fruit are safe to eat after chemical treatment (cabbage being treated for whitefly, above), keep to the interval recommended on the product label.

frequency or time may prove ineffective, damaging or simply wasteful and so it is essential to follow the manufacturer's instructions carefully.

RULES FOR SAFETY

When applying chemicals there are a number of general rules which should be followed. Chemicals should be used only when necessary and must always be applied according to the instructions. Avoid contact with your eyes and skin, and also inhalation. Although there is no stated need for protective clothing, wearing rubber gloves when handling concentrates is advisable. Different products should not be mixed together unless recommended.

Keep children and animals away and never eat, drink or smoke while applying chemicals. Do not spray or dust in windy, hot, sunny or completely calm weather, to minimize the threat of chemical drift affecting adjacent plants. Spray in the late afternoon, early evening or early morning. Consider carefully any potential risks involved to children if using composts containing chemicals.

EQUIPMENT FOR APPLICATION

You may need to invest in some special equipment. This should all be labelled clearly and kept separately, so that it is used for no other purpose

---

HAZARD SYMBOLS

OXIDIZING AGENT

HARMFUL OR IRRITANT

Garden chemicals have been formulated so that their contents will not cause any harm, as long as they are used correctly. The symbols shown here are standard throughout the European Union and will be found on some garden products. The only garden chemical with the oxidising agent symbol is the weedkiller sodium chlorate. It indicates that there can be fire risk in its use.

MIXING AND APPLYING CHEMICAL SPRAYS

1 Measuring. Measure out the required amount of concentrate accurately. Avoid making up more than necessary and always read the instructions carefully. Wear rubber gloves and take care not to spill any of the chemical.

2 Mixing. Part-fill the sprayer with water and then carefully pour the concentrate in. Rinse out the measurer with water and add this to the sprayer. Top up to the required level with water and close all containers.

3 Applying. If possible spray both surfaces of the leaves. This is particularly important if contact action sprays are used. Spray at dusk to minimize risk to pollinating insects and to help prevent scorching.

(*see below*). Use completely different equipment again for the application of weedkillers, as even trace amounts of these can cause serious damage to garden plants. A small sprayer is suitable for most chemicals unless you have a very large garden. One that produces a fine spray is needed for controlling most pests and diseases. Insecticide or fungicide drenches or weedkillers can be applied using watering cans fitted with dribble bars.

It is important to clean all equipment regularly and, when rinsing it out after use, to spray all washings onto empty land or suitable plants. Never dispose of any dilute or concentrated chemicals by pouring them down a lavatory or drain, into watercourses, or onto land where they may contaminate underground water.

APPLYING CHEMICALS IN ENCLOSED SPACES

When chemicals such as sprays or dusts are used in enclosed spaces, such as greenhouses or conservatories, particular care is needed to avoid contact or inhalation. Temperatures are often higher and ventilation much lower than outside in the garden. Greenhouses or conservatories which open directly into the house pose further difficulties and wherever possible, doors or windows into the house should be closed tightly.

The risk of scorching the plants being treated is also greater, due both to the raised temperatures and to the magnifying effect that glass may have on the sun's rays. To minimize problems, timing of application, in order to avoid the hotter, brighter parts of the day, is particularly important.

STORING CHEMICALS

It is very important to store garden chemicals safely, since even small amounts can sometimes be harmful to people, animals and garden plants. Below are tips to help you store chemicals correctly. Rules for safe storage should also be applied to the equipment with which you apply chemicals. It is worth marking items such as watering cans clearly so that they are not confused with equipment for general garden use.

STORAGE TIPS
• Store chemicals in a safe place, preferably a locked cupboard inaccessible to children or animals.
• Keep chemicals in their original containers, with their tops tightly closed.
• Make sure all containers are clearly labelled and kept separately.
• Clear out cupboards annually, so that you do not accumulate old and obsolete products. To dispose of them, wrap them well and contact your local council. Check for and get rid of leaking products similarly.

SAFETY HAZARDS
• Do not transfer any chemical out of its original container into another one with an incorrect label or no label at all.
• Only buy sufficient chemicals for one to two years' use and do not store them for a long period of time. Chemicals become less effective once they have been opened and may go out of date.
• Do not keep surplus quantities of made-up solutions and sprays. Dispose of them and rinse out equipment directly you have finished treating plants.
• If you need to get rid of old chemicals, do not simply add them to household rubbish nor pour them down household drains. Dispose of them safely (*see left*) without removing them from their original containers.

GARDEN HEALTH AND PROBLEMS

CHEMICAL-FREE CONTROLS

RELYING SOLELY on chemicals to control pests and diseases can bring long-term problems in the form of pesticide resistance, harm to non-pest animals and consequent damage to the environment. Integrated control is a more environmentally friendly approach to tackling pests and diseases. It includes recognizing the role that natural enemies play in reducing pest numbers and devising complementary control measures that enable pest predators and parasites to survive. This may involve more careful monitoring of pest and disease infestations to determine whether chemical treatments are needed and, if so, when they should be applied to achieve the best effect. This approach can significantly reduce the number of spray applications given compared with a standard spray programme.

Good garden practices (*see pp.92–96*), such as removing badly infested or diseased plants, ensuring plants have adequate water and nutrients, growing plants and cultivars suitable for the local soil and climate conditions and selecting cultivars with pest or disease resistance where available, are all part of an integrated control approach to limiting pest and disease problems, as is the use of biological controls (*see pp.90–91*). Picking off pests

COMPANION PLANTING
Companion plants were once thought to deter pests from crops. It is more likely, however, that their bright flowers attract pest predators, initiating a form of biological control (see p.90).

A BEER TRAP FOR WASPS
Half-fill an old jar with beer and seal it with some paper and an elastic band. Make a small hole in the paper and, in their attempts to reach the beer, the wasps will drown.

by hand at an early stage can prevent many problems from escalating. Pests that are most active at night, such as slugs and snails, are best gathered up with the aid of a torch on damp evenings. On plants grown in or around garden ponds, you should always pick off any pests by hand, since it is dangerous to fish and other pond life to apply chemicals in or near water.

There is also a number of traps and barriers that either prevent pests reaching their intended goal, or draw them to a more tempting destination. Some, such as beer traps for slugs and wasps, eliminate the pests; others are monitors, giving an indication of when controls will be most effective.

PHEROMONE TRAPS
Pheromones are volatile chemicals produced by some insects as a means of communicating with others of the same species. They play an important role in maintaining the organization of social insects such as ants, bees and wasps. Pheromones are also used by some insects, usually females, as a means of attracting males for mating purposes.

The pheromones of some pest species, such as codling moth, plum moth and carnation tortrix, have been identified and the chemicals synthesized for use in pheromone traps, which are sold in kit form (*see facing page*). Once the trap is in position

HOW TO ASSEMBLE A PHEROMONE TRAP

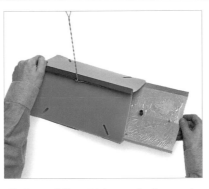

1 Kit contents. A self-assembly pheromone trap kit will include a fold-together, open-sided box with hanging wire, some sticky sheets and some synthetic pheromone lures.

2 Assembling. Make up the box and place a sticky sheet in the bottom of it. Put a pheromone lure on top of this sheet. The sheet and pheromone should be replaced after six weeks.

3 Hanging. Hang the trap by its wire on the tree, at about head height and within the central canopy. It should be put in position at the expected beginning of the moth's flight period.

male moths fly into it, expecting to find a virgin female, but instead get caught on the sticky sheet. The main function of these traps is to record the moths' main flight and egg-laying period accurately so that, if necessary, insecticides can be applied at the right time to control the newly-hatched larvae. On isolated apple or plum trees pheromone traps may catch sufficient male moths to reduce the mating success of females, and so limit the number of fertile eggs that are laid.

A STICKY TRAP
This trap is used to catch winged insects, who are attracted by the yellow colour of the strips. The strips are covered in non-drying glue and so trap the insects when they fly onto them.

GREASE BANDS AND STICKY TRAPS

Some insects cannot fly and it is possible to prevent them from gaining access to their host plants by placing a sticky barrier in their path. Grease bands are mainly applied to the trunks of fruit trees to trap winter moth and other related species that have wingless females. These emerge from pupae in the soil during autumn to spring and crawl up the trunk to lay eggs on the twigs. Grease bands applied to the trunk and tree stake in mid- to late autumn will reduce the number of eggs laid.

Other crawling insects that can be deterred by sticky or greasy barriers are earwigs, ants and adult vine weevils. A band around the base or the rim of containers can be very effective. Smearing grease around stems stops earwigs climbing up into flowers such as chrysanthemums and dahlias. Flea beetles can be made to jump on to a sticky trap: grease a piece of wood or card, and draw it, grease side down, along the tops of plants, being careful not to touch them with it, but agitating the foliage with your hand or a stick. Flea beetles jump when disturbed, and so will be caught on the trap.

Sticky traps are also useful in greenhouses for catching winged insects, such as whiteflies, thrips and fungus gnats. The best traps are made up of yellow plastic strips that are covered in non-drying glue. They are suspended above the plants and when insects alight they are unable to escape. Such traps may not give adequate control on their own, but they can give an early warning of the presence of pests. It is often easier to see whitefly on a yellow trap than to find them by searching on the underside of leaves. Early detection of an infestation allows the pest to be controlled, either by introducing biological controls or by using insecticides, before the plants are damaged.

GARDEN HEALTH AND PROBLEMS

OTHER TRAPS

Other chemical-free traps can be used to protect flowers from earwigs, and roots from wireworms. Grapefruit and beer traps can be used against slugs and snails. The earwig trap uses an inverted flower pot filled with dry grass and placed over a cane amongst susceptible plants. Earwigs feed at night and will crawl into the trap to shelter from the daylight. You can then remove the pot and pick out the earwigs. To limit wireworm damage, fix a potato on a stick and bury it: the wireworms will move in, preferring it to plant roots. The potato with its pest inhabitants can be regularly removed and renewed.

Hollowed-out grapefruit halves can be propped up on stones on the ground in flower-beds; slugs and snails are attracted to the smell, crawl in and remain there. The grapefruit skin can be picked up in the morning, complete with the slugs, which can be disposed of by dropping them in hot water. A jar half-full of beer or milk can be almost buried in the ground, where the smell will attract slugs and snails, which fall in and drown. The addition of some porridge oats makes it an even more effective bait. (For a beer trap for wasps, *see p.86.*)

BARRIERS

There are various barriers that can be constructed in the garden to deter pests and provide a non-chemical alternative to control. They range in size from fruit cages (*see below*) and large glass cloches to individual plastic cloches (*see facing page*).

To protect smaller fruit bushes, such as strawberries or currants, against bird and animal damage, use mesh netting to construct a low frame, making sure that the base is secured with bricks to prevent the pests getting in underneath. Apple and pear trusses can be individually enclosed in bags

EARWIG TRAP
Earwigs are a particular nuisance on chrysanthemums because they crawl to the top of stems and eat the buds and petals. This earwig trap is an easy and effective control.

made from muslin or old tights. Although time-consuming this can be effective and should be done when the fruit is almost ripe – the time it is most likely to be attacked by wasps or birds.

A variety of traps can be used in a greenhouse (*see p.79*); these include putting legs of tables or any other staging in saucers or pots of water. This will prevent insects climbing up onto the plants.

Some animals, such as hedgehogs and frogs, are highly beneficial in a garden (*see pp.60–63*) and should be encouraged. However, others, such as deer, rabbits and moles, can cause extensive damage. Fences can be used against rabbits and deer; wire netting buried at least 15cm below

FRUIT CAGE

PROTECTIVE FLEECE

PROTECTIVE COVERS
The fruit cage (far left) is made from wire or plastic netting supported by metal posts. This type of cage can help protect small fruit trees or bushes from bird and animal attack. The fleece (left) protects and insulates plants and seedlings, while allowing light and air to reach them. This tunnel is constructed by using strong wire hoops to support a sheet of horticultural fleece.

ground with 75cm above should deter rabbits. Deer will need a much taller barrier (*see p.123*). Mice (*see p.149*) and moles (*see p.150*) need traps.

HORTICULTURAL FLEECES

In recent years several types of horticultural fleece have become available. These are finely spun light-weight materials, usually white or green in colour and made of polypropylene or other plastics, that can be laid over developing vegetable crops. These fleeces give the young plants valuable protection against wind and cold, and also exclude pests such as pigeons, cabbage butterflies, carrot fly, cabbage root fly and flea beetles.

When using fleece to give protection against pests it is essential also to adopt a crop rotation system (*see pp.92–96*). If, for example, carrots are sown where carrots or parsnips were grown the previous year and then covered with a fleece, adult carrot flies may emerge from pupae buried in the soil and will then be trapped under the covering with the crop.

BRASSICA COLLARS

Transplants of brassicas and other related vegetables, such as swedes, are prone to attack by cabbage root fly. A ready-made collar, or one made using a piece of carpet, underlay or cardboard, is placed around the stem of the young plant. This prevents the female flies from laying their eggs in the soil; instead, they will lay them on the collar, where they will dry up before hatching.

HOT WATER TREATMENT

This form of pest control is mainly carried out by commercial growers, but can be attempted by gardeners. It is used against certain bulb pests, such

AN INDIVIDUAL CLOCHE
A cut-off plastic bottle placed over young plants in early spring makes an ideal slug deterrent and also acts as a mini-cloche, enabling soil around the roots to warm up more quickly.

as narcissus eelworm and bulb fly, and also against eelworms on some herbaceous plants such as chrysanthemum. The technique involves immersing dormant bulbs or plant stools in water held at a temperature that is high enough to kill the pest without harming the plant. For details of hot water treatments, see chrysanthemum eelworm (*p.118*) and narcissus eelworm (*p.151*).

Care must be taken to ensure that the temperature is kept constant throughout the period of treatment, since too much heat damages the plants while insufficient heat allows the pest to survive. After treatment, bulbs are planted out in fresh soil, while herbaceous plants are potted up to produce cuttings.

REPELLENTS

Repellents are used principally against birds and mammals, where the aim is to deter feeding or encourage them to move elsewhere, rather than to cause them any harm.

Humming tapes are used to protect plants against pigeons and other birds, while devices that produce ultrasonic sounds are available for frightening away moles and cats. Other means of keeping pets and pest animals away from plants rely on using irritant substances, such as pepper dust, or compounds which give plants unpleasant tastes or smells. Repellents may need frequent replenishment to maintain the effect. It is also advisable to change the type of repellent in use to avoid the target animals becoming familiar with the sound, smell or taste and losing their fear of it.

Collar about 10cm in diameter

Brassica transplant

BRASSICA COLLAR
Prevent female cabbage root flies from laying their eggs in the soil around the stems of vulnerable young brassicas by placing a disc or collar around the stem base.

BIOLOGICAL CONTROLS

THIS APPROACH to control involves enlisting the help of pests' natural enemies – predators, parasites and fungal, bacterial or viral diseases. These natural enemies limit pest populations, but are not always successful in stopping them from damaging plants. This is often because pests breed more rapidly in the spring and early summer so that heavy infestations are already present by the time beneficial insects become numerous.

Biological control occurs naturally in gardens; ladybirds eat aphids, for example, and thrushes attack snails. Today, however, the term "biological control" is more commonly taken to refer to the use of natural enemies that are introduced under suitable conditions to give effective control. This approach has more success in greenhouses where specific predators and parasites can be purchased for use against most of the major pests. There are few successful biological controls for plant diseases, but in some parts of the world insects, mites and fungal diseases are being used to control alien plants that have become weeds.

BENEFITS OF BIOLOGICAL CONTROL

Biological controls may often be more suitable to use than pesticides. When used on food plants, such as tomato and cucumber, for example, there is

Thrush needs to find a stone to break snail's shell

USEFUL BIRD
While a few birds are a nuisance in the garden, the song thrush has a useful role, feeding on snails. Conveniently placed flat stones in or near beds and borders provide them with anvils on which to smash the shells.

no problem with pesticide residues on the edible parts. Similarly, the risk of chemicals causing spray damage is avoided.

Predators and parasites can seek out pests living in places that are difficult to reach with pesticides. Once released, natural enemies can be left to get on with controlling the pests; chemical controls often require repeat applications, particularly during the summer when high temperatures allow pests to breed rapidly. Biological controls are selective, controlling particular pests, with no harmful effects on non-target species. Some pests, such as glasshouse whitefly and red spider mite, may acquire resistance to pesticides, making biological control the only effective treatment.

ENCOURAGING PEST PREDATORS

The mixture of flowers and other plants found in a typical garden is helpful in providing, directly and indirectly, food and shelter for a wide range of beneficial insects, birds and pest-eating mammals such as bats, shrews and hedgehogs. Installing a pond can also help as it provides a breeding site for frogs and toads. In the autumn many of these pest predators will seek out sheltered places where they can survive the winter. A thorough tidying-up of the garden at the end of the summer may be pleasing to the eye, but could result in the loss of many potential allies in the battle against pests.

The choice of chemicals used on garden plants can also have a significant effect on natural populations of pest predators. Some pesticides are more selective than others for pests and their use limits the harm caused to non-pest animals. Organic sprays, for example, give good control of aphids, but, because of their short persistence, have limited effect against most beneficial insects and mites.

A COLOURFUL GARDEN
Bright flowers, particularly in shades of purple and yellow, will attract many types of beneficial insects, such as hoverflies, whose larvae feed on greenfly and other aphids.

BUYING AND USING BIOLOGICAL CONTROLS

The predators, parasites and pathogenic agents that are commercially available are intended for use against specific pests. It is therefore essential that pests are accurately identified. The various biological controls are compatible with each other and can be used in combination; however, they must not be used together with any chemical control, or immediately afterwards. Predators and parasites are very susceptible to most insecticides and must be regarded as an alternative to chemical control rather than an additional treatment. The instructions for use should give a safe interval which must elapse between the application of chemicals and of the biological control.

All of the biological controls are living animals, which should be released onto the infested plants, or the soil or potting compost, as soon as they have been obtained. They can sometimes be bought from garden centres, but more usually are sold by mail-order firms who advertise in gardening magazines and can generally be contacted through their websites.

APPLYING BIOLOGICAL CONTROLS

Pathogenic nematodes are watered into the soil or potting compost. The nematodes must have moist, well-drained soil to enable them to survive and move between the soil particles. The vine weevil nematode and the slug nematode can work in a soil temperature of 5°C (41°F) upwards, but nematodes used against chafer grubs and leatherjackets need soil temperatures of 14–21°C (57–70°F). It is best

PREDATORY MIDGE
Aphids can be among the most damaging and common of pests. The larvae of a tiny fly, Aphidoletes aphidomyza, *though only up to 3mm long, will kill and eat up to 80 aphids each.*

to apply nematodes in the evening to avoid rapid drying out of the soil surface. The other predators and parasites that are available to gardeners (*see the table below*) need time to breed before they are numerous enough to overcome pests. It is therefore necessary to introduce these biological controls before pest infestations become heavy. Although they can survive lower temperatures, these predators and parasites need warm daytime temperatures, usually at least 21°C (70°F), and good light, if they are to be effective. Because of this their period of use is likely to be between early spring and mid-autumn. The bacterium *Bacillus thuringiensis* is no longer available to amateur gardeners for caterpillar control.

TYPES OF BIOLOGICAL CONTROL

Pest	Name of biological control	Type of predator or parasite	How it works
APHIDS	Aphidoletes aphidomyza	Predatory midge	Larvae eat aphids.
	Aphidius species	Parasitic wasps	Larvae develop inside the aphids' bodies.
CHAFER GRUBS	Heterorhabditis megidis	Pathogenic nematode	Infects chafer grubs with a fatal disease.
FUNGUS GNATS (SCIARID FLIES)	Hypoaspis miles	Predatory mite	Feeds on the flies' larvae.
GLASSHOUSE RED SPIDER MITE	Phytoseiulus persimilis	Predatory mite	Eats the eggs, nymphs and adult spider mites.
GLASSHOUSE WHITEFLY	Encarsia formosa	Parasitic wasp	Attacks the pest's scale-like nymphal stages.
LEATHERJACKETS	Steinernema feltiae	Pathogenic nematode	Infects leatherjackets with a fatal disease.
MEALYBUGS	Cryptolaemus montrouzieri	Predatory ladybird beetle	Both the adult and larval ladybirds eat the eggs and active stages in the mealybug's life-cycle.
SLUGS	Phasmarhabditis hermaphrodita	Pathogenic nematode	Infects slugs with a fatal disease.
VINE WEEVIL	Steinernema kraussei	Pathogenic nematode	Infects vine weevil grubs with a fatal disease.
WESTERN FLOWER THRIPS	Amblyseius species	Predatory mites	Eat the immature thrip nymphs.

G A R D E N   H E A L T H   A N D   P R O B L E M S

PREVENTIVE MEASURES

WELL GROWN, vigorous plants are not immune to pests and diseases, but they are less likely to be badly damaged. Attention should be paid to selecting plants and cultivars that are likely to do well under the local soil and climatic conditions. It is worthwhile paying more for good quality seeds, bulbs and plants from reputable suppliers. Cheap lots from dubious sources may be in poor condition, or carrying pests and diseases that may be difficult to eliminate once introduced into a garden or greenhouse.

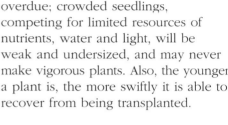

RESISTANT CULTIVAR
This calabrese, 'Trixie', has been bred for its resistance to clubroot, a devastating soil-borne disease.

RESISTANT CULTIVARS

Many cultivars have been developed in recent years to withstand common strains of fungal or bacterial disease, or susceptibility to pest attack. These include some fruits and vegetables that virtually eliminate the risk of several serious problems in the kitchen garden. Resistant cultivars suggested in the *A–Z of Pests, Diseases and Disorders, pp.97–194*, are to be recommended for general garden use. It is always advisable to check seed and plant catalogues regularly as the availability of resistant cultivars changes.

CARE OF YOUNG PLANTS

Take care to ensure that planting or sowing is done properly; sowing too early or late (when soil may be cold and wet), or planting pot-grown plants too deeply or with a pot-bound root system, will result in plants that will always struggle and fail to make good growth. When growing your own plants, ensure that you raise strong, healthy specimens by following procedures correctly. It is important, for example, to thin, prick out and pot on seedlings (*see facing page*) as soon as necessary, rather than neglecting them until the task is well overdue; crowded seedlings, competing for limited resources of nutrients, water and light, will be weak and undersized, and may never make vigorous plants. Also, the younger a plant is, the more swiftly it is able to recover from being transplanted.

MAINTAINING HEALTHY PLANTS

Plant problems can arise at any time of the year but are more prevalent during the spring to autumn season, their peak growing period. It is a good idea to make regular tours of inspection in order to detect problems before they become serious. It is important to look for plants which are under-performing; this may show as slow growth, often with symptoms of discoloration, distortion or damage that should be investigated as soon as possible.

Plants showing signs of attack by pests or diseases should not be ignored, since it is likely that other plants may also become affected. Treatment with a suitable chemical is most often completely effective when the problem is detected and dealt with early on. In a group of plants, removal of the

CHOOSING HEALTHY PLANTS AND BULBS

Vigorous growth

SELECTING HEALTHY BEDDING PLANTS
Choose plants that are free of any sign of pests or disease. Look for compact growth, with a good leaf colour and a well-developed root system. Avoid those with yellowing leaves, weak growth or premature flowering.

BUYING BULBS
Since there is a good deal of variety in the quality and size of bulbs available to the gardener, they should be closely examined before buying. Bulbs should be as fresh and firm as possible (see bottom right*) and without signs of mould or discoloured patches. The bulb shown top right has both diseased tissue and damaged outer scales, so it would not be a good choice. Undersized bulbs may not flower in their first year, so always try to choose the largest bulbs available of the chosen cultivar.*

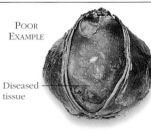

POOR EXAMPLE

Diseased tissue

GOOD EXAMPLE

GOOD PLANTING

1 Sowing. Sowing into a seed tray is a good way of raising large numbers of seedlings, but they will soon need more space to allow them to develop. Over-crowding will restrict their growth.

2 Pricking out. Select the stronger seedlings and carefully remove them from the tray. Handle them by gently holding the seed leaves rather than the stem, which is easily bruised.

3 Transplanting. Space the seedlings out and plant out or put into pots once they are well established. Avoid damping off by using fresh compost, rather than garden soil, and tap water.

affected one helps stop the problem spreading. As plants grow and develop, thin out any which have become overcrowded. This improves their vigour and allows air to circulate around stems and leaves, thus reducing damp air conditions that favour germination and infection by fungal spores.

FEEDING PLANTS

Most plants will benefit from supplementary feeding, to increase yields of fruits and vegetables and to increase the quality and quantity of flowers. Organic fertilizers include garden compost, farmyard manure, bone meal, dried blood and seaweed extracts. These all provide a variety of nutrients, although not necessarily in the right balance required for optimum plant growth.

Man-made fertilizers can be "single nutrient" compounds, such as super phosphate (supplying phosphorus) or ammonium sulphate (nitrogen), but more widely used are general fertilizers which provide all the major nutrients. These are available as granules for adding to the soil, or in soluble form for watering in or spraying on the leaves (foliar feeding). The slow-release fertilizer pellets now available will give container-grown plants the nutrients they need throughout the growing season.

Feeding is generally done during the spring and early summer, when plants are making rapid growth. Fertilizers applied in autumn or winter may be largely wasted; dormant plants do not need feeding, and nutrients added then may be washed away by rain before spring.

WATERING

Watering is always crucial in a garden, particularly at critical stages in a plant's development, such as the seedling stage and when flowers and fruits are being produced. In dry periods, it is important to water before plants show signs of drought stress, and also to water at regular intervals.

Watering is best done in the evening so that less water is lost by evaporation, and water should be directed into the soil, around the roots, not on the plant itself. Do not water plants in sun as the plant can be scorched. Covering the soil surface with a mulch will help retain moisture. Rain butts can be used to collect water from the roof. This source of water is suitable for most plants but is best avoided for seedlings and cuttings, as it can harbour fungal diseases that cause damping off and stem rots.

SUPPORTING CLIMBING PLANTS
Plants that cannot support their stems need tying in to keep them off the ground, where growth will be vulnerable to insects and slugs. Even self-clinging climbers need help when young.

DRAINAGE

Waterlogged soil can be just as damaging as dry earth to plants. A heavy soil, such as clay, has only small spaces between the soil particles, making it moisture-retentive and difficult for water to drain through. Improve heavy soil texture by digging in plenty of coarse organic matter, such as compost, leaf mould or manure, to improve aeration.

Thin soils over chalk are very alkaline. Water drains through easily, as do soil nutrients, so that acidic types of organic matter such as manures are needed to give bulk and counteract the alkalinity. Sandy soil, too, is very light, needing plenty of mulching and feeding, but it is easy to work, and warms up quickly in spring.

WEEDING

Good weed control is important. Weeds compete with cultivated plants and if they gain dominance, make them weak and more vulnerable to problems. Most weeds can be removed by hand or by hoeing, but in lawns and garden paths using a suitable chemical weedkiller may be more convenient. Also, like cultivated plants, weeds may attract common pests, viruses and fungal diseases, and are a source of reinfestation after the cultivated plants have been sprayed. They also help pests and diseases survive in the absence of cultivated host plants and thus diminish the value of crop rotation.

REMOVING DEAD PLANT MATERIAL
Unless features such as winter seed heads are required, dead or dying flowers and foliage, together with their stalks or stems, should be removed, as they can become sites for rots and fungal growth, which may spread to living plant material.

REMOVING SUCKERS TO KEEP PLANTS STRONG
Rose suckers grow from the roots of the plant, since roses are grafted onto a rootstock. Gently clear away soil round the sucker to its union with the root, and cut or pull it off.

REMOVING OTHER UNWANTED GROWTH

To keep plants strong, suckering growth from the roots needs removing, particularly on grafted plants such as roses and most fruit trees. These are usually grafted onto rootstocks which will produce sucker growth of inferior quality to, but usually of greater vigour than, the named cultivar. If possible, the suckers should be traced back to the roots from which they arise and cut off at source. Cutting them off at soil level simply encourages more suckers. Avoid digging the soil around roses, cherries and other plants prone to suckering as damage to the roots can stimulate the growth of suckers.

DEAD, DISEASED AND DAMAGED GROWTH

Although the routine pruning of plants that require it is usually best carried out at a particular time of year, it is, with very few exceptions, safer to remove any dead, diseased or damaged growth from a plant as soon as it is noticed, rather than waiting for the usual pruning time. Branches that have been broken by storms or other causes should be tidied up with a clean cut to speed callusing. Branches infected with fungal diseases, such as coral spot, canker or silver leaf, should be cut out well below the signs of infection, to prevent the further spread of these diseases.

ROUTINE PRUNING

Not all plants require pruning but many woody plants do benefit from some form of thinning or reduction in size. This may be done on an annual basis – as with apple trees, in order to promote the development of spurs that will produce flower buds and hence fruits. Other plants may need infrequent pruning, with the intention of doing no more than

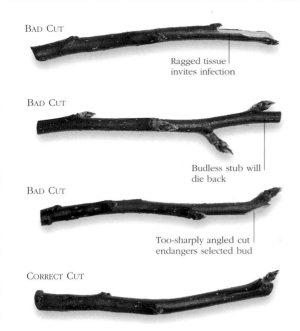

BAD CUT

Ragged tissue
invites infection

BAD CUT

Budless stub will
die back

BAD CUT

Too-sharply angled cut
endangers selected bud

CORRECT CUT

GOOD AND BAD PRUNING CUTS
*Make sharp pruning cuts close to a strong bud, with the cut
sloping away from the bud. Avoid making ragged cuts or
leaving stubs above the bud. Use a sharp knife or secateurs.*

removing the occasional branch or shoot that is
growing too far in the wrong direction. Always
prune to correct situations where branches cross
each other, since in chafing together they may
create wounds that are open to infection.

Pruning also provides an opportunity to thin
shrubs and trees with a dense system of branches
to provide a more open centre. This improves
growth and the development of flowers and fruits
by allowing sunshine in. There is also a better flow
of air amongst the foliage and branches which will

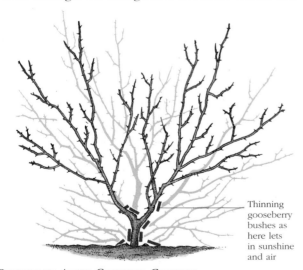

Thinning
gooseberry
bushes as
here lets
in sunshine
and air

PRUNING TO AVOID CROWDED GROWTH
*Cut out some of the older stems in order to encourage new
growth. Thin the stems if overcrowded to create a more open
centre. This will help reduce the incidence of diseases.*

help deter the germination of fungal spores. With
flowering shrubs the removal of about one-third of
the older stems each year will revitalize the plants
by stimulating the development of young shoots.

MAKING CORRECT CUTS

Bad pruning can cause problems, rather than curing
or preventing them, so it is important to make cuts
correctly with clean, sharp tools. Invest in good
quality secateurs to prune stems up to the thickness
of a pencil; anything larger is better tackled with
loppers or a pruning saw. Shoots should be cut
cleanly back to a strong bud with a slanting cut
(*see left*). Never leave a stump of stem above the
bud as this will encourage dieback.

Do not cut a branch off absolutely flush with the
trunk or branch from which it arises. There will be
a slight thickening or collar at the base of the
branch and the cut should be made just outside this
point. If the collar is also removed, the growth of
scar tissue over the pruning wound is impaired and
can result in a large exposed area of inner wood.

It is sometimes necessary to remove large
branches from trees. Before doing so, it is generally
wise to seek the advice
of a professional
arboriculturist; major
work such as this can
result in an unstable or
unbalanced tree.

Large branches are
heavy, and it is often
advisable for safety
reasons to remove them
in smaller pieces instead
of making a single cut.
This also reduces the
danger of the branch
dropping as it is being
cut and tearing a strip of
bark from the trunk. This

CROSSING BRANCHES
*Crossing branches rub
against each other, wearing
away the bark and creating
entry points for disease.*

risk can be further reduced by initially making a cut
on the underside of the branch before completing
the task by sawing down from the top.

WOUND PAINTS

The use of proprietary wound paints on large
pruning cuts is less fashionable today than in the
past. There is some evidence to show that they
can restrict the natural healing process that occurs
around correctly-made pruning cuts. However,
where trees and shrubs are known to be
susceptible to fresh wound-infecting diseases,
such as silver leaf or coral spot, it may be helpful
to use a wound paint.

COMPOST HEAP
Some types of infested or diseased material must never be composted, but many pathogens may be destroyed by the heat generated in the centre of the heap. Never leave rotten apples on top of compost like this, but bury them deep within the heap.

RAKING UP OLD LEAVES
Dead leaves harbour slugs and encourage grey mould (Botrytis) and other diseases. Keep the garden tidy by collecting plant debris and putting it on the compost heap.

DISPOSAL OF DEBRIS AND PRUNINGS

At the end of the growing or cropping season, tidy up plants and rake up fallen leaves to remove hiding-places for pests. Composting garden debris is an excellent way of returning nutrients and organic matter to the soil, but there are some basic rules that must be followed when composting unhealthy plant material. Most pests and diseases cannot survive in the heat generated by compost heaps, but it is advisable to cover infested plants with compost or with a layer of grass clippings or weeds. Plants with root pests or diseases should never be put onto compost heaps but burned or bagged up for removal with the domestic rubbish, as should prunings that come from plants affected by canker, silver leaf and coral spot.

REPLANTING AND CROP ROTATION

Growing the same or closely-related plants in the same piece of ground in successive years can lead to a build-up of problems. Vegetables, bedding plants and bulbs are, however, particularly prone to some pests and diseases, and changing the growing site is not guaranteed to protect them. Many insects have winged adults which will have little difficulty in locating their host plant's new position. Similarly, many diseases are spread by wind-blown spores. There are, however, some less mobile problems, such as eelworms and root rots, which can be avoided by adopting a crop rotation system.

There are other good reasons for changing the growing sites of plants. Roses and apples suffer from replant disorders. If replacement plants are put in the same place from which an apple or rose has recently been removed, it is likely that the new plant will fail to thrive. Micro-organisms associated with the roots of the old plant seem to have an antagonistic effect on the new plant. Crop rotation in the vegetable garden also helps make better use of the available nutrients, as plants differ in their nutritional requirements. When devising a crop rotation system it is advisable to group related plants together in order to get maximum advantage when next year the planting sites are changed round. A simple rotation for the vegetable garden involves a three-year cycle; at the end of this time each of the three groups will have been grown for one year in a different area.

SIMPLE CROP ROTATION

GROUP 1	GROUP 2	GROUP 3
Cabbage	Potato	Beans
Brussels sprouts	Tomato	Peas
Cauliflower	Carrot	Onions/Shallots
Kale	Parsnip	Leeks
Broccoli	Celery	Lettuce
Turnip	Parsley	Sweetcorn
Swede	Beetroot	Marrow
Radish	Spinach beet	Cucumber

FIRST YEAR	SECOND YEAR	THIRD YEAR
GROUP 1	GROUP 2	GROUP 3
GROUP 2	GROUP 3	GROUP 1
GROUP 3	GROUP 1	GROUP 2

A BED OF GROUP 3 CROPS, INCLUDING PEAS AND BEANS

A–Z of Pests, Diseases and Disorders

This dictionary section contains entries for all the most common pests, diseases and disorders likely to be found in the garden. Some occur in varying forms on many plants, and have a general entry with cross-references. Many others are largely host-specific, and will be found under plant names. Suggested controls and preventive measures give a choice between chemical and non-chemical methods wherever appropriate.

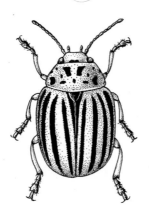

ACER GALL MITES

SYMPTOMS In early summer, small, raised, reddish pimples appear on the upper leaf surfaces of maples and sycamores (*see p.17*), or a felt like growth of yellowish white or purplish-red hairs develops on leaf undersides.
CAUSE Microscopic mites that suck sap and secrete chemicals into the foliage that induce the abnormal growths. *Aceria cephaloneus* causes pimple galls and *Aceria eriobius* is responsible for felt galls.
CONTROL None of the chemicals available to gardeners is effective, but no control is needed, because apart from creating the galls, these mites have no harmful effect on their hosts. On small plants, light infestations may be controlled by removing affected leaves.

ACORN GALL WASP

SYMPTOMS In late summer, oak acorns develop as sticky, ridged, yellowish-green structures (*see p.52*) instead of the familiar nut and cup. Later, the acorn (or knopper) galls become brown and woody and drop from the tree. In some years nearly all acorns on an oak tree are affected.
CAUSE A gall wasp, *Andricus quercuscalicis*, lays single eggs in developing acorns in early summer. The egg hatches into a small, white, legless grub which secretes chemicals that cause the developing acorn to be converted into a gall. The grub pupates inside the gall, and the adult wasp emerges

the following year.
CONTROL No control is available; the tree itself is unharmed by gall wasps.

ADELGIDS

Only conifers, particularly pines (*Pinus*), spruce (*Picea*), larch (*Larix*), Douglas fir (*Pseudotsuga menziesii*) and silver fir (*Abies*), are attacked by adelgids. These black aphid-like insects feed by sucking sap from the foliage, stems and bark. Some species, particularly on spruce, cause galls at the shoot tips which they inhabit, but on other conifers they move about, although invariably covered by a fluffy, white waxy substance secreted from their bodies. Heavy infestations may cause some yellowing of the foliage, but generally the conifers will continue to make satisfactory growth. Control measures are limited, but are detailed in individual entries.
• *see also* DOUGLAS FIR ADELGID *p.124;* LARCH ADELGID *p.143;* PINE ADELGID *p.161;* PINEAPPLE GALL ADELGID *p.162.*

ALDER SUCKER

SYMPTOMS Between mid-spring and early summer a fluffy, white waxy substance covering small, greenish-black insects appears on the foliage and shoot tips of alder trees (*see p.32*).
CAUSE The woolly wax is secreted from the bodies of the immature nymphs of *Psylla alni*. The adult stage is a winged aphid-like insect that does not produce this substance. Both nymphs and adults are sap-feeders but cause little harm.
CONTROL Usually unnecessary. However, if the infestation is causing concern small trees

could be sprayed with bifenthrin or pyrethrum.

ALGAE

SYMPTOMS Powdery deposits develop on the plant's surface; these are commonly green or grey-green and can be rubbed off, leaving no discoloration on the plant tissue beneath. The plant's growth and development are not affected.
CAUSE Harmless microscopic, non-flowering plants which develop in films of moisture on plant surfaces. They are particularly common on plants growing in damp situations. Leaves and trunks or stems are most commonly affected, especially those on the north-facing side of the plant.
CONTROL No control is needed as the plant is not damaged. If deemed to be unsightly, algal deposits can usually be washed off with a sponge and soapy water. Improve air circulation, and reduce dampness to prevent regrowth.

ALLIUM LEAF MINER

SYMPTOMS White legless maggots up to 10mm long tunnel through the leaves and stems of leeks, onions, shallots and other related plants during spring and autumn. The dark brown barrel-shaped pupae are 2–3mm long and can be found by peeling back the outer leaves of leeks in summer and winter. The adult flies feed on sap from the leaves and cause a row of white dots along the leaf.
CAUSE Maggots of a leaf-mining fly, *Phytomyza gymnostoma*, which has two generations a year. This pest was first detected in Britain

in 2003 and is spreading in the Midlands region.
CONTROL There is no pesticide available for use on onions, leeks, and other edible host plants. The female flies can be prevented from laying eggs by practising crop rotation and covering new crops with horticultural fleece in spring and autumn. Crops sown or planted after April will avoid the first generation of flies.
• *see also* LEEK MOTH *p.145.*

ALTERNARIA LEAF SPOT

SYMPTOMS Grey or brown spots appear on affected foliage. These frequently show concentric rings of tissue dieback with central, raised pycnidia, or fruiting bodies, of the fungus. In extreme conditions the spots may join together, causing extensive areas of leaf to be killed. In the majority of cases, however, the damage remains as distinct near-circular spots. Areas of pale or variegated foliage are likely to be the most readily affected, when the leaf may wither and die back. Plants commonly affected include potatoes, carrots and brassicas, and also cinerarias.
CAUSE Various species of the fungus *Alternaria*. The spores are air-borne or carried by rain or water splash.
CONTROL Remove and dispose of affected leaves as soon as they appear. This should help to limit spread.
• *see also* BRASSICA DARK LEAF SPOT *p.111;* POTATO EARLY BLIGHT *p.165.*

AMERICAN GOOSEBERRY MILDEW

SYMPTOMS Whitish-grey powdery fungal growth

appears on the upper surface of gooseberry foliage (*see* **p.25**), stems and fruits (*see* **p.49**). Severely affected leaves, especially the young growth towards the shoot tips, may become distorted and die off. Shoot tips may also die back. The mildew on the skin of affected berries turns pale brown or buff in colour as it ages, and can be scraped off. Infected fruit is edible, but unsightly, turning brown when cooked. Blackcurrants may also be affected.
CAUSE The fungus *Sphaerotheca mors-uvae*, which is encouraged by stagnant air around the branches and by excessive use of high-nitrogen fertilizers. It overwinters on the branches and within the buds.
CONTROL Prune out any branches that have been affected as soon as they are seen; also prune to thin growth and improve air circulation within the crown of the bush. Avoid excessive use of nitrogenous fertilizers; a general fertilizer is usually the preferable choice. Consider growing cultivars that show resistance to this infection, for example 'Invicta' or 'Greenfinch'.

ANEMONE RUST
See PLUM RUST *p.163*.

ANEMONE SMUT
SYMPTOMS Anemone stems and leaves show dark, raised blisters, which burst to release dark brown to black spores. The plant may show some dieback, but is rarely killed. Anemone smut may also affect globe flowers (*Trollius*).
CAUSE The fungus *Urocystis anemones*, which can overwinter in infected plant

debris and on other plant material on the soil surface. It is easily spread by rain splash.
CONTROL Remove affected parts promptly, before the blisters erupt. Destroy severely affected plants and choose a fresh site for growing new anemones or globe flowers.

ANTHRACNOSE
Various fungi cause patches or spots of discoloration to appear on infected plant tissue. Leaf, stem or, in beans, pod tissue may be killed, and in severe cases the whole plant may die back. Precise symptoms vary with the host plant and the particular fungus involved.
• *see also* BEAN ANTHRACNOSE *p.105;* PLANE ANTHRACNOSE *p.162;* WILLOW ANTHRACNOSE *p.192*.

ANTIRRHINUM RUST
SYMPTOMS Concentric rings or circles of pustules full of dark brown spores appear on the underside of snapdragon (*Antirrhinum*) leaves (*see* **p.15**). A chlorotic (yellow) "halo" develops around each patch of spores, and the upper leaf surface may also become yellowed. Affected leaves die, and if left unchecked, the infection may spread to stems and flower buds. Severely infected plants fail to flower and may be killed.
CAUSE The fungus *Puccinia antirrhini*, which may be seed-borne and which can overwinter on snap-dragons if provided with sufficient warmth to survive the winter.
CONTROL Remove and dispose of severely infected plants promptly. Otherwise, spray with a fungicide

containing penconazole or myclobutanil. Remove all snapdragons at the end of the season and do not grow them on that site for at least two years. Try growing cultivars which show a degree of resistance to the infection, such as the Tahiti series, 'Monarch', and 'Royal Carpet Mixed'.

ANTS
Several species of ants occur in gardens, including the black ant, *Lasius niger*, the yellow meadow ant, *Lasius flavus,* and red ants of various *Myrmica* species. They are often seen on plants infested with aphids and other sap-feeding insects. Ants collect the sugary honeydew excreted by these insects and may protect aphids by driving off predators. Fine heaps of soil appear on lawns (*see* **p.59**) and in flower-beds, where soil excavated from the underground nests the worker ants build for their queen ants is deposited on the surface. This can interfere with mowing and partly bury low-growing plants.
CONTROL Ants can be reduced in numbers but they cannot be eliminated from gardens. They are generally a nuisance rather than a significant pest. Many queen ants that fly into gardens are in fact killed and eaten by ants from existing nests before they can establish new colonies. Rigorous attempts to eliminate nests may increase the survival rate of queens flying in from nearby gardens, and lead to even more nests in the near future. Control measures should be confined to nests which are causing real problems. In gardens, treat nest sites with dusts

containing bendiocarb, permethrin or pyrethrum, or spray with cypermethrin, deltamethrin, or plant oils. Ants in buildings can be killed with baits containing borax, bendiocarb, or fipronil, or nest entrances and ant pathways can be sprayed with aerosol lacquers that leave persistent deposits of permethrin + tetramethrin or deltamethrin.

APHIDS
Many species of aphid occur in gardens and few plants are never attacked. The more important aphid pests are described in more detail in individual entries. Generally, affected plants show reduced growth with varying degrees of leaf distortion. The upper leaf surface becomes sticky with the sugary honeydew excreted by aphids; a black sooty mould will grow on the honeydew, especially under damp conditions. Aphids may be green, pink, yellow, black, greyish-white or brown, and are usually about 2mm in length, although some conifer aphids can be up to 6mm long. All nymphs shed their outer skins several times as they grow larger. These whitish cast skins accumulate on the upper surface of leaves below those where the aphids are feeding. Some virus diseases, especially of strawberries and raspberries, are spread on the mouth-parts of aphids as they move from one plant to another.
CONTROL Control on specific plants may be detailed in individual entries (*see below*). Use acetamiprid, thiacloprid, thiamethoxam, or bifenthrin on plants in active growth. Organic insecticides include pyrethrum, fatty acids and plant oils.

• *see also* BEECH WOOLLY
APHID *p.106*; BLACK BEAN APHID
p.107; BLACKFLY *p.108*; CEDAR
APHID *p.116*; CHERRY BLACKFLY
p.117; CONIFER APHIDS *p.120*;
CURRANT BLISTER APHID *p.121*;
CYPRESS APHID *p.122*; GREEN
SPRUCE APHID *p.135*;
HONEYSUCKLE APHID *p.139*;
JUNIPER APHID *p.142*; LETTUCE
ROOT APHID *p.146*; LUPIN APHID
p.147; MEALY CABBAGE APHID
p.148; MEALY PLUM APHID *p.148*;
MELON COTTON APHID *p.148*;
PEACH APHID *p.157*; PEAR
BEDSTRAW APHID *p.158*; PLUM
LEAF-CURLING APHID *p.162*; ROOT
APHIDS *p.171*; ROSE APHIDS
p.172; ROSY APPLE APHID *p.174*;
WATER LILY APHID *p.190*; WILLOW
BARK APHID *p.192;* WOOLLY
APHID *p.194*.

APPLE AND PEAR CANKER

SYMPTOMS Areas of bark
sink inwards (*see **p.36***),
usually starting near a
wound or bud. The bark
becomes discoloured and
then shrinks and cracks,
forming concentric rings
of flaky bark. The branch
may become swollen
around the cankered area.
As the canker enlarges, the
shoot may be girdled and so
foliage and growth above it
starts to die back. Stems and
trunks may be attacked and
very occasionally semi-
mature fruits are infected
and show rotting. In summer
raised, white fungal pustules
develop on the canker, and
in winter raised, red fruiting
bodies develop. Apples and
pears are frequently affected;
poplars (*Populus*), beeches
(*Fagus*), willows (*Salix*),
hawthorn (*Crataegus*) and
Sorbus may also succumb.
CAUSE The fungus *Nectria
galligena*, which is spread
mainly during the spring as
wind-borne spores enter
through wounded areas such
as leaf scars, pruning cuts,
scab infection, frost crack,
attack by woolly aphid.
CONTROL Prune out whole
spurs or branches where
possible. On larger branch
or trunk infections, carefully
pare away all infected and
damaged bark and wood,
cutting back to completely
clean tissue, and treat the
wound with a proprietary
wound paint. Dispose of
prunings and parings
carefully. Spray affected
apple trees with Bordeaux
mixture to help suppress
canker. Improve general
growing conditions, in
particular soil drainage, since
unsuitable conditions such
as wet soil may aggravate
the disease. Avoid growing
apples known to be
particularly susceptible, such
as 'Cox's Orange Pippin',
'James Grieve', 'Worcester
Pearmain' and 'Spartan'.
The following cultivars
seem to show a degree
of resistance under most
circumstances: 'Laxton's
Superb', 'Newton Wonder',
'Bramley's Seedling', and
'Lane's Prince Albert'.

APPLE BITTER PIT

SYMPTOMS The skin of the
apple fruit may develop
sunken brown spots, each
less than 1mm in diameter
(*see **p.48***). The flesh is
spoiled by numerous pale
brown spots or freckles.
Affected apples may have
an unpleasant, slightly bitter
taste. Large fruits are
especially susceptible, as are
fruits produced by heavy-
cropping trees. Symptoms
may appear while the fruits
are on the tree, but often
only develop during storage.
CAUSE Calcium deficiency in
the fruit. This results in the
collapse of small groups of
cells within the flesh, so that
brown flecks develop.
Calcium levels in the soil
may be perfectly adequate,
but in dry conditions, the
tree cannot take up the
amount it needs. Large
fruits and heavy-cropping
trees are more prone to this
disorder because of their
greater demand for calcium.
Too high a concentration
of magnesium or calcium
in fruits may also be a
cause of bitter pit.
CONTROL Keep apple trees
well watered and mulched
to maintain soil moisture
levels. Maintain even growth
by feeding with a balanced
fertilizer; avoid excessive
use of high-nitrogen feeds.
Spray developing fruits
with calcium nitrate solution
from early summer until
early autumn. The cultivars
'Bramley's Seedling',
'Discovery' and 'Crispin'
may be damaged by this
chemical, so care should be
taken when treating them.

APPLE BLOSSOM WEEVIL

SYMPTOMS Apple flower
buds start to develop but
the petals fail to open and
turn brown (*see **p.46***). If
these "capped" blossoms
are pulled apart, a creamy-
white grub, up to 7mm long,
or the pupal stage, will be
seen. Damaged blossoms
cannot produce fruits.
CAUSE A weevil, *Anthonomus
pomorum*, which lays eggs
in the unopened buds
during mid- and late spring.
The adult weevil is brown
with whitish markings on
its upper surface and is
4–6mm long.
CONTROL Light infestations
have little impact on the
yield of fruits. Spraying with
bifenthrin when flowers are
at the green cluster stage
is only worthwhile if heavy
attacks have occurred in
previous years.

APPLE CAPSID

SYMPTOMS Ripe apples have
raised yellowish-brown corky
bumps or scabby areas on
the skin (*see **p.47***). Leaves at
shoot tips may be distorted,
with many small holes.
CAUSE Apple capsid,
Plesiocoris rugicollis, which
is a bright green sap-feeding
insect up to 7mm long. It
injects a toxic saliva into
the plant as it feeds on the
foliage and immature fruits
during early summer. This
kills some plant cells and,
as the developing leaves
grow and expand, these
dead areas tear into small
holes. Fruits respond to
the capsid's saliva by
developing raised bumps
or scabby areas.
CONTROL The fruit damage
is only skin-deep and light
attacks can be tolerated. If
damage has been extensive
in previous years, spray
with bifenthrin or pyrethrum
at petal fall.

APPLE FROST DAMAGE

SYMPTOMS Russeting of the
skin, usually on the most
exposed face of the fruit.
The more protected faces are
usually unharmed. Damaged
areas may develop a corky,
roughened appearance
and the fruit may become
deformed and uneven in
growth. Flowers and young
foliage subjected to frost
wither and die off rapidly.
CAUSE Frost on young,
unhardened plant tissues.
This is exacerbated if the
plant is growing in a position
where, following overnight
frosts, it is then subjected
to rapid warming by early
morning sun.

CONTROL Rarely a recurrent problem, unless the tree is in an unsuitable position. Seasonal use of horticultural fleece or similar material draped over trees will prevent injury.

APPLE FRUIT SPLIT

SYMPTOMS Shallow cracks occur, usually only just penetrating into the flesh and most commonly running round the stalk end of the fruit (see **p.48**). These cracks may heal over and a corky layer forms, but if they occur when the fruit is more mature, the fruit usually succumbs to secondary infections.
CAUSE An erratic supply of water is the most common cause. The splitting occurs when the tree puts on a sudden spurt of growth, when moisture becomes available after a period of drought.
CONTROL Keep trees regularly watered and apply a mulch to moist soil to try to retain soil moisture.

APPLE LEAF MINER

SYMPTOMS Silvery-white or brown lines appear on the foliage of apple and cherry trees and ornamental *Malus* and *Prunus* (see **p.19**), particularly in late summer. If the leaf miners have finished feeding and are pupating, hammock-like cocoons of white silk will be attached to the undersides of leaves or on the bark.
CAUSE Small green caterpillars of a tiny moth, *Lyonetia clerkella*, live within the leaves, and eat out meandering mines or tunnels, which show as white or brown lines.
CONTROL Although leaf miners are sometimes abundant, this pest is active too late in the growing season to have much effect on the tree's health or ability to produce fruit, so its presence can be tolerated. There are currently no insecticides suitable for use against leaf miner.

APPLE POWDERY MILDEW

SYMPTOMS Leaves are covered in a powdery white fungal growth (see **p.25**). Young leaves breaking from buds which were infected in the previous year often show the symptoms. If attacked early on, the foliage rarely reaches full size, but is distorted and dies prematurely.
CAUSE The fungus *Podosphaera leucotricha*, which overwinters in the apple buds or on the young stem growth (as silvery white patches on the bark). This fungus may also occasionally attack pears, quinces and medlars, but does not cause serious damage. Disease spread and development is most rapid in hot summers and if there are heavy dews at night.
CONTROL Keep apples well watered, as a dry soil seems to encourage the development of powdery mildew. Apply a mulch to the moist soil in order to preserve soil moisture. Prune out any severely infected growth and mildewed stems. Prune established trees to improve air circulation within the crown. Spray with a suitable formulation of myclobutanil. Grow cultivars that show show good resistance, such as 'Blenheim Orange', 'Discovery', 'Greensleeves' or 'Worcester Pearmain'.

APPLE SAWFLY

SYMPTOMS Apple fruitlets fall off in early to mid-summer with a maggot hole in the side (see **p.48**). This hole is filled with excrement produced by a white caterpillar-like larva, or maggot, with a brown head. Mature fruits may be misshapen, with a broad ribbon-like scar on the skin, often running up from the eye end (opposite the stalk) to the middle of the fruit (see **p.47**).
CAUSE A sawfly, *Hoplocampa testudinea*, lays eggs in the fruitlets at flowering time. After hatching, the larva bores into the fruitlet and at first tunnels just beneath the skin before boring into the core. Fruitlets damaged in this way drop off, and when fully fed the larvae leave them (creating the exit hole) and go into the soil to pupate. However, if the larva dies before it can bore into the core, the fruit will reach maturity. The early feeding beneath the skin causes fruits to be misshapen with the scar mentioned above.
CONTROL Pick off and destroy any fruitlets showing signs of damage before the larvae have a chance to escape into the soil. Spray at petal fall with bifenthrin if sawfly has affected a significant proportion of the crop in previous years.

APPLE SCAB

SYMPTOMS Blackish-brown, scabby patches develop on the fruits, with similar but more greenish-grey spots on the leaves (see **p.16**). In severe cases the fruit may be almost entirely covered with scabby patches, and be small and misshapen. Infected fruits may crack or split and so become infected with secondary organisms such as brown rot. Affected leaves fall early and may also show extensive infection on the petiole (leaf stalk). A very similar disease affects poplars.
CAUSE The fungus *Venturia inaequalis*, which overwinters on scabby patches on the young stems and also on fallen infected leaves. Scab can occur on ornamental and edible *Malus* species and cultivars. It is usually most prevalent in damp seasons and on trees with crowded branches.
CONTROL Rake up and dispose of any affected leaves as soon as they fall. Prune out cracked or scabby shoots to limit the overwintering places of the fungus. Spray with a fungicide containing mancozeb. Grow apple cultivars that show resistance to this infection, for example 'Discovery', 'Lane's Prince Albert', 'Gavin' or 'Sunset'.

APPLE SCALD

SYMPTOMS Reddish or brown areas develop on the skin of the fruit (see **p.47**), on surfaces that are subjected to the full impact of the sun. In extreme cases the flesh beneath may be damaged, but the discoloration is usually limited to the skin.
CAUSE The scorching or scalding is caused by strong sunlight on the tender skin of young apples.
CONTROL In temperate zones, scald is generally only seen in unusually hot, bright summers, so controls are not really justified.

APPLE SOOTY BLOTCH

SYMPTOMS Sooty, greenish-black spots develop on the

fruit skin, usually on apples but occasionally on plums. They do not penetrate into the flesh, however, and can be scraped off. The fruit's development is not affected and it is not distorted.

CAUSE The fungus *Gloeodes pomigena*, which overwinters on young stems and which is most common in trees growing in shade, and in cool, damp conditions.

CONTROL Although common in some seasons, it is rarely widespread and usually only occasional. No specific control measures are recommended, but trees sprayed for apple scab (*see p.101*) will only rarely develop sooty blotch.

APPLE SUCKER

SYMPTOMS Flattened, pale green insects, up to 2mm long, occur on apple blossom trusses in the spring (*see p.43*). Heavy infestations kill the flowers and prevent fruit setting.

CAUSE Sap-feeding, aphid-like insects, *Psylla mali*, that hatch from overwintering eggs as apples come into leaf. Most of the damage is caused by the immature nymphs during the blossom period. The adult stage resembles a winged aphid and is present after petal fall.

CONTROL Spray with bifenthrin or pyrethrum before flowering, at the green bud stage.

APPLE WATER CORE

SYMPTOMS This disorder is also known as "glassiness" because affected fruits develop a glassy, almost wet appearance to the flesh. This is usually most apparent close to the core, but may occur throughout.

Symptoms may decrease when apples are stored, but more often affected fruits become brown and start to disintegrate.

CAUSE Can be encouraged by unsuitable cultural conditions, including excessive pruning and thinning, high temperature, high light intensity and extensive damage or removal of bark on the trunk or branches.

CONTROL Avoid excessive thinning or pruning. Other factors are usually impossible to control.

AQUILEGIA SAWFLY

SYMPTOMS Pale green caterpillar-like larvae, up to 18mm long, feed on the leaf margins of aquilegias. They can quickly defoliate plants, leaving just the leaf stalks and flower stems. Several generations occur between mid-spring and early autumn.

CAUSE A sawfly, *Pristiphora aquilegiae*, which lays small batches of eggs on the undersides of leaves. The adults have black bodies 6mm long, two pairs of transparent wings and reddish-brown legs.

CONTROL Watch out for signs of leaf damage on aquilegias, especially in the spring and early summer, when defoliation will have a more severe effect on the plant's health. Either remove the larvae by hand or spray with thiacloprid, pyrethrum or bifenthrin.

ARABIS MOSAIC VIRUS

SYMPTOMS Yellow mosaic-like flecking and discoloration on leaves, not only of arabis but of many plants. Infected

plants are often stunted and distorted, and commonly fail to reach maturity.

CAUSE Arabis mosaic virus, which is one of the most common and widespread viruses infecting garden plants, both edible and ornamental.

CONTROL There is no control measure available. Affected plants should be removed and disposed of promptly.

ARBUTUS LEAF SPOT

SYMPTOMS Numerous small spots develop on the leaves of strawberry trees (*Arbutus*). Each has a dry, grey-white central area bordered by a dark purple band (*see p.14*). The vigour of the tree is not affected.

CAUSE The fungus *Septoria unidonis*, which is common and widespread.

CONTROL No control is available, nor is one really required. Fallen leaves should be cleared up.

ARUNCUS SAWFLY

SYMPTOMS Pale green caterpillar-like larvae, up to 20mm long, feed on the edges or central areas of the leaves of goatsbeard (*Aruncus dioicus*). Several generations occur between late spring and early autumn and plants can be stripped of foliage.

CAUSE A sawfly, *Nematus spiraeae*, sometimes known as the spiraea sawfly because *Aruncus* was at one time classified as a *Spiraea* species. This sawfly does not attack any of the plants currently known as *Spiraea*. The adults have black and yellow bodies 6mm long, two pairs of transparent wings and legs that are mainly yellow.

CONTROL Watch out for signs of leaf damage, especially in the spring and early summer, when defoliation will have a more severe effect on the plant's health. Either remove the larvae by hand or spray with thiacloprid, pyrethrum or bifenthrin.

ASH CATKIN GALL MITE

SYMPTOMS Instead of producing the usual ash seeds or keys, the flowers become converted into lumpy, brown, woody galls. These persist on the tree and can be a distinctive feature on infested ash trees in the autumn and winter after leaf fall.

CAUSE Microscopic gall mites, *Aceria fraxinivorus*, which suck sap from flowers in the spring, and secrete chemicals that convert the flowers into galls. Individual trees growing close together can vary considerably in the extent to which they are affected, indicating that some trees are more susceptible than others.

CONTROL Apart from reducing the number of seeds produced by ash trees, this mite has no adverse effect on their growth. There is no effective chemical control available to amateur gardeners.

ASH SUCKER

SYMPTOMS In early summer the margins of ash leaves become purplish-red, thickened and curled over. On the undersides of the curled leaves will be some flattened aphid-like insects, which secrete a white waxy material from their bodies.

CAUSE The ash sucker, *Psyllopsis fraxini*, is an

aphid-like, sap-feeding insect. The damage is caused by the wingless immature nymphs; the adults, which resemble winged aphids, are present during the summer but have no noticeable effect on the tree.

CONTROL Tall trees cannot be treated and spraying is not usually required, even on saplings. Although the distorted leaves may cause concern and look unsightly, ash suckers do not have a serious impact on the tree's health or vigour.

ASPARAGUS BEETLE

SYMPTOMS Asparagus plants lose their foliage as adult beetles and their larvae eat the leaves and outer bark from the stems (*see* **p.40**). The damage to the bark causes stems to dry up and turn yellowish-brown above the point where bark has been removed from all round the stem. The adult beetles are 6–8mm long, black with six yellow blotches on their wing cases, and a reddish thorax. They emerge from the soil in late spring and begin laying eggs on the stems and foliage. The larval stage is creamy-black in colour, up to 10mm long, with three pairs of legs towards the head end. There are two generations between late spring and early autumn.

CAUSE Both the adults and larvae of a leaf beetle, *Crioceris asparagi*.

CONTROL Burn old stems at the end of the year to destroy some of the over-wintering beetles. Watch for signs of infestation from late spring onwards. Hand-removal of adults and larvae is feasible on small asparagus beds; otherwise spray with pyrethrum. Asparagus

flowers attract bees, so when plants are in flower, spray at dusk if insecticides are needed.

ASTER WILT

SYMPTOMS Rapid wilting of the plant, followed by death. The stem base is blackened and shrinks inwards. A faint, furry, pale pink or white fungal growth may develop on the affected areas.

CAUSE The fungus *Fusarium oxysporum* f. sp. *callistephi*, or fusarium wilt. It is most common in heavy or wet soils but can attack asters in any soil, and even in seed trays. The fungus is usually introduced on infected seedlings or plants, but may also be present in the soil. It may also be seed-borne.

CONTROL Promptly remove and dispose of infected plants and the soil in the vicinity of their roots. Use a fresh site for future plantings of asters and only plant perfectly healthy-looking plants. Discard any with dead roots or damaged stem bases. Avoid planting asters on heavy soils, or improve soil texture before planting. Grow china asters that show resistance, such as 'Duchess', 'Ostrich Plume', Milady Mixed, Gala Mixed, 'Violet Cutter', 'Starlight' or 'Super Sinensis'.

AZALEA GALL

SYMPTOMS Pale green, waxy growths develop in place of foliage, or occasionally in place of petals. Sometimes they have a pinkish-red colour. The galls later turn white as a powdery spore layer develops on their surface, and at the end of the summer they

wither and turn brown. The infected plant continues to grow and flower normally. Symptoms may occasionally also develop on rhododendrons.

CAUSE The fungus *Exobasidium vaccinii*, which is common on outdoor and indoor Indian hybrid azaleas (*R. indicum*). Spores are carried on air currents and by insects, and months may elapse between infection and the appearance of symptoms.

CONTROL Remove galls promptly, before the spore layer develops.

AZALEA WHITEFLY

SYMPTOMS White-winged insects, 2mm long, readily fly up from the undersides of the leaves of evergreen azaleas during the summer. The immature nymphs are flat, oval, whitish-green, scale-like creatures that live on the undersides of leaves. They are present for most of the year, as this pest overwinters in the nymphal stage. Both adults and nymphs excrete a sugary substance called honeydew that makes the upper leaf surfaces sticky and allows the growth of sooty moulds.

CAUSE A whitefly, *Pealius azaleae*, which feeds by sucking sap from the leaves of evergreen azaleas.

CONTROL The adult stage, which is reached in mid-summer, is more susceptible to insecticides than the eggs or nymphs. When adult whitefly are seen, spray the underside of the foliage thoroughly with pyrethrum, fatty acids or plant oils, or with an insecticide containing bifenthrin, thiacloprid, or acetamiprid.

BACTERIA

SYMPTOMS The symptoms of bacterial infection vary with both the plant part and the type of plant infected, as well as the precise bacterium involved. Leaves usually become spotted, often with a yellow "halo" around the edge of the leaf spot. Fleshier plant parts such as stems, rhizomes, tubers, roots and bulbs may discolour and disintegrate rapidly, often becoming slimy in texture and giving rise to unpleasant odours. Specific bacterial infections, such as bacterial canker and fireblight, have individual entries in this section.

CAUSE Various species of bacteria may be involved in general infections, probably the most common being species of *Erwinia*. Many bacterial infections occur initially on injured tissues but then spread into other areas of the host plant. Bacteria may be spread by rain splash, by insects and other animals, and also on pruning tools; some, such as crown gall, may spread through soil water movement.

CONTROL The precise method of control varies, but prompt removal of all infected areas and adjacent plant tissue is essential. Copper-based fungicides can be used to control bacterial canker of stone fruits.

• *see also* BACTERIAL CANKER *p.104*; BACTERIAL SOFT ROT *p.104*; BEAN HALO BLIGHT *p.105*; FIREBLIGHT *p.128*; LEAF SPOT (BACTERIAL) *p.144*.

BACTERIAL CANKER

SYMPTOMS Clearly defined areas of bark flatten and sink inwards (see *p.35*). Amber-coloured, resin-like ooze may appear, closely associated with injured bark. Only *Prunus* are commonly affected; symptoms are most obvious on cherry trees, far less so on plums. Buds at the tips of affected branches may fail to open, or leaves may appear but later wither and die back, due to the ringing effect produced by the expanding canker. Leaves on affected limbs are often yellowed and small in size. Foliage may develop shothole symptoms (see *p.176*).
CAUSE The bacterium *Pseudomonas morsprunorum* (on plum and sweet cherry) and *Pseudomonas syringae* (on peach, apricot, cherry and plum). Most infections occur during wet, windy weather in autumn, or in damp conditions in spring, when young stems and leaves are readily infected. Spring infections usually start on the leaves and then spread to the wood. Most wood infections occur through injured bark, such as that caused by pruning, frost crack or leaf fall. Infection during summer is rare.
CONTROL Prune out infected areas during the summer months. Spray infected trees with a copper-based fungicide, such as Bordeaux mixture, or with copper oxychloride. Sprays should be applied once at the end of summer, once in early autumn and once more in mid-autumn. Consider growing cultivars which show a degree of resistance, such as the cherries 'Merton Glory', 'Merton Premier', 'Merla' and 'Merpet', or plum cultivars 'Marjorie's Seedling' and 'Warwickshire Drooper'.

BACTERIAL LEAF SPOT

See LEAF SPOT (BACTERIAL) p.144.

BACTERIAL SOFT ROT

SYMPTOMS Roots, tubers, rhizomes and fruits are generally the most susceptible areas of host plants. Discoloured areas develop and, as the tissue beneath deteriorates, the injured area may sink inwards. Infection usually spreads rapidly and can prove fatal or seriously damaging if its progress is not halted.
CAUSE Various bacteria may be involved, most commonly *Erwinia atroseptica* and its various pathovars or strains. The bacteria usually enter via wounds and are easily spread on pruning tools, by handling or by insects and other animals.
CONTROL Control is rarely possible as the infection spreads so rapidly. It may, however, be possible to limit the damage caused by very prompt removal of infected areas together with the adjacent tissue. Disinfect pruning tools by using a garden disinfectant.
• *see also* BRASSICA BACTERIAL SOFT ROT *p.111*.

BANDED PALM THRIPS

SYMPTOMS Indoor plants, especially those which have relatively tough leaves, such as *Ficus*, *Schefflera*, *Monstera* and palms, develop a silvery-brown discoloration of the upper leaf surface (see *p.22*). Close examination will reveal narrow-bodied insects up to 2mm long, either creamy-yellow in colour or banded black and white.
CAUSE Sap-feeding thrips, *Parthenothrips dracaenae*, which are creamy-yellow as immature nymphs. The adults are darker in colour, with dark bands on pale wings. The wings are folded over the thrips' bodies when not in use, which gives the adults their striped appearance.
CONTROL Spray the foliage thoroughly with acetamiprid, thiacloprid, thiamethoxam, pyrethrum, or bifenthrin. Two or three applications at fortnightly intervals may be needed to break the pest's life cycle. Damaged leaves will not regain their normal colour but new growth will stay clean once thrips have been controlled.

BARK BEETLES

SYMPTOMS Many woody plants are affected by bark beetles, especially trees or shrubs already in poor health, that have dead limbs, or have recently died. The larvae feed between the bark and the wood and eat out a series of channels that can be seen if the bark is peeled away (see *p.33*). The tunnelling pattern is often distinctive for a particular species of bark beetle; it consists of a maternal tunnel made by the adult female, in which eggs were laid, and larval tunnels which radiate away from the original tunnel. Fully-fed grubs pupate at the ends of the tunnels and adult beetles chew their way out through the bark.
CAUSE Many different species of bark beetle can occur on conifers and deciduous trees and shrubs. Most are harmless, as they are secondary pests on plants that are already dying from other causes. The most notorious bark beetle is *Scolytus scolytus*, the elm bark beetle. This breeds in dying elm trees, including those infected with Dutch elm disease (see *p.125*). When adult beetles emerge in early summer they have spores of the fungal disease on their bodies and can carry it to other elms in the area.
CONTROL Prune out and burn affected branches, which will be weak or dying, and check to see if another problem, such as honey fungus (see *p.138*), is present. Insecticides are not effective against bark beetles, since for much of the year they are protected under the bark.

BASAL ROT

See FOOT AND ROOT ROTS p.129.

BAY SUCKER

SYMPTOMS During the summer, leaves at the shoot tips of sweet bay (*Laurus nobilis*) develop thickened, pale yellow margins which curl over (see *p.26*). Often just one half of the leaf is affected, and later the damaged parts dry up and turn brown. Immature insects feed underneath the curled leaf margins but emerge when ready to moult to the adult stage. These fully-grown nymphs are grey, flattened insects which secrete a fluffy, white waxy material from their bodies. The adults are winged, aphid-like insects, 2mm in length, which may be seen on the shoot tips during summer.

CAUSE A sap-feeding insect, *Trioza alacris*, which has two or three generations between late spring and mid-autumn.

CONTROL Deal with light infestations by picking off any leaves that have been affected. Thiacloprid can be used when leaf curling begins.

BEAN ANTHRACNOSE

SYMPTOMS Sunken, brown longitudinal marks develop on the stems of affected dwarf or runner beans. Leaf veins may develop a red coloration and may subsequently turn brown, wither and then die off. A reddish-brown spotting develops on infected pods and, when the weather becomes wet, a pink-coloured slimy fungal growth may develop.

CAUSE The fungus *Colletotrichum lindemuthianum*, which is usually seed-borne. The fungal spores are spread from young infected seedlings onto adjacent plants either by rain or by irrigation splash.

CONTROL Remove infected plants immediately, and do not save their seed. Grow cultivars showing some resistance, such as 'Aramis' or 'Rido Kenyan'.

BEAN CHOCOLATE SPOT

SYMPTOMS Chocolate-brown spots develop on the upper surfaces of broad bean leaves (*see p.12*). Chocolate-brown streaks may also appear on the stems, pods and flowers, and the seed coat may be affected in a

similar way. Adjacent spots or streaks may coalesce to form large blotches. Plants that have been severely infected may die off, and even a very mild infection may reduce yields.

CAUSE The fungus *Botrytis fabae*, which is encouraged by damp or humid air. The fungus may overwinter on the remains of infected plants and may also be seed-borne.

CONTROL Grow beans on a properly drained site, avoiding the use of high nitrogen fertilizers (which will only increase the production of soft, infection-prone growth). Apply sulphate of potash so that growth will harden slightly. Space plants wider apart than normal – at least 23cm – to improve air circulation, and to ensure good weed control is maintained between the rows. This should help to minimize humidity and damp air around the plants.

BEAN HALO BLIGHT

SYMPTOMS Small, angular spots appear on the leaves of dwarf French or runner beans, at first appearing water-soaked, and then darkening, each one surrounded by a bright yellow "halo". Leaves then develop yellowing between the veins (interveinal chlorosis); the entire affected leaf may yellow and die. Growth is affected and yields are reduced. If pods or stems become infected, they develop grey, apparently water-soaked patches.

CAUSE The bacterium *Pseudomonas phaseolica*, which attacks dwarf French and runner beans. Water splash (from rain or overhead watering)

spreads the bacteria. Initial infection usually starts from infected seed.

CONTROL Pick off affected leaves promptly, and avoid overhead watering. Remove and burn all affected plants at the end of the season, discarding their seed. Consider growing the dwarf French bean cultivars 'Forum' or 'Red Rum', which both show some resistance to this infection.

BEAN RUST

SYMPTOMS Numerous dark brown pustules develop on the underside of the leaves of French and runner beans. Occasionally stems and pods are infected. Rarely, the white cluster cup form of the fungus appears beneath the leaves. Severely infected leaves wither and in extreme cases cropping is reduced.

CAUSE The fungus *Uromyces appendiculatus*, which is most common in warm, damp summers.

CONTROL Remove affected leaves promptly.

BEAN SEED BEETLES

SYMPTOMS Holes (from which adult beetles have emerged) are seen in dry beans and peas kept for seed (*see p.51*). Several plants are vulnerable, but in gardens it is mainly broad beans that are affected. While the beetle grub is present within the seed, a pale circular patch on the seed coat indicates where it has eaten a cavity in the seed's cotyledon.

CAUSE Several species of seed beetle, up to 3mm long, can attack pea and bean seeds. On broad beans it is usually *Bruchus rufimanus*. This lays eggs in the bean

pods as the seeds are forming. The small grubs go unnoticed in bean seeds that are cooked for eating, but their presence becomes obvious in beans kept for next year's seed when the adult beetles emerge.

CONTROL There is no effective treatment available to amateur gardeners. Feeding damage by the larvae is to the seed's food reserve (known as the cotyledons) rather than the embryo, so damaged seeds can still germinate.

BEAN SEED FLY

SYMPTOMS Germinating seeds of runner and French beans are eaten by white, legless fly larvae. Seedlings are sometimes killed before the shoots emerge from the soil, but more usually ragged leaves and stems damaged by the soil-dwelling larvae appear above soil level (*see p.51*). Such plants will usually survive although they may make slow growth initially, especially if the growing points have been killed, causing sideshoots to develop.

CAUSE Maggots of a fly, *Delia platura*, which is similar in appearance to a house fly.

CONTROL Slow-germinating seeds are most at risk, so avoid sowing under adverse conditions, such as when the soil is cold or very wet. The female flies are attracted to soil that contains fresh organic matter, so apply this in autumn rather than spring. To avoid bean seed fly problems, beans can be germinated in pots or seed trays, and transplanted after the first true leaves have expanded. There are no suitable insecticides for bean seed fly.

BEECH BARK DISEASE

SYMPTOMS Usually seen in association with beech bark scale (*see below*). Small areas of bark (a few centimetres in diameter) die and may appear as slight depressions (*see **p.35***). They may then exude a black, tar-like substance formed from the tree's sap. As the infection spreads, larger areas of bark are killed and the leaves become yellowed. Leaves produced subsequent to the infection are often unusually small and yellow. Tiny red fungal bodies the size of pinheads develop on affected bark. Trees decline in vigour and may be killed, often snapping off part way up the trunk.
CAUSE The fungus *Nectria coccinea*, which infects beech trees via wounds in the bark created by beech bark scale. Young beeches are only rarely infected.
CONTROL Fell severely infected trees, to minimise spread and because the weakened state of their wood may render them dangerous.

BEECH BARK SCALE

SYMPTOMS A white powdery substance appears in crevices in the bark of beech trees, especially on the trunk and larger branches (*see **p.35***). Small orange insects up to 1mm long live underneath this waxy exudation.
CAUSE A sap-feeding scale insect, *Cryptococcus fagisuga*, also called the felted beech coccus. It has little direct effect on the tree, but it is believed that heavy infestations make trees more susceptible to a form of canker known as beech bark disease (*see above*).

CONTROL There are no effective insecticides for this scale insect.

BEECH GALL MIDGE

SYMPTOMS In late summer, yellowish-brown, cylindrical structures, or galls, which may be hairy or smooth, develop on the upper surfaces of beech leaves. These galls, each containing a single fly larva, readily break off and drop to the ground in the autumn.
CAUSE Small flies – gall midges – lay eggs in leaves in mid-summer. The hairy galls are caused by *Hartigiola annulipes* and in some years are abundant; the larger smooth-surfaced galls are caused by *Mikiola fagi*, which is uncommon in Britain but can be frequent elsewhere in Europe.
CONTROL Apart from creating galls, these midges have no effect on the health of beech trees, so their presence can be tolerated.

BEECH LEAF-MINING WEEVIL

SYMPTOMS In spring, beech tree foliage develops many small, round holes eaten out by the adult weevils. These are black, 2mm long, and jump like flea beetles when disturbed. The larval stage feeds as a leaf miner, making a brown linear mine from the centre of the leaf, where the egg was laid, to the leaf tip. A large, dried-up brown area develops at the leaf tip where most feeding occurs (*see **p.24***).
CAUSE A weevil, *Rhynchaenus fagi*, which in some years is so abundant that nearly every leaf is affected by larvae. This can be unsightly but the tree is

able to withstand attack.
CONTROL Nothing can be done on mature trees because of the problem of spraying large specimens. Clipping beech hedges will remove many of the damaged leaves and new growth produced after late spring will not be attacked. Chemical treatment is generally not worthwhile.

BEECH WOOLLY APHID

SYMPTOMS Pale yellow insects covered with a fluffy, white waxy material (*see **p.32***) form dense colonies on the shoot tips and undersides of beech leaves during late spring to mid-summer, after which their numbers decline. The foliage becomes sticky with the aphids' sugary excrement, called honeydew, and this encourages the growth of sooty moulds. Winged aphids are present in mid-summer and fly up when beech hedges are clipped. Because they are white, they are sometimes mistaken for whiteflies.
CAUSE A sap-feeding aphid, *Phyllaphis fagi*, a specific pest of beech trees.
CONTROL Tall trees cannot be sprayed effectively and so the aphid must be tolerated. On hedges look for signs of aphid activity in late spring and spray with thiacloprid, acetamiprid, bifenthrin, or pyrethrum before heavy infestations develop.

BEEFSTEAK FUNGUS

See FISTULINA HEPATICA *p.128*.

BEES

See BURROWING BEES *p.113*; LEAF-CUTTING BEES *p.144*; NECTAR ROBBING *p.152*.

BEET LEAF MINER

SYMPTOMS Leaves of beetroot, Swiss chard, and spinach beet develop large brown areas (*see **p.17***) where internal tissues have been eaten by white maggots.
CAUSE Larvae of the leaf-mining fly, *Pegomya hyoscyami*, which has two generations during summer.
CONTROL With light infestations, remove affected leaves or crush the larvae inside the leaf mines. There are no insecticides approved for leaf miner control on beetroot and spinach.

BEETLES

In terms of named species, beetles are the most numerous insects in the world. They are characterized by their modified fore wings, hard opaque structures (the elytra) that form a covering for the abdomen. Some beetles, such as ladybirds (*see p.143*) and ground beetles (*see p.136*), are beneficial as predators of garden pests. There are also a number of beetles which attack various parts of plants.
Root damage – *see* CHAFER GRUBS *p.116*; TURNIP GALL WEEVIL *p.186*; VINE WEEVIL *p.188*; WIREWORMS *p.193*.
Stem damage – *see* BARK BEETLES *p.104*; SHOT-HOLE BORERS *p.177*.
Foliage damage – *see* ASPARAGUS BEETLE *p.103*; BEECH LEAF-MINING WEEVIL *p.106*, COLORADO BEETLE *p.119*; FIGWORT WEEVIL *p.128*; FLEA BEETLES *p.128*; LEAF WEEVILS *p.145*; LILY BEETLE *p.147*; MINT BEETLE *p.149*; PEA AND BEAN WEEVIL *p.156*; ROSEMARY BEETLE, *p.173*; VIBURNUM BEETLE *p.187*; VINE WEEVIL *p.188*; WATER

LILY BEETLE *p.190*; WILLOW LEAF BEETLES *p.192*.

Flower damage – *see* APPLE BLOSSOM WEEVIL *p.100*; POLLEN BEETLES *p.163*.

Seed and fruit damage – *see* BEAN SEED BEETLES *p.105*; NUT WEEVIL *p.153*; RASPBERRY BEETLE *p.168*; STRAWBERRY SEED BEETLE *p.181*.

BEGONIA POWDERY MILDEW

SYMPTOMS Greyish-white, powdery fungal growth appears in distinct spots or patches on the upper leaf surface (*see* **p.23**). Infected leaves either become dry and brown or, more commonly with the fleshy-leaved types, yellowed and soggy. Stems and flower-heads may also be affected. The disease is more common on indoor begonias; plants grown in the open are rarely affected.
CAUSE The fungus *Microsphaera begoniae*, which is encouraged by dry conditions around the plant's roots, and by moist air around the foliage.
CONTROL Pick off affected leaves immediately. Keep plants adequately moist, and improve air circulation to help reduce humidity. Spray with a suitable fungicide such as myclobutanil, penconazole or sulphur. Unless the disease is noticed and tackled early, control is very difficult.

BERBERIS SAWFLY

SYMPTOMS Sawfly larvae up to 18mm long eat the leaves on barberry shrubs (*Berberis* spp.) and *Mahonia* spp. The caterpillar-like larvae have black heads and white bodies marked with small black dots and larger yellow blotches. The adult sawflies have metallic bluish black bodies 7–10mm long and the wings are black. There are two generations of larvae between late spring and early autumn when plants can be stripped of leaves. The fully fed larvae go into the soil to pupate.
CAUSE A sawfly, *Arge berberidis*, which has become more widespread in Europe in recent years. It has been present in Britain since 2000. The most susceptible barberries are *Berberis thunbergii* and *B. vulgaris*.
CONTROL The larvae can complete their feeding and cause serious damage in as little as two to three weeks so check regularly for the presence of larvae. If seen, spray with bifenthrin, thiacloprid, or pyrethrum. Spray at dusk if the plants are in flower to avoid harming bees.

BIG BUD MITES

SYMPTOMS Buds become abnormally enlarged and rounded, and fail to develop into shoots and leaves (*see* **p.41**). Each big bud will contain many hundreds of microscopic, elongate white mites. Dried-up big buds can be found at any time of year, but they become swollen over the winter months.
CAUSE Several species of gall mite cause this problem. The mites most frequently found on garden plants are *Cecidophyopsis ribis* on blackcurrant, *C. psilaspis* on yew and *Phytoptus avellanae* on hazel.
CONTROL Big bud mites on yew and hazel will kill some buds, but generally not enough to affect the plants' growth, so they can be tolerated. Blackcurrant big bud mite is more serious, since apart from causing a loss of vigour, it can spread a virus-like disease called currant reversion (*see p.122*). The chemicals available to amateur gardeners will not control big bud mites. Removing big buds from lightly infested bushes in late winter will limit infestations. Scrap badly affected bushes after fruiting and replace in the autumn. 'Ben Hope' is a blackcurrant cultivar that is resistant to big bud mite.

BIRCH WITCHES' BROOMS

SYMPTOMS Extensively branched twigs develop, clustered closely together to form "witches' brooms" scattered throughout the crown of the tree. The leaves on the brooms turn yellow and bear a layer of fungal spores on their lower surface. During the winter, the growths may, at first, be mistaken for birds' nests and can spoil the outline of the tree or hedge, but during the summer they appear as dense, leafy areas and so are less obtrusive. The growths are particularly susceptible to frost and so may be killed during the winter.
CAUSE The fungus *Taphrina betulina*, which is spread by water splash and on air currents. *Betula pendula* and *B. pubescens* are most frequently attacked. Similar witches' brooms may develop on birch as a result of a mycoplasma infection.
CONTROL Prune out the growths, if desired. Spread is very slow, but removal helps to limit the problem. No chemical control is available.

BIRD DAMAGE ON FRUITS

SYMPTOMS Apples, pears and plums show bird pecks on the ripening fruit (*see p.47*). Wasps often extend the damage. Smaller fruits such as strawberries, cherries and currants are eaten whole.
CAUSE Various birds, including blackbirds, thrushes and starlings.
CONTROL Small trees and bushes can be netted or grown in a fruit cage. On large trees, bags made from muslin or old nylon tights drawn over the better fruit trusses will protect them from birds.

BITTER PIT

See APPLE BITTER PIT *p.100*.

BITTER ROT

See GLEOSPORIUM ROT *p.134*.

BLACK BEAN APHID

SYMPTOMS Dense colonies of black insects, up to 2mm long, mass on the shoot tips and undersides of leaves (*see* **p.32**) of broad beans and other plants, including globe artichokes, nasturtiums, dahlias and poppies, during the summer. Affected plants are weakened and bean pods fail to develop. The same pest may occur on viburnum and mock orange (*Philadelphus*) during the spring, causing leaf curling at the shoot tips.
CAUSE A sap-feeding aphid, *Aphis fabae*.
CONTROL Inspect susceptible plants at regular intervals and spray immediately aphids are seen, either with bifenthrin, acetamiprid, or thiacloprid. The last two cannot be used on beans or artichoke; bifenthrin

cannot be used on artichoke. The organic insecticides, pyrethrum, fatty acids and plant oils are also effective as long as they are applied thoroughly before heavy infestations have had a chance to develop. They can be used on edible plants but, because of their short persistence, may need more frequent use.

BLACK LEG OF CUTTINGS

SYMPTOMS Before or shortly after roots start to form, the base of an affected cutting darkens and atrophies; the upper parts start to discolour and then die off.
CAUSE Various soil- and water-borne fungi, in particular those responsible for causing damping off (*see p.123*) in seedlings. They are usually introduced on dirty tools, pots or trays, in unsterilized compost or by the use of non-mains water.
CONTROL Maintain good hygiene when taking cuttings. Use only clean or sterilized materials and tools, and mains water. Using a rooting powder will stimulate rapid rooting, which may in turn help to minimise black leg.

BLACKBERRY CANE SPOT

SYMPTOMS Silvery-grey, lens-shaped spots with purple margins appear on the canes. Pinprick-sized, black pycnidia, or fungal fruiting bodies, develop in the centre of each lesion. In severe cases canes may die back.
CAUSE The fungus *Elsinoe veneta*.
CONTROL Prune out severely infected canes.

BLACKCURRANT GALL MIDGE

SYMPTOMS Leaves of blackcurrants remain small and crumpled (*see p.26*), due to whitish maggots up to 2mm long feeding on leaves as they develop at shoot tips. Affected leaves dry up and die. Three generations occur during the summer, with symptoms first appearing in late spring at flowering time. Shoot tips may be killed, causing side-shoots to grow.
CAUSE The leaf distortion is induced by chemicals secreted by the larvae of a tiny fly, *Dasineura tetensi*. While the maggots pupate, the shoot may produce a couple of normal leaves before the next generation affects more new leaves.
CONTROL The first generation in late spring is potentially the most harmful. There are no effective insecticides available for controlling this pest. Some blackcurrant cultivars, such as 'Ben Connan' and 'Ben Sarek', are resistant.

BLACKFLY

Blackfly is a common name given to several species of black aphids.
• *See also* APHIDS *p.99;* BLACK BEAN APHID *p.107;* CHERRY BLACKFLY *p.117.*

BLEEDING CANKER

SYMPTOMS Yellow or brown substances ooze from patches of dead or dying bark on the trunks or large limbs of trees. The ooze may dry on the bark, leaving crusty deposits. Affected bark often cracks, allowing other decay organisms to enter. Stems completely ringed by the lesions will die back. Horse chestnuts, limes, silver birch and apple trees are particularly prone to attack.
CAUSE Various species of the fungus *Phytophthora* (*see p.161*).
CONTROL There is no cure for this fungus. An inspection should be made by an experienced arboriculturist to determine the tree's health and safety.

BLIND SHOOTS

SYMPTOMS Occasionally, flowering shoots of roses and other flowering shrubs develop normally, but fail to develop a terminal flower bud (*see p.40*). The plant is perfectly healthy, and other shoots flower normally.
CAUSE Unknown; possibly a cultural disorder or adverse weather conditions.
CONTROL Prune back the blind shoot to a strong bud or pair of buds. Normal flowering shoots should grow from these buds.

BLINDNESS OF BULBS

SYMPTOMS Bulbs produce healthy, normal foliage, but fail to flower. Either no buds are formed at all, or buds form, but are dry and virtually empty of petals (*see p.41*).
CAUSE Occasionally due to large bulb fly infestation or basal rot (on *Narcissus*), but more often due to cultural problems. An inadequate supply of moisture and nutrients to the bulb causes it to decline in size and fail to produce buds, or a full complement of petals. Excessively crowded clumps (which are competing for moisture and nutrients), naturalized bulbs which have not been fed and bulbs growing in dry situations or suffering from drought are also likely to be affected. Narcissi are frequently affected, especially the multi-headed or double forms.
CONTROL Divide and replant congested clumps, preferably into a fresh, well-fertilized site. Ensure an adequate supply of water during dry weather. Feed established clumps in the spring with a complete fertilizer and treat with a foliar feed whilst in leaf. Do not tie up or cut off bulb foliage for at least six weeks after flowering is complete.

BLOSSOM END ROT

SYMPTOMS A sunken patch develops at the blossom end (base) of developing fruits. The base of the fruit becomes tough and leathery and darkens to brown or black as cells collapse. Not all fruits on a truss will necessarily be affected, and some trusses on a plant may remain perfectly healthy. Tomatoes, especially those growing in containers, and occasionally sweet peppers are affected.
CAUSE Dry conditions around the plant's roots, which prevent the plant from taking up sufficient calcium. The low calcium content of the fruits causes cells to collapse and discolour. This disorder is also encouraged by a very acidic growing medium.
CONTROL Ensure an adequate, regular supply of moisture at all times, especially on fast-growing and heavy-cropping plants. If blossom end rot does develop, pick off affected fruits and improve the watering routine. Grow small-fruiting tomatoes, as these are rarely affected. Grow plants in open ground or in larger containers where

the soil or compost is less prone to drying out.

BLOSSOM WILT

SYMPTOMS Blossom withers and dies shortly after appearing. Dead trusses of blossom remain on the tree. Infection may then spread to adjacent leaves which wilt, turn brown and die, remaining attached to the branch. Pinprick-sized, raised, buff-coloured fungal pustules develop on infected areas, and localized dieback may occur (*see* **p.41**). *Malus*, *Pyrus* and *Prunus*, including cherry, plum, apple, apricot and peach, may be attacked. Occasionally *Amelanchier* is affected.
CAUSE The fungus *Sclerotinia laxa* (on all except apple) and *S. laxa* f. *laxa* (on apple) which may overwinter as cankers on infected stems, or pustules on infected flowers or foliage. The spores are wind-borne and spread is most rapid during damp weather.
CONTROL If feasible, prune out infected flower trusses, preferably before the infection spreads to the foliage or into the spur.

BLUE MOULD ON BULBS

SYMPTOMS Bulbs in store are most likely to develop symptoms, but they may also occur while bulbs are in the ground. Reddish-brown lesions develop on the side of the affected bulb or corm, and pink fungal bodies may subsequently develop beneath the skin. A pale, bluish-green fungal growth appears on top of the lesion as it increases in size (*see* **p.54**). On bulbous irises this may be accompanied by rapid softening and rotting.
CAUSE Various species of the *Penicillium* fungi. These fungi are all encouraged by over-warm, moist storage conditions and poor air circulation. The fungus is believed to enter the bulbs whilst they are in the ground, but the symptoms will only develop if storage conditions are unsuitable.
CONTROL Avoid purchasing infected bulbs or those which are damaged, as these too are very susceptible. Store bulbs for as short a time as possible, and in cool but frost-free and well-ventilated conditions.

BLUEBELL RUST

SYMPTOMS Dark brown fungal pustules develop in elliptical, often concentric, formations on the leaves. Flower stems may also be attacked. Leaves yellow and wither prematurely and if infections are recurrent and severe, flowering may be somewhat reduced. Scillas and grape hyacinths (*Muscari*) may also be affected.
CAUSE The fungus *Uromyces muscari*. Infection is encouraged by damp growing conditions.
CONTROL Rarely necessary, but remove infected foliage to limit the spread of the infection.

BLUSHING BRACKET

See DAEDALEOPSIS CONFRAGOSA *p.122.*

BOLTING

SYMPTOMS Affected vegetable plants produce flowers and seeds prematurely, before the crop has developed fully or been harvested (*see* **p.44**).
CAUSE Various factors may be involved; these may include exposure to low temperatures at a certain growth stage (often induced by a cold, late onset of spring) which then induces flowering, or by excessively dry soil conditions.
CONTROL Where feasible, avoid growing early varieties of susceptible crops, as these are usually more prone to the problem. Check seed catalogues for cultivars showing resistance to bolting.

BORON DEFICIENCY

SYMPTOMS Symptoms vary depending on the crop affected, but root vegetables such as beetroot, turnip, carrot, radish and swede are most susceptible. Cauliflower, cabbage, celery, pears and strawberries occasionally develop symptoms.
Beetroot – Roughened, cankered areas develop on the sides of the roots. In severe cases, the subsequent rotting spreads to cause a heart rot, which shows as brown ring development in the inner root tissues and the crown. Foliage is unusually small and may die.
Cabbage – Leaf distortion and hollow areas in the stems.
Carrot – Splitting to expose the greyish and sometimes split central core, as well as yellow and pink discoloration of the older foliage and splitting of the leaf stems.
Cauliflower – This shows curd browning, poor curd development and roughening of the main stem, leaf stalks and midribs.
Celery – Transverse cracks appear in the outer surface of the leaf stalk, after which the inner tissues turn reddish brown. Leaves may occasionally show distortion.
Pears – Distorted fruits with brown, hardened flecks in the flesh, often combined with slight shoot dieback.
Radishes – Splitting and dull skin, often with woody flesh.
Strawberries – Small leaves and general stunting, with puckering and yellowing at the leaf tips. The berries are small, pale in colour and with a characteristic "waist" close to the calyx.
Swede and turnip – These show "brown heart", which can be seen when the roots are cut across to reveal grey or brown discoloured areas, often in concentric rings, in the lower parts of the root.
CAUSE This disorder can often occur if the soil has been limed excessively or if it has been allowed to become very dry. Boron is readily leached out of the soil, so heavy rainfall may encourage this deficiency on light soils.
CONTROL For vegetables, apply about 35g per 20m² of borax before sowing or planting (mixing the borax with horticultural sand to make an even application possible). For pear trees, spray at petal fall with 70g borax in 22 litres of water (2½oz in 5 gallons), together with a wetting agent.

BOTRYTIS

See GREY MOULD *p.135.*

BOX BLIGHT (CYLINDROCLADIUM SP.)

SYMPTOMS In autumn or winter brownish-black spots develop on the leaves, which gradually coalesce and may cover the entire leaf. On the lower surface patches of greyish-white fungal growth may be visible. The fungus may develop in the stems, causing black streaking on the bark. Defoliation and then dieback occurs. During the growing season the infected box plant will re-grow from the live parts. Although the appearance of the plant may be badly spoiled, there have, as yet, been no recorded cases of this fungus actually killing a plant. This infection has only been recognized in the United Kingdom since the mid-1990s and so far it has only been recorded on species of *Buxus*. It is known to attack *Buxus sempervirens*, in particular 'Suffruticosa', the native *Buxus sempervirens, B. microphylla* and *B. sinica*.
CAUSE The species of the fungus *Cylindrocladium*, which is encouraged by high humidity and rain. The spores are probably carried by rain- or water-splash, although animals such as birds may also carry it. When box plants are kept tightly clipped, perhaps when used for topiary, the fungus is all the more likely to cause extensive problems because air circulation around the foliage is restricted.
CONTROL As the fungus can form resting bodies in infected plant material, which then falls to the ground and remains viable, it seems likely that the fungus may be carried when plants, or soil on contaminated boots or gardening tools are moved from one location to another. It is essential to try to minimize this risk. As there are no fungicides available to gardeners to control this pathogen, infected plants should be removed promptly. Removal of infected stems is unlikely to prove adequate. If infected leaves have fallen then these should be removed before box plants are re-planted in the area. It is essential to ensure that any new box plants are completely healthy; buy from a reputable source and keep new plants heeled in a damp spot for about three weeks – infected plants should show symptoms within about 10 days. Consider striking your own cuttings from plants known to be disease free.

BOX BLIGHT (VOLUTELLA)

SYMPTOMS Infected leaves turn brown and then fall. The fungus overwinters on these leaves and in the spring produces spores which are then splashed on to the leaves and stems of the plant. The fungus enters via natural stem cracks and pruning or clipping wounds. Masses of pinkish spores are produced on the lower surface of infected leaves. Large patches of dead stems may develop, spoiling the plant's appearance.
CAUSE The fungus *Volutella buxi*, which may attack *Buxus sempervirens* and other species of box, and is encouraged by wet weather. The symptoms it causes are less severe than those caused by *Cylindrocladium* sp.
CONTROL Remove and dispose of all infected plant parts promptly, ideally during dry weather to minimize spread of the spores. There are no chemicals available to control this fungus.

BOX RED SPIDER MITE

SYMPTOMS A light whitish flecking develops on the upper leaf surface of box (*Buxus* species).
CAUSE A tiny sap-feeding mite, *Eurytetranychus buxi*, that lives and feeds on the underside of the leaves. The mite overwinters on its host plants as eggs. These hatch in the spring and most of the feeding damage occurs during spring to early summer, although the leaf flecking remains visible throughout the year.
CONTROL The damage is generally not sufficient to affect the plants' growth, so this is a pest that can be tolerated. If control measures are required, the plants could be sprayed with bifenthrin in the spring.

BOX SUCKER

SYMPTOMS The shoots of box are stunted, with closely-spaced cupped leaves that resemble cabbages (*see p.26*). Droplets of liquid with a white, waxy coating can be tipped out of the shoot tips during late spring. This is excrement produced by the sucker's nymphal stages.
CAUSE The immature nymphs of a sap-feeding insect, *Psylla buxi*, which attack the new growth in spring. The nymphs are flattened, pale green insects up to 2mm long. Winged adults are present in mid-summer, but they do not affect the plant's growth.
CONTROL If the box plant has grown to the desired height and is being clipped to keep it in shape, something that restricts shoot extension is not really a problem. Protect young plants that need to grow larger by spraying with bifenthrin, acetamiprid, or thiacloprid in mid-spring.

BRACKET FUNGUS

SYMPTOMS Fungal outgrowths develop on large limbs or the main trunk of affected trees (*see pp.37, 38*). They may appear throughout the tree, but are most common on the trunk, particularly towards the base. Some may be found growing from the buttress roots, or even attached to roots at some distance from the trunk. Some are annual, but most are perennial and quite woody in texture. Most have a tough leathery or woody upper surface, with spores being produced from pores on the undersurface. Most bracket fungi cause damage to trees – usually, extensive deterioration of the structural woody tissues. Infected trees often become dangerous because they are weakened structurally, and may suffer wind snap or wind throw.
CAUSE Various fungi, some of the most common being *Fistulina hepatica*, species of *Ganoderma, Innonotus hispidus, Laetiporus sulphureus, Meripilus giganteus, Phaeolus schweinitzi, Phellinus pini, Piptoporus betulinus*, and *Sparrisis crispa*.
CONTROL Precise identification is essential, together with examination of the tree and consideration of its environment. As infected trees may be hazardous, employ an arboriculturist to examine the tree. Removal of the fungal brackets may limit

spread, but will not cure the problem as the deterioration continues within.

BRASSICA BACTERIAL SOFT ROT

SYMPTOMS Patches of foliage or areas within the heart of the plant, often towards the stem base, discolour. Rapid disintegration of the tissues follows, usually accompanied by a foul smell.
CAUSE Various bacteria may be involved, but *Erwinia carotovora* is the most common. The bacteria usually gain entry through injured areas of the plant, when it has been attacked by pests such as slugs, snails and caterpillars.
CONTROL Avoid initial injury, where possible. It is rarely possible to save an infected plant because the soft rot is generally only noticed when in a fairly advanced state, but prompt removal of affected parts when the first symptoms occur may prevent its development.

BRASSICA DARK LEAF SPOT

SYMPTOMS Numerous round, brown to grey spots develop on leaves (*see* **p.13**), each with concentric rings of pinprick-sized fungal bodies. Infected areas of the leaf are killed and may drop away, leaving holes and a rather ragged-looking leaf.
CAUSE The fungus *Alternaria brassicola*, which is encouraged by wet weather and poor air circulation, and thrives on plants which are grown rather soft.
CONTROL Remove infected leaves promptly. Burn debris at the end of the season. Improve air circulation by spacing plants well and

weeding between the rows. Avoid overhead watering and excessive use of high-nitrogen fertilizers.

BRASSICA DOWNY MILDEW

SYMPTOMS Yellow blotches appear on upper leaf surfaces, each corresponding with a patch of greyish-white fungal growth on the lower leaf surface. Seedlings and young plants are more prone to attack than mature plants, but all stages can be susceptible. Severe infection may cause extensive leaf yellowing and the death of seedlings.
CAUSE The fungus *Peronospora parasitica*, which is encouraged by mild, damp weather and humid conditions. It has a wide host range, attacking many cruciferous plants, including some ornamentals and weeds. The spores can be spread by wind to nearby plants, and infection of young plants may also occur through the roots from infection within the soil.
CONTROL Improve air circulation by wide plant spacing and regular weeding. Remove affected leaves as seen.

BRASSICA LIGHT LEAF SPOT

SYMPTOMS Discoloured circles of leaf tissue develop, becoming covered in fluffy, off-white fungal growth. In extreme cases, spots merge, causing extensive damage.
CAUSE The fungus *Cylindro-sporium concentricum*.
CONTROL Remove affected leaves promptly, and dispose of all debris at the end of the season. Rotate crops, or at least allow a two-year break between brassica crops.

BRASSICA RING SPOT

SYMPTOMS Purple-brown spots with yellow edges develop on leaves, the older, outer leaves being most susceptible. Well-defined concentric ringing of the tissue is usually visible, and the spot may be covered in numerous, black, raised, pin-prick-sized, fungal fruiting bodies. This disease may occur on all brassicas but is most common and damaging on Brussels sprouts, where it may also attack the buttons.
CAUSE The fungus *Mycosphaerella brassicicola*, which is encouraged by damp weather conditions.
CONTROL Remove severely affected leaves and clear up all plant debris at the end of the season.

BRASSICA WIRE STEM

SYMPTOMS Leaves on brassica seedlings, especially those of cauliflower and broccoli, become yellowed and narrowed, or may die back.
CAUSE A deficiency of molybdenum, which is most common in acid soils.
CONTROL Lime the soil to reduce acidity, and water affected plants or seed rows with sodium molybdate.

BROOM GALL MITE

SYMPTOMS During early summer, buds on broom (*Cytisus*) fail to develop into normal shoots and leaves. They proliferate instead into green cauliflower-like growths which are covered in silvery hairs (*see* **p.39**). Later in the summer, these galled buds dry up and become greyish-brown. They remain on stems over winter,

and will usually persist on the plant for several years. Infestations increase in successive years until the shrub's flowering is significantly reduced.
CAUSE A microscopic gall mite, *Aceria genistae*, which lives and feeds within the affected buds.
CONTROL None of the insecticides available to amateur gardeners is effective against this mite. Remove badly infested plants and replant with clean stock in the autumn. Limit light infestations by picking off galled buds.

BROWN CORE OF PRIMULAS

SYMPTOMS Primula and polyanthus plants show signs of poor growth, and leaves may be yellowed. Plants wilt readily, especially during warm weather, but even when there is plenty of soil moisture available. Plants wither and die, usually within a single season. The root system deteriorates, leaving a reduced number of roots and little fine root growth. If roots are cut lengthwise, a distinct reddish-brown core of discoloration is seen running down them.
CAUSE The fungus *Phytophthora primulae*, which is soil-borne and builds up in the soil when susceptible plants are grown on the same site for a number of years in succession.
CONTROL Remove infected plants promptly. There is no soil treatment available, but growing fresh plants on a new site is usually successful. Take care not to spread soil from around infected plants to other parts of the garden.

BROWN ROT

SYMPTOMS A soft, brown area develops on the fruit skin of tree fruits (*see* **p.47**) and also in the flesh beneath when it starts to deteriorate. The whole fruit is ultimately affected. Raised creamy-white pustules appear on the rotted areas, in roughly concentric rings (*see* **p.49**). Infected fruits either fall off the tree, or may become "mummified" and dry and remain attached to the branches. Most tree fruits can be affected, especially apple, pear, plum, peach and nectarine.
CAUSE The fungi *Sclerotinia fructigena* (on apples) and *S. laxa* (other host plants), which gain entry through injured skin. Bird pecks, codling moth exit holes, frost crack, cracking due to irregular growth and fungal scab infections are all common causes of the initial injury. The spores are readily spread by insects or birds, by rain splash and by fruits coming into contact with a source of infection.
CONTROL Avoid, or minimize, possible causes of injury to fruits by taking appropriate control or preventive measures. Remove infected fruits promptly, as well as fallen ones, and dispose of them well away from the garden, or burn them. Prune out all mummified fruits, together with a short section of the spur to which they are attached.

BROWN SCALE

SYMPTOMS Brown, shell-like, convex objects, which can be up to 5mm in length, are attached to the stems of many ornamental shrubs and fruit bushes (*see* **p.36**), especially those growing against walls or in other sheltered places. The honeydew excreted by the insects makes the foliage sticky and allows the growth of sooty moulds.
CAUSE A sap-feeding insect, *Parthenolecanium corni*, which usually has one generation a year. Its eggs are laid under the scale.
CONTROL The newly-hatched scale nymphs are the most vulnerable stage in the life cycle, feeding on the stems and undersides of leaves in early to mid-summer. Spray ornamental plants in mid-summer with thiacloprid or acetamiprid. Fruit trees and bushes can be sprayed with fatty acids and plant oils.

BROWN-TAIL MOTH

SYMPTOMS Greyish-black, hairy caterpillars, up to 40mm in length, feed together in colonies under silk webbing on hawthorn, cherry, apple, blackthorn, blackberry, rose and other trees and shrubs. The caterpillars have red and white markings on their bodies and a pair of orange warts on the upper surface at the rear end. They overwinter as small larvae inside dense, white silk webs spun at shoot tips.
CAUSE Caterpillars of a moth, *Euproctis chrysorrhoea*, which during late spring and early summer cause severe defoliation. In Britain, the moth is mainly found on the south and east coasts of England, but sometimes occurs further inland.
CONTROL Prune out the overwintering silk nests after leaf fall. Wear gloves to avoid contact with the hairs, as some people develop an intensely irritating rash when the caterpillar hairs come into contact with their skin. From mid-spring to early summer spray infested plants with an insecticide such as bifenthrin or carefully prune out infested shoots. Young caterpillars are more susceptible to sprays than the older larvae.

BRYOBIA MITES

SYMPTOMS Black-bodied mites, up to 1mm in length, with four pairs of pinkish legs, occur on the upper surface of leaves on ivies, primulas, polyanthus, gooseberry and other plants during late winter to early summer. Leaves develop a fine, pale mottling on the upper surface.
CAUSE One of several species of sap-feeding mites, *Bryobia* species.
CONTROL Spray thoroughly with bifenthrin when bryobia mites are seen. Heavy infestations may need three applications at two-week intervals.

BUD DROP

SYMPTOMS Buds form, usually to full size, but drop while still apparently healthy. The plant is otherwise healthy.
CAUSE A period of dry weather while the flower buds are forming. On very susceptible plants, such as camellia and rhododendron, just a few consecutive days of drought at the end of summer when the buds are forming are enough to do damage. Plants growing in containers are particularly vulnerable because they are more prone to dry conditions around the roots. Feeding too late in the growing season (after early summer) may also induce bud drop the following spring.
CONTROL Water thoroughly and regularly and never let the soil or compost dry out completely. Apply a deep mulch around the root area.

Feed no later than early to mid-summer and avoid excessive feeding.

BUD, DRY

SYMPTOMS Buds form, but may be undersized and are usually dry and brown; the plant is otherwise healthy.
CAUSE As with bud drop (*see above*), a period of dry weather while the flower buds are forming. On very susceptible plants, such as rhododendron, just a few consecutive days of drought at the end of summer are enough. Plants growing in containers are particularly vulnerable because they are more prone to dry conditions around the roots.
CONTROL Water thoroughly and regularly and do not let the soil dry out completely. Apply a deep mulch around the root area.

BUFF-TIP MOTH

SYMPTOMS Colonies of black and yellow chequered caterpillars, up to 50mm in length, feed together on leaves of deciduous trees and shrubs, including lime, oak, hornbeam, birch, hazel, rose and cherry, from mid- to late summer. Unlike most other gregarious moths, buff-tip larvae produce no silk webbing.
CAUSE Caterpillars of a moth, *Phalera bucephala*, which defoliate branches on their host plants.
CONTROL Buff-tip moth is only likely to harm young trees or shrubs. If necessary, prune out infested shoots or spray with an insecticide such as bifenthrin.

BULB SCALE MITE

SYMPTOMS Daffodil and amaryllis bulbs grown

indoors for early flowering are much more likely to show symptoms of this pest than similar bulbs planted out in gardens. The whitish-brown mites are less than 1mm long and barely visible to the naked eye. Their feeding causes stunted and distorted growth. Flower stems and leaves have saw-tooth scarring along the edges and show a sickle-shaped curvature.

CAUSE Sap-feeding mites, *Steneotarsonemus laticeps*, which live in the neck area of dormant bulbs. These should not be confused with *Rhizoglyphus* species mites, which are larger, broader, pearly-white mites often found feeding on rotting bulbs. These are not pests, as they feed on damage initiated by some other cause.

CONTROL None of the chemicals available to amateur gardeners will control bulb scale mite. Infested bulbs should be thrown away.

BULLFINCHES

Bullfinches (*Pyrrhula pyrrhula*) feed on wildflower seeds for most of the year, but turn to flower buds when the seed supply runs out during winter. During late autumn to spring they will peck out the flower buds of plum (*see* **p.46**), pear, cherry, apple, gooseberry, almond and forsythia. The outer bud scales are discarded and are easily seen when the ground below the plant is covered by snow or frost. Buds at the shoot tips often survive and so, in the spring, shoots which are mostly devoid of flowers will have clusters of blossom at the tips.

CONTROL Only netting will give complete protection; bird-repellent sprays are unreliable and may need frequent applications.

BURROWING BEES

SYMPTOMS Small, conical heaps of fine soil appear in lawns or on bare soil during late spring and summer (*see* **p.58**). They are distinguished from ant nests by the hole at the top of the heap where the bee enters its nest. Each nest consists of a vertical shaft in the soil; this shaft has several side-chambers which the bee provisions with nectar and pollen as food for its grubs.

CAUSE Many different species of solitary bees nest in this way in gardens. A common species is a ginger-brown bee, *Andrena fulva*, which is active in late spring. Although all burrowing bees are solitary, with each female constructing her own nest, the nests may be clustered together and the same piece of ground may be used by the bees in successive years.

CONTROL Control measures are not necessary as burrowing bees cause no real damage to lawns, and they are of benefit as pollinating insects for fruit trees and bushes. The females have stings, but lack aggressive instincts and do not attack humans.

CABBAGE MOTH

SYMPTOMS All types of brassica are damaged during the summer by yellowish-brown or green caterpillars, up to 42mm in length. They eat holes in the foliage (*see* **p.30**), especially the heart leaves of cabbages. Cabbage moth caterpillars also enter onion leaves, making holes and causing them to flop over.

CAUSE Caterpillars of a moth, *Mamestra brassicae*, which has one generation a year.

CONTROL Check brassicas regularly in summer and take control measures as soon as caterpillars are seen; they are difficult to deal with once they grow larger and go into the heart leaves. Use an insecticide containing pyrethrum or bifenthrin against the young larvae.
• *see also* CABBAGE WHITE BUTTERFLY (*below*).

CABBAGE ROOT FLY

SYMPTOMS All brassicas and allied vegetables such as turnip, swede and radish are attacked. Some related ornamentals, including stocks, alyssum and wallflowers, are sometimes damaged. Affected plants grow slowly and tend to wilt on sunny days. Brassicas are particularly vulnerable as young plants, especially shortly after transplanting, when they can be killed. White, legless maggots, up to 9mm in length, eat the finer roots until just a rotting stump is left. The maggots also bore into the swollen roots of radish, swede and turnip.

CAUSE Maggots of a fly, *Delia radicum*, which has three generations from mid-spring to early autumn, hatching from eggs laid in the soil.

CONTROL Established plants usually tolerate cabbage fly attacks: it is seedlings and transplants that need protection. There are no insecticides for use against cabbage root fly. Infestations can be reduced without using insecticides by placing squares or discs made of carpet underlay, cardboard or roofing felt, about 10cm in diameter, on the soil around the stems of transplants. The female flies deposit their eggs on the discs, instead of in the soil; they then dry up before hatching. Growing brassicas, etc., under fleece will exclude the adult flies.

CABBAGE WHITE BUTTERFLIES

SYMPTOMS Holes are eaten in the leaves of brassicas and some ornamental plants such as nasturtium. Caterpillars of the large cabbage white butterfly are yellow with black markings. They have hairy bodies, up to 40mm long, and feed mainly on the outer leaves of brassicas. Small cabbage white butterfly larvae are up to 25mm long and feed mainly in the hearts of cabbages and other brassicas. They are pale green with a velvety appearance, unlike the smooth and largely hairless caterpillars of cabbage moth (*see above*).

CAUSE Large cabbage white, *Pieris brassicae*, and small cabbage white, *P. rapae*, which each have two generations a year, with caterpillars occurring from spring to early autumn.

CONTROL Check brassicas regularly from early summer to early autumn and take control measures as soon as caterpillars are seen. They are difficult to deal with once they have grown larger and have gone into the heart leaves. Use an insecticide such as pyrethrum or bifenthrin against the young larvae.

CABBAGE WHITEFLY

SYMPTOMS Brassicas, such as Brussels sprout, kale, cabbage, cauliflower and broccoli, have 2mm long, white-winged insects on the undersides of the leaves (*see **p.32***). The adults are active throughout the year and readily fly up when disturbed. The immature nymphs are flat, oval, scale-like insects that remain immobile on the lower leaf surface. Heavy infestations make the foliage sticky with the whiteflies' sugary excrement, and sooty moulds will grow on this substance.
CAUSE The adults and nymphs of a sap-feeding insect, *Aleyrodes proletella*. This insect has several generations during the summer, and overwinters as an adult whitefly.
CONTROL Light infestations can be tolerated, but if large numbers of whitefly are present and the foliage is becoming soiled with sooty moulds, spray with bifenthrin, pyrethrum, fatty acids or plant oils. Several treatments at weekly intervals may be necessary with the organic insecticides.

CACTUS CORKY SCAB

SYMPTOMS Buff to tan corky-brown markings develop on the skin and may become sunken (*see **p.22***). Central areas often become grey or dark brown. Adjacent areas of the plant usually retain their normal green colour. Cacti and many succulents can be affected, especially *Opuntia* and *Epiphyllum*.
CAUSE Unsuitable growing conditions, in particular excessively high humidity or light levels.
CONTROL Improve growing conditions by moving the plant to a slightly cooler position and improving ventilation to reduce humidity. Take care not to alter conditions too dramatically or rapidly or the plant may suffer even more.

CALCIUM DEFICIENCY

SYMPTOMS These vary according to the plant and plant part affected. The most common symptoms are blossom end rot and bitter pit. Other symptoms may include: browning of the internal tissues of Brussels sprouts' buttons and the heads of cabbage; cavity spot of carrots; stunting and blackening of central leaves on celery; tip burn of lettuce, and the production of rolled leaves and spindly shoots on potato. On other plants the most common symptom is poor growth of young shoots and leaves, often combined with some curling and shoot or leaf deterioration.
CAUSE A deficiency of calcium, either in the soil or compost, or because of an inadequate uptake of calcium due to a low moisture content of the soil. This disorder is most common on acidic soils or growing media.
CONTROL Ensure an adequate supply of moisture and encourage soil moisture retention by mulching, where appropriate. Where possible on acid soils lime the soil to raise the pH level.
• see also APPLE BITTER PIT p.100; BLOSSOM END ROT p.108; CARROT CAVITY SPOT p.115.

CAMELLIA GALL

SYMPTOMS Large, pale green swellings, often forked or hand-like, develop in place of the leaves (*see p.26*). Each gall may measure several centimetres in width and up to about 20cm in length. The surface soon becomes covered in a layer of spores, giving the gall a creamy-white appearance. Small shrivelled pieces of leaf may protrude from the tips of the galls. There are rarely more than one or two galls per plant. The overall vigour and flowering of the camellia is not affected.
CAUSE The fungus *Exobasidium camelliae*, which seems to be encouraged by damp weather. The spores are probably spread by insects and by rain splash.
CONTROL Remove galls, preferably before the spore layer develops.

CAMELLIA PETAL BLIGHT

SYMPTOMS Brown patches develop on the flowers and spread to the base. Infected flowers fall prematurely and a hard, black fungal resting body (sclerotium) forms at the base of the petals. These remain dormant beneath the tree and later produce structures which produce spores that land on petals of the next season's blooms. The sclerotia can remain viable for up to five years.
CAUSE The fungus *Ciborinia camelliae,* which is spread on air currents. Soil contaminated with the sclerotia can also spread the fungus.
CONTROL Remove fallen flowers promptly. Avoid moving contaminated soil; check treads of boots, etc., too.

CAMELLIA YELLOW MOTTLE

SYMPTOMS Irregular blotching and mottling of camellia leaves, with bright yellow or near-white markings on the dark green surface (*see **p.22***). Occasionally whole branches may bear completely white or strongly chlorotic leaves. The overall vigour of the plant seems not to be affected.
CAUSE An organism known as camellia yellow mottle, which is virus-like in its structure and behaviour. This organism is systemic within the plant, so propagation should not be carried out using material from any part of an affected plant.
CONTROL Prune out all affected stems.

CANKER

SYMPTOMS The precise symptoms of the various cankers (see references below) vary with both the host plant and the causal organism. The bark is usually raised and often roughened and may split to reveal the tissues beneath. In some cases ooze may develop around the cankered areas. As cankers enlarge, they often ring the infected stem and so growth above the canker deteriorates and may die.
CAUSE Various fungal and bacterial organisms, most of which have a fairly restricted and specific host range.
CONTROL Remove affected areas as soon as possible and, where possible, consider treating with an appropriate chemical. Improve growing conditions.
• see also APPLE AND PEAR CANKER p.100; BACTERIAL CANKER p.104; MULBERRY CANKER p.150; NECTRIA CANKER p.152; SEIRIDIUM

CANKER *p.176*; PHOMOPSIS CANKER *p.160*; POPLAR CANKER *p.164*.

CAPSID BUGS

SYMPTOMS Leaves at the shoot tips of many plants, such as fuchsias, *Caryopteris*, roses, hydrangeas, currants, forsythias, dahlias and chrysanthemums, are misshapen, with many small holes (*see* **p.29**). Flowers of dahlia and chrysanthemum open unevenly and those of fuchsia (*see* **p.46**) abort at an early stage.

CAUSE Pale green insects, such as the common green capsid, *Lygocoris pabulinus*, and tarnished plant bug, *Lygus rugulipennis*, both of which are about 6mm long when adult. They suck sap from the shoot tips, and secrete a toxic saliva that kills some of the plant cells. Later, as leaves develop from the damaged shoots, these dead areas tear into many small holes.

CONTROL Damage occurs any time between late spring and late summer. Check vulnerable plants during the summer and spray with thiacloprid or bifenthrin as soon as capsid bugs or signs of damage are seen.

• see also APPLE CAPSID *p.100*.

CARNATION TORTRIX MOTH

SYMPTOMS Pale green caterpillars up to 18mm long, with brown heads, feed between two leaves bound together with silk webbing, or under a curled leaf margin similarly held in place by silk. The caterpillars graze away the inner surfaces of their hiding places, causing the remaining damaged tissues to dry up and turn brown (*see* **p.20**). This pest occurs on a wide range of shrubs and herbaceous plants, both in gardens and in greenhouses.

CAUSE Caterpillars of a small, brown-winged moth, *Cacoecimorpha pronubana*. This moth has several generations a year, and its caterpillars can be found at any season on plants growing indoors or in sheltered places.

CONTROL The fact that the caterpillars are concealed within their hiding places makes them difficult to reach with insecticides. Some control can be achieved by looking for bound-up leaves and squeezing them to crush the caterpillars and pupae. If plants are too heavily infested for this to be feasible, spray with bifenthrin. A pheromone trap can be used to capture adult male moths in greenhouses.

CARROT CAVITY SPOT

SYMPTOMS Lens-shaped spots of discoloration develop on carrot roots. Cracks and associated craters develop beneath. Secondary organisms may then invade and cause rapid deterioration of the root.

CAUSE The precise cause is uncertain, but a deficiency of calcium is believed to be the main cause. This deficiency causes groups of cells to collapse, so causing discoloration and crater formation.

CONTROL Ensure an adequate supply of moisture. Where feasible on acid soils, lime the soil to raise the pH level to about 6.5.

CARROT FLY

SYMPTOMS The roots of carrots and related plants, such as parsley, parsnip and celery, are tunnelled by slender, creamy-yellow maggots up to 9mm long. These cause rusty brown lines on the outside of the tap roots where tunnels close to the surface have collapsed (*see* **p.51**). Damaged roots are susceptible to secondary rots, which can affect their storage during the winter.

CAUSE Larvae of carrot fly, *Psila rosae*, which has two or three generations between late spring and late autumn. Carrot is the favourite host.

CONTROL Seedlings of carrots sown after late spring will miss the first generation of larvae; carrots harvested before late summer will miss the second generation. Growing susceptible plants on a fresh site under horticultural fleece prevents female flies getting close enough to lay eggs. There are currently no insecticides approved for carrot fly control. Some carrots, such as 'Sytan', 'Resistafly' and 'Flyaway', are less susceptible to damage than most other cultivars.

CARROT MOTLEY DWARF VIRUS

SYMPTOMS Foliage of carrots and parsley develops distinct yellow and pink hues (*see* **p.25**), and is stunted and sometimes twisted. Leaf stalks may show pronounced twisting. Affected plants are stunted, and carrots form little if any edible root. Root tips and root hairs die back. Similar foliage discoloration may be seen when these plants are suffering from a deficiency of magnesium, but the other symptoms described are not seen.

CAUSE Carrot motley dwarf virus is usually found together with carrot red leaf virus, and it is these two viruses, in combination, which are responsible for the symptoms. The viruses are both spread by aphids.

CONTROL Dig up and dispose of infected carrots. Attempt thorough aphid control to minimize the risk of infection development and spread.

CATERPILLARS

Caterpillars are the larval stage of butterflies and moths. There is much variety in their colour, size and degree of hairiness, but they generally have an elongate tubular body shape with a distinct head capsule. They have three pairs of jointed legs at the head end and two to five pairs of clasping legs (prolegs) on their abdomen. Larvae of sawflies look similar but have more than five (usually seven) pairs of prolegs. Most pest species feed on leaves, sometimes gregariously under silk webbing. Others live in soil and feed on roots, or bore into stems, or feed inside leaves as leaf miners, or feed on fruits, berries and seeds.

Root damage – *see* CUTWORMS *p.122*, SWIFT MOTHS *p.182*.

Stem borers – *see* CURRANT CLEARWING MOTH *p.122*; LEOPARD MOTH *p.146*.

Leaf miners – *see* APPLE LEAF MINER *p.101*; HORSE CHESTNUT LEAF MINER *p.139*; LABURNUM LEAF MINERS *p.142*; LEEK MOTH *p.145*; LILAC LEAF MINER *p.147*; PYRACANTHA LEAF MINER *p.167*.

Leaf damage – *see* CABBAGE WHITE BUTTERFLY *p.113*; CABBAGE MOTH *p.113*; CARNATION TORTRIX MOTH *p.115*; CHINA MARK MOTH *p.117*; DELPHINIUM MOTH *p.123*; ELEPHANT HAWK MOTH *p.126*; MULLEIN MOTH *p.151*;

VAPOURER MOTH *p.187*; WINTER MOTH *p.193*; YELLOW-TAIL MOTH *p.194*. **Gregarious leaf feeders, often with silk webbing** – *see* BROWN-TAIL MOTH *p.112*; BUFF-TIP MOTH *p.112*; COTONEASTER WEBBER MOTH *p.120*; GYPSY MOTH *p.136*; JUNIPER WEBBER MOTH *p.142*; LACKEY MOTH *p.143*; OAK PROCESSIONARY MOTH *p.153*; SMALL ERMINE MOTHS *p.178*. **Fruit, seed and berry damage** – *see* CODLING MOTH *p.119*; PEA MOTH *p.157*; PLUM MOTH *p.162*; TOMATO MOTH *p.184*. Other species can sometimes be removed by hand, especially on small plants. Many caterpillars feed after dark and are easier to find by torchlight on mild evenings. An alternative treatment is to use an insecticide, such as pyrethrum or bifenthrin, when signs of caterpillar feeding are seen.

CATS

Despite their value as pets, cats can be a nuisance in gardens. They soil flower borders and vegetable plots and may damage plants as they attempt to bury their droppings. Their urine can scorch foliage, and the bark of trees and shrubs is damaged when cats use them as scratching posts.
CONTROL In localities where cats are numerous, it is difficult to stop them being a problem. Proprietary animal repellent sprays and powders often fail to give reliable long-term protection. Devices which emit ultrasonic sounds that are virtually inaudible to humans but sound loud to cats can encourage them to go elsewhere. However, as cats become used to the noise some may turn a deaf ear to it. Cats looking for

toilet areas are attracted to loose dry soil or mulch-covered flower beds. Close planting helps to cover up the soil surface with vegetation, making it less accessible.

CAVITY DECAY

SYMPTOMS Once exposed to the elements and open to infection by fungal spores, damaged areas on trees may soon deteriorate and form a cavity. The decay may then spread further into the trunk or limb, which may result in sufficient structural damage to render the tree dangerous.
CAUSE Initial injury may result from storm damage, poor pruning or other physical injury.
CONTROL Prevent cavity formation wherever possible. When cavities develop, the rotting wood should not be removed and the healthy wood exposed: this was once believed to be the best method of dealing with wood rotting, but it has since been determined that this action simply promotes the breaching of the barriers which trees form in response to injury, and more extensive rotting subsequently develops. Trees with cavities should be examined regularly to determine their condition and safety.

CEDAR APHID

SYMPTOMS The foliage and stems of cedars become thickly coated with a black, sooty mould that grows on the sugary honeydew excreted by the aphids (*see p.32*). Heavy infestations cause the foliage to turn yellow in colour and drop off in mid-summer.
CAUSE A greyish-brown, sap-feeding aphid, *Cedrobium*

lapportei, up to 2mm in length, which forms dense colonies at the leaf bases in late spring to early summer.
CONTROL Spray small trees with bifenthrin, acetamiprid, or thiacloprid when aphids or sooty moulds are detected. Damage to large trees has to be tolerated but, fortunately, infestations resulting in defoliation are infrequent occurrences.

CELERY LEAF MINER

SYMPTOMS The leaves of celery and related plants such as parsnip, parsley and lovage develop brown dried-up patches (*see p.19*). This is where the interior of the leaf has been eaten by one or more white maggots up to 7mm long. Heavy attacks give the plants a scorched appearance, with celery plants producing thin, bitter-tasting stems.
CAUSE Larvae of a leaf-mining fly, *Euleia heraclei*; it has two generations a year, with damage occurring from late spring to late summer.
CONTROL With light infestations, pick off the damaged portions of the leaves. There are no suitable insecticides for celery leaf miner.

CELERY LEAF SPOT

SYMPTOMS Pinprick-sized spots develop on the leaf surface (*see p.12*), then even smaller black fungal fruiting bodies develop on the spots. The infection may spread to the leaf stalks, and severely affected leaves die off.
CAUSE The fungus *Septoria apiicola*, which is spread by handling, by insects and by water splash. In dull, damp weather, or on damp, misty nights, the spores are spread

very readily. The infection may also be seed-borne and the fungus may overwinter on infected crop debris.
CONTROL Grow plants with care to ensure that they are vigorous and so better able to compensate for any losses. Always use disease-free seed which has been treated against leaf spot, from a reliable company. Spray young plants with a copper-based fungicide. Practise crop rotation (*see pp.92–96*) to minimize the risk of disease carry-over from year to year. Destroy all infected plant debris at the end of the season.

CENTIPEDES

Centipedes are yellowish-brown animals with many body segments, each of which has one pair of legs. This distinguishes them from the superficially similar millipedes which have two pairs of legs per segment. Centipedes also have longer antennae on their heads, and are more active animals, which run away if disturbed. They are predators of small soil-dwelling animals and may help control soil pests.

CHAFER GRUBS

SYMPTOMS Plants may suddenly wilt. Those species of chafers that eat grass roots cause yellow patches in lawns (*see p.58*). Heavy infestations attract foxes, badgers, rooks, magpies and crows, which scratch out the loosened turf to feed on the grubs during late summer to spring.
CAUSE Plump, white-bodied grubs with brown heads and three pairs of legs that live in the soil and feed on roots (*see p.53*). Their bodies are

curved like the letter C. They can kill small plants and gnaw cavities in root vegetables. Several species of chafers occur in gardens. The largest species is the cock chafer or maybug, *Melolontha melolontha*, which has larvae up to 40mm long. It mainly occurs in cultivated soil such as flower-beds and vegetable gardens. Lawns are attacked by the garden chafer, *Phyllopertha horticola* and Welsh chafer, *Hoplia philanthus*. The latter are particularly associated with sandy soils and are smaller, with grubs up to 20mm long. When fully fed, the grubs pupate in the soil and emerge as adult beetles in late spring to early summer.

CONTROL Chafer grubs are difficult to kill with insecticides once they have grown large enough to cause damage. Infestations in flower borders and vegetable gardens are generally light, and it is feasible to search for the grubs in the soil near plants which have suddenly wilted. Problems in lawns are sporadic, but can be devastating when they do occur. A pathogenic nematode, *Heterorhabditis megidis*, can be watered into lawns in the summer while the soil is moist and in the temperature range 14–20°C (57–68°F). The nematodes enter the bodies of the grubs and kill them by releasing bacteria that cause a fatal infection. Lawns can also be treated with imidacloprid in early summer.

CHERRY BLACKFLY

SYMPTOMS Leaves at the tips of shoots become curled in late spring, and dense colonies of black aphids are visible on the leaf undersides

(see **p.24**). By mid-summer some of the affected leaves will turn brown and dry up. **CAUSE** A sap-feeding aphid, *Myzus cerasi*, which overwinters as eggs on fruiting cherries and some ornamental species such as *Prunus avium* and *P. padus*. The Japanese flowering cherries are generally not attacked. **CONTROL** Look for aphids on new growth from mid-spring onwards and spray ornamental cherries with acetamiprid, thiacloprid or bifenthrin before leaf curling has become extensive. Only thiacloprid and organic sprays, such as pyrethrum, fatty acids or plant oils, can be used on fruiting cherries.

CHERRY LEAF SCORCH

SYMPTOMS In the winter infected trees stand out because their branches remain covered with dead, brown, shrivelled leaves. The buds and stems, however, remain alive. Close examination of the lower surface of infected leaves with a hand lens reveals tiny reddish-brown blisters. These produce the fungal spores and so, once the spores have been shed, small indentations are all that is left on the leaf. In the summer yellow-margined brown blotches develop on the leaves, which then brown completely and die. **CAUSE** The fungus *Apiognomonia erythrostoma*. The spores from the over-wintering dead leaves infect new foliage in the spring. Little is known about the precise host range, but it is most widely seen on *Prunus avium* and its cultivars. **CONTROL** Despite the

premature leaf browning, this disease seems to have few, if any, serious adverse effects. No fungicidal controls are available, but it may be worth removing infected leaves (perhaps with a garden vacuum). Infected leaves must not be composted.

CHERRY LEAF SPOT

SYMPTOMS In the summer the leaves develop purplish spots, often followed by shot-hole symptoms or yellow mottling. The lower leaf surface may be marked with white flecks. Leaf fall is common, and both twigs and fruit crop levels and quality may be affected. **CAUSE** The fungus *Blumeriella jaapii*, which is widespread in much of Europe but has only recently been reported in the UK. The fungus overwinters on fallen leaves then produces spores which infect new foliage in the spring. Rain and high humidity encourage spread of the infection. **CONTROL** Rake up and dispose of fallen leaves; do not compost them.

CHICKEN OF THE WOODS

See LAETIPORUS SULPHUREUS *p.143.*

CHINA ASTER FUSARIUM WILT

See ASTER WILT *p.103.*

CHINA MARK MOTH

SYMPTOMS Water lily pads and other floating leaves of pond plants have oval holes up to 15mm in length (see **p.30**). The removed leaf pieces are spun together with silk by a small

caterpillar to form a protective case around it. The caterpillars become fully fed in summer. **CAUSE** Caterpillars of the china mark moth, *Elophila nymphaeata*. The adult moths have brown wings with white blotches and they rest on pond-side vegetation. **CONTROL** Insecticides cannot be used in ponds as they will kill fish and other pond life. Search for the caterpillars' leaf cases, which are often on the undersides of floating leaves, and remove them by hand.

CHIRONOMID MIDGES

SYMPTOMS The leaves of water lilies and other plants with floating foliage, such as water hawthorn, frog-bit and *Potamogetum* species, have a nibbled appearance around the edges. The foliage goes yellow and rots. Slender, transparent larvae, up to 10mm long, can be found feeding in damaged tissues. **CAUSE** Maggots of certain chironomid flies (*Cricotopus* species). These seem to be a particular problem in new ponds or recently cleaned established pools. Usually, once the pond has settled down and regained its complement of predatory insects, the midge population is kept at a non-damaging level. **CONTROL** Insecticides cannot be used because of the danger to fish and other pond life. Remove damaged leaves and introduce fish if there are none in the pond.

CHLOROSIS

SYMPTOMS Yellowing of plant tissue, most commonly in the foliage. The loss or deterioration of the green pigment (chlorophyll) may also allow other pigments, which are usually masked by the green, to be seen. In addition to yellowing, red or orange discoloration may also develop.

CAUSE Many quite unrelated factors may be involved. Most common are deficiencies of iron and manganese (so-called lime-induced chlorosis), nitrogen and magnesium. Waterlogging, low temperatures, virus infection or weedkiller contamination are also fairly commonly responsible. The precise patterning of the chlorotic symptoms and their distribution within the plant may help to determine which factor is responsible.

CONTROL Take appropriate action to remedy any mineral deficiencies, sources of contamination or cultural problems.

CHOCOLATE SPOT

See BEAN CHOCOLATE SPOT p.105.

CHRYSANTHEMUM EELWORM

SYMPTOMS Foliage on chrysanthemums, Japanese anemones, penstemons and many other herbaceous plants progressively turns brown and dries up from the base of the plants upwards. Spread of the microscopic eelworms causing the damage within the leaf is restricted by the larger leaf veins, therefore recently infested leaves have brown wedges or islands separated from green healthy parts by the leaf veins. Eventually the entire leaf will turn brown or black.

CAUSES Microscopic nematodes or eelworms, *Aphelenchoides ritzemabosi*, which feed within the foliage. Although present on suitable host plants all year, infestations develop most rapidly during damp weather in late summer and autumn as the foliage starts to age.

CONTROL None of the insecticides available to amateur gardeners is effective. Destroy badly affected plants. Dormant chrysanthemum stools dipped in hot water will produce shoots for cuttings that are free of eelworm. The old stems are cut down and soil washed from the stool before immersing it for five minutes in water held at 46°C (115°F). The amount of heat is critical, as too much will damage the plant, while too little will allow eelworms to survive.

CHRYSANTHEMUM LEAF MINER

SYMPTOMS Sinuous white or brown lines develop in the leaves of chrysanthemums (see *p.25*), gerberas, cinerarias, marguerites, pyrethrum and other related composite flowers. If an infested leaf is held up to the light, a grub or its pupa may be seen at the end of its tunnel. Small, pale, slightly raised round scars appear on the upper surface of leaves at shoot tips where the adult flies have been feeding.

CAUSE Maggots of a leaf-mining fly, *Phytomyza sygenesiae*, which has two generations during the summer on garden plants, but can breed continuously throughout the year in greenhouses.

CONTROL Deal with light infestations by picking off mined leaves, or crush the grubs at the ends of their tunnels. There are currently no insecticides available to amateur gardeners for controlling leaf miner.

CHRYSANTHEMUM PETAL BLIGHT

SYMPTOMS Water-soaked lesions or smaller, brown spots develop on the petals of chrysanthemums (see *p.45*) and occasionally other members of the Compositae family, and on anemones. Infected flowers deteriorate rapidly.

CAUSE The fungus *Itersonilia perplexans*, which is encouraged by wet weather and by humid conditions. It is less common than ray blight (see *below*).

CONTROL Remove infected plant parts or whole plants immediately. Improve air circulation and attempt to lower humidity.

CHRYSANTHEMUM RAY BLIGHT

SYMPTOMS Red-brown spots develop on the petals; as the spots increase in number and size, the petals are killed off and the whole flower head starts to decay. Rotting of shoot tips and brown spots on the foliage may also be seen. Occasionally infections occur at the stem base, causing the whole plant to be stunted and discoloured, and eventually to die. Similar symptoms may be caused by chrysanthemum petal blight (see *above*) or grey mould (see *p.135*).

CAUSE The fungus *Didymella chrysanthemi*, which can overwinter on infected plants and plant debris. Conditions of high humidity, especially common in glasshouses, encourage disease development and spread.

CONTROL Remove and dispose of infected plants immediately. Decrease humidity by opening vents and windows to improve air circulation.

CHRYSANTHEMUM WHITE RUST

SYMPTOMS Pale buff or cream-coloured raised, warty pustules develop on the lower leaf surface (see *p.14*). The upper surface is marked with yellow pits, each pit corresponding with a pustule. As pustules age they become pale to tan-brown, and leaves may become distorted and die.

CAUSE The fungus *Puccinia horiana*, which is encouraged by moist, humid atmospheres. It may persist from year to year on over-wintering plants and stools, and can infect indoor and outdoor chrysanthemums.

CONTROL Remove infected leaves promptly. Spray all chrysanthemums in the vicinity with a suitable fungicide such as penconazole or myclobutanil, giving at least three applications. Destroy severely infected plants and do not propagate from them. Improve air circulation to decrease humidity.

CLEMATIS GREEN PETAL

SYMPTOMS The flowers of clematis open to reveal a distinct green coloration. This is especially clear on pale-flowered or white-petalled clematis. It is usually only the first flush of flowers that is affected.

CAUSE The most common causes are adverse growing or cultural conditions at a critical stage in the flower's development. Green flowers frequently develop a few weeks after there has been an unusually cold spell of weather. If cold damage is responsible, subsequent flushes of flowers are perfectly normal. However, if the problem lasts throughout the season, and reappears the following year, then the development of green pigments in the petals could be due to the presence of a mycoplasma infection.

CONTROL Protect plants against excessively cold conditions with horticultural fleece or similar material. If the plant has a mycoplasma infection it should be destroyed, as there is no cure available.

CLEMATIS WILT

SYMPTOMS Leaves and shoots wilt and die back. Wilting usually starts at the shoot tips, leaf stalks blacken where they join the leaf blade and young foliage wilts. New, healthy shoots may appear from below soil level or from nodes beneath the wilted area. A small spot of discoloration may develop under the lowest pair of wilting leaves. Waterlogging, graft failure, insect or slug damage may cause wilting too, but in these cases affected clematis do not show the other symptoms.
CAUSE The fungus *Ascochyta clematidina*.
CONTROL Prune out affected stems, cutting right back into healthy tissue, even cutting below soil level if necessary. Mildly infected and otherwise vigorous plants should reshoot.

CLICK BEETLES

Click beetles are elongate brown beetles, 8–20mm long. If a click beetle is placed on its back it will right itself by jumping into the air with an audible clicking noise, hence its common name. Click beetles cause no damage as adult insects, but several species are the adult stage of soil pests known as wireworms (*see p.193*).

CLUBROOT

SYMPTOMS Roots become swollen, and the whole root system may become distorted (*see **p.55***). The foliage is pale, chlorotic or pinkish in colour and wilts readily, especially during hot weather, but may even wilt when the soil is moist. Plants fail completely or produce a poor crop. Members of the Cruciferae may be attacked, in particular brassicas. Certain ornamentals, including wallflowers, stocks and candytuft, and also some weeds, including shepherd's purse, charlock and wild radish, are susceptible.
CAUSE The slime mould *Plasmodiophora brassicae*, which is soil-borne and can remain viable in the soil for in excess of 20 years, even in the absence of susceptible host plants. Infection occurs via the root hairs. When swollen roots disintegrate, they release many spores. The pathogen is easily introduced into a garden, on infected plants, on soil adhering to boots, tools and wheelbarrows and in soil, manure or garden compost. Acidic and water-logged soils are most prone to clubroot.
CONTROL Improve soil drainage and lime the soil in order to deter the pathogen. Burn all infected plants immediately, certainly before the roots start to disintegrate. Avoid spreading the spores around the garden on tools or boots, or by moving soil that is potentially infective. Raise your own plants, rather than buying them in; if this is not possible, make sure that you only purchase plants from a reputable source and check them carefully before planting. Practise good weed control. If you have no choice but to grow crops of brassicas on infested soil, raise plants in proprietary compost in individual pots that are at least 5cm in diameter. Plant out only once plenty of roots have formed. Choose resistant vegetable cultivars such as calabrese 'Trixie', Chinese cabbage 'Harmony', kale 'Tall Green Curled' and swede 'Marian'.

CODLING MOTH

SYMPTOMS In mid- to late summer, fruits of apple and pear are tunnelled by small white caterpillars with brown heads (*see p.47*). By the time the fruit is ripe, the caterpillars have usually finished feeding and have sought overwintering sites under loose flakes of bark. The caterpillar feeds in the core and makes an exit tunnel leading either to the eye end or elsewhere on the fruit surface.
CAUSE Caterpillars of a small moth, *Cydia pomonella*, which is generally more widespread and troublesome than the other apple fruit pest, apple sawfly (*see p.101*). The latter attacks apples at the fruitlet stage while codling moth damages the ripening fruits in late summer.
CONTROL The appropriate insecticide is bifenthrin, which is applied in early summer with a second treatment three weeks later. Pheromone traps (*see p.86*) are available for detecting when adult codling moths are active and likely to be laying eggs. Accurate timing of insecticides is important, since it is necessary to kill the young caterpillars before they have bored into fruits. Pheromone traps may catch enough male moths to reduce the mating success of females on isolated apple trees and so may reduce the number of maggoty apples.

COLLAR ROT

See PHYTOPHTHORA *p.161*.

COLLETOTRICHUM

SYMPTOMS Leaf and occasionally stem spotting, developing into lesions that may become sunken. Stems may become ringed, causing the whole plant, or parts of it, to die back.
CAUSE Various species of the fungus *Colletotrichum*, some of which are believed to be seed-borne, and which can all be spread by water splash.
CONTROL Remove affected plant parts promptly, and entire plant if stem infections are present. Do not save seed from infected plants.

COLORADO BEETLE

SYMPTOMS The foliage of potato, tomato, aubergine and pepper plants is eaten by pale yellowish-orange beetles, up to 10mm in length, with five black stripes on each wing case. Rotund, orange-red grubs with black heads and two rows of black spots along the sides of the bodies also attack leaves and can cause extensive defoliation.
CAUSE A leaf beetle, *Leptinotarsa decemlineata*,

which originates from North America but is now found throughout Europe, except in Britain and Ireland.

CONTROL Strains of the beetle which have gained resistance to some insecticides can occur. If Colorado beetles are found in Britain, they must be reported to the Department for Environment, Food and Rural Affairs, which will organize the necessary measures to eradicate them. Elsewhere, infested plants can be sprayed with an insecticide approved for this purpose when the adults or larvae are seen.

CONIFER APHIDS

SYMPTOMS Colonies of greyish-black or brown insects develop on the shoots, branches and sometimes trunks of conifers. Spruce (*Picea*), pine (*Pinus*), juniper (*Juniperus*), cypress (*Cupressus*) and Leyland cypress (x *Cupressocyparis leylandii*) are liable to be attacked by the aphids which, at 3–6mm long, are significantly larger than most aphids on other plants. The foliage and stems become sticky with honeydew excreted by the aphids from late spring to mid-summer. A black, sooty mould grows on the honeydew, and heavy infestations can cause foliage to turn yellow and dry up.

CAUSE Several aphid species, mainly *Cinara* species.

CONTROL Only small trees and hedges can be treated effectively. Spray with thiacloprid, acetamiprid, or bifenthrin in mid- to late spring before heavy infestations build up.

• *see also* CEDAR APHID *p.116*; CYPRESS APHID *p.122*; GREEN SPRUCE APHID *p.135*; JUNIPER APHIDS *p.142*.

CONIFER RED SPIDER MITE

SYMPTOMS Foliage becomes finely mottled and yellowish-brown in mid-summer, and may fall prematurely.

CAUSE Careful examination with a hand lens will show tiny yellowish-green mites with dark markings crawling amongst the foliage on fine silk webbing. Orange-red spherical eggs will also be found on the leaves and stems. Spruce is the most susceptible conifer, especially *Picea albertiana*, but this mite also attacks juniper, thuya, cedar and cypress.

CONTROL Watch for signs of infestation from late spring onwards. If necessary, spray thoroughly with bifenthrin on two or three occasions at weekly intervals.

CONIFER WOOLLY APHIDS

See ADELGIDS *p.98*.

CONIFERS, NEEDLE RUSTS OF

See NEEDLE RUSTS OF CONIFERS *p.152*.

COPPER DEFICIENCY

SYMPTOMS Blue-green or slightly yellow discoloration of the foliage, occasionally followed by dieback. Symptoms are not easy to recognize or differentiate from other problems, and are fairly rare, being most likely on peaty soils.

CAUSE A deficiency or poor availability of copper in soil.

CONTROL Treatment with a complete fertilizer may help to rectify the problem, but other possible causes for the symptoms shown should be investigated too.

CORAL SPOT

SYMPTOMS Bright, coral to orange raised pustules appear on dead, woody stems and branches (*see **p.41***). The pustules may only appear after the stem has been dead for several weeks. Many different plants may be affected, but currant bushes (*Ribes*), elaeagnus, magnolias and maples (*Acer*) are very susceptible. Dieback occurs, and if the infection spreads right down into the crown, the whole plant may be killed.

CAUSE The fungus *Nectria cinnabarina*, which may also live on dead logs, twigs, woody debris and old pea sticks. Spores are produced all year, and water, either as rain or irrigation splash, is the main method of dispersal. The fungus enters a plant via a wound or colonizes a dead snag left by physical injury or poor pruning.

CONTROL Prune out all dead or dying stems promptly, cutting well back into sound, healthy wood. Remove wood bearing the pustules. Dispose of garden debris that may be harbouring the disease. Do not compost or pass through a shredder for use as mulch.

COTONEASTER WEBBER MOTHS

SYMPTOMS In early summer, the stems of cotoneaster are covered with silk webbing spun by numerous dark brown caterpillars up to 12mm long (*see **p.37***). The leaves dry up and turn brown where they have been grazed by the caterpillars. Less extensive webbing and feeding damage also occurs in late summer, caused by the newly-hatched caterpillars before they hibernate for the winter and resume their activities in spring.

CAUSE Caterpillars of two moths, *Scythropia crataegella*, which also attacks hawthorn, and *Numonia suavella*. Both have one generation a year and their caterpillars pupate amongst the silk webbing.

CONTROL Prune out light infestations, otherwise spray with an insecticide such as bifenthrin or pyrethrum when webbing is seen.

CRANE FLIES

Crane flies, or daddy-long-legs, have soil-dwelling larvae – leatherjackets – that feed on roots and the stem bases of small plants. The adults do not feed.

• *see also* LEATHERJACKETS *p.145*.

CROWN GALL

SYMPTOMS Rounded swellings or galls – which are off-white when young and harden and turn brown with age – develop on woody or herbaceous plants. Galls are very variable in size, ranging from less than one centimetre to more than 30cm in diameter. They usually develop in small groups. Crown gall may appear on the roots or stem base (the crown of the plant). Less frequently, symptoms of an aerial form may show at some distance up the stem or trunk (*see **p.38***). Galls on woody plants are persistent, those on herbaceous plants, are softer and more short-lived. They soon disintegrate, often allowing entry of secondary organisms in the process. The plant's overall vigour is rarely affected. However, if galls ring the stem or trunk, or cause extensive disruption of water and nutrient flow, the whole plant may decline. If stem or root splitting occurs due to the erupting out of the galls, secondary organisms may gain entry and cause dieback.

CAUSE The bacterium *Agrobacterium tumifaciens*, which is soil-borne, and capable of living saprophytically (without a living host) in the soil. The bacteria gain entry via wounds and then stimulate rapid proliferation of cells, so forming the galls. Gall growth is greatest when the plant is growing rapidly, so infections that occur at the very end of the growing season may only produce symptoms the following year.

CONTROL Improve soil drainage. Check all plants carefully before planting. Keep root and stem injury to a minimum. Remove infected plants and grow root crops (except beet) on the site for at least a year to minimize the risk of subsequent infections, as these crops are not susceptible to crown gall.

CROWN ROT

SYMPTOMS Deterioration and rotting of the tissues at the crown (the junction of the stems and roots) of the plant. Herbaceous plants are very prone, but other woodier hosts, such as wisterias, may also succumb. It may then cause deterioration of the stems and foliage and the whole plant can die.

CAUSE Various soil- and water-borne fungi and bacteria, often several in combination.

CONTROL If dealt with very promptly it may be possible to save an infected plant. Remove affected areas, cutting well back into healthy tissue.

CUCKOO SPIT

SYMPTOMS Blobs of white frothy liquid appear on the stems of many plants in early summer (*see* **pp.39, 43**). A yellowish-green insect lives in the froth where it feeds by sucking sap.

CAUSE The immature nymph stage of froghoppers, such as *Philaenus spumarius*. Once they have reached the adult stage in early to mid-summer, they stop producing froth and become less noticeable.

CONTROL Spraying is often unnecessary, since cuckoo spit is only a temporary nuisance on plants, and the nymphs cause little harm unless they are feeding at the shoot tips, in which case they may cause some distorted growth. Picking off the pests by hand is usually sufficient to control most cases of infestation.

CUCUMBER FOOT AND ROOT ROT

See FOOT AND ROOT ROTS *p.129.*

CUCUMBER GUMMOSIS

SYMPTOMS Discrete sunken grey-black spots, often with a dense, furry, grey fungal growth, develop on the fruit. A viscous liquid may ooze from these lesions, and severely infected fruits may split. Stem and leaf lesions may also develop but are less common.

CAUSE The fungus *Cladosporium cucumerinum*, which is encouraged by cool, damp conditions.

CONTROL Remove infected areas promptly. Improve ventilation and, if feasible, raise temperatures. Grow resistant cultivars such as 'Birgit', 'Femspot', 'Jazzer', 'Petita' or 'Tyria'.

CUCUMBER MOSAIC VIRUS

SYMPTOMS Plants are stunted and show distinct yellow mosaic patterning (*see* **p.21**), usually on stunted, deformed foliage (*see* **p.25**). Flowering is reduced or non-existent, and plants may even die completely. If fruits are produced on cucurbits (such as marrows, courgettes, or cucumbers), they are small, pitted and strangely dark green with bright yellow blotches. Not only is their texture hard but the fruits will be inedible. On ornamental plants, flower breaking (changes in colour on clearly defined areas on the petals) may occur.

CAUSE The cucumber mosaic virus, which has an extremely wide host range, including many common garden plants, food crops and weeds. Virus particles are readily spread by handling and by sap-feeding pests, in particular aphids.

CONTROL There are no cures available and infected plants should be destroyed. Practise aphid control and minimize the risk of infection spreading. Remove weeds. Avoid unnecessary handling of plants and always handle healthy plants before touching infected or suspect ones. After handling infected plants, ensure that you wash your hands thoroughly. Grow resistant varieties where available, such as aubergine 'Bonica'; courgette 'Defender', 'Supremo' or 'Tarmino'; cucumber 'Bush Champion', 'Crispy Salad', 'Jazzer', 'Petita', or 'Pioneer'; marrow 'Badger Cross' or 'Tiger Cross'.

CURRANT AND GOOSEBERRY LEAF SPOT

SYMPTOMS In early summer, very small dark brown or black spots develop on the leaves of currants, goose-berries and, occasionally, ornamental currants (*Ribes*). The spots may enlarge and join together, and the infected leaf turn brown. Affected leaves fall prematurely and the bush may be extensively defoliated by mid-summer. In extreme cases, fruits shrivel.

CAUSE The fungus *Drepanopezizsa ribis*. It produces spores from the leaf spots that are spread by wind and rain. The fungus overwinters on the leaf spots on fallen leaves.

CONTROL Regularly remove and dispose of fallen leaves to prevent the fungus overwintering. Spray with a suitable fungicide such as mancozeb, ensuring that the undersides of the leaves are treated thoroughly. Feed and maintain plants well, as this appears to make them less prone to attack.

CURRANT BLISTER APHID

SYMPTOMS Leaves at the shoot tips become crumpled, with raised areas which have small pale yellow insects on the undersides. The affected leaves are yellowish-green or have a reddish coloration (*see* **p.27**), especially on redcurrants.

CAUSE Sap-feeding aphids, *Cryptomyzus ribis*, which are active on the foliage between bud burst and mid-summer. After that they develop wings and migrate to the wildflower known as hedge woundwort (*Stachys sylvatica*). There is a return migration in autumn when overwintering eggs are laid on currant bushes.

CONTROL Although unsightly, currants are not greatly harmed by this pest. To prevent infestations, spray with pyrethrum, thiacloprid, fatty acids or plant oils between bud burst and flowering.

CURRANT CLEARWING MOTH

SYMPTOMS Red- and black-currant stems are tunnelled by white caterpillars, up to 15mm long, with brown heads (*see p.33*). Their feeding activities sometimes cause shoots to die but more usually the presence of a caterpillar is detected when a weakened stem snaps. This is likely to happen during windy weather, especially when stems are heavy with berries. A blackened tunnel can be seen in the centre of attacked stems.

CAUSE Caterpillars of a clear-wing moth, *Synanthedon tipuliformis*. The name clearwing is derived from the fact that the moth's wings are largely devoid of coloured scales and are therefore transparent, except at the wing tips. This feature, combined with a black and yellow-striped abdomen, gives this day-flying moth a wasp-like appearance.

CONTROL There is no effective chemical treatment, as it is not possible to reach caterpillars inside stems. Fortunately, this pest does not usually damage all of the stems on a bush. During the winter, tunnelled stems can sometimes be detected by gently bending them, causing them to snap at the weak point. These stems should be cut back to sound wood and the prunings burnt to destroy the caterpillars.

CURRANT REVERSION

SYMPTOMS Plants produce leaves with slight yellowing, and unusually small main veins. The leaves are also smaller and have fewer lobes than healthy ones. Flowering and ultimately cropping of the plant is also reduced.

CAUSE The blackcurrant reversion virus, which is spread largely by big bud mite (*see p.107*) and may also be transmitted by grafting. Symptoms of mite infestation, such as swollen buds, may also be present.

CONTROL Remove infected plants, as the problem will not decline. Such plants will be a source of infection for other, healthy, bushes. Always buy blackcurrants from a reputable source and ensure that they are certified free from pests and diseases on purchase.

CUSHION SCALE

SYMPTOMS Leaves of hollies, camellias, rhododendrons, evergreen azaleas, *Trachelospermum* and *Euonymus* develop sooty mould on the upper surfaces, especially during winter. Flat, oval, yellow-brown scales are attached to the undersides of the leaves, and in early summer the females deposit eggs under broad, white, waxy bands (*see p.15*).

CAUSE Sap-feeding scale insects, *Chloropulvinaria floccifera*.

CONTROL Spray the undersides of the leaves thoroughly with thiacloprid or acetamiprid in mid-summer to control the more vulnerable newly-hatched scale nymphs.

CUTWORMS

SYMPTOMS Small plants in flower beds and vegetable gardens wilt and die. Examination of the plants shows that the roots just below soil level have been severed, or the outer bark has been gnawed away. Root vegetables, such as potatoes and carrots, may have cavities eaten in them. Caterpillars, often creamy-brown or sometimes greenish-brown in colour and up to 40mm long, are often found in the soil near recently damaged plants (*see p.53*). At night these caterpillars will feed above soil level on the foliage of low-growing plants.

CAUSE The soil-dwelling caterpillars of several moths, including large yellow under-wing (*Noctua pronuba*), lesser yellow underwing (*N. comes*), turnip moth (*Agrotis segetum*), heart and dart moth (*A. exclamationis*) and garden dart (*Euxoa nigricans*).

CONTROL Cutworms tend to work their way along row crops such as lettuce, so it is worthwhile sifting the soil near a damaged plant to find the culprit. These caterpillars are more troublesome in weedy plots, so good cultivation does help. There are currently no insecticides available for controlling cutworms.

CYPRESS APHID

SYMPTOMS Conifers, such as *Cupressus* species, *Chamaecyparis lawsoniana*, x *C. leylandii* and *Thuja*, develop sooty mould on the foliage and stems. This grows on the sticky honeydew excreted by the aphids, which are brownish-black and up to 4mm long. Heavy attacks during the early summer cause shoots to dry up and turn yellow-brown in mid-summer. On hedges, the damage is generally worse on the lower parts, with some green growth left at the top.

CAUSE A sap-feeding aphid, *Cinara cupressivora*, which is active during summer.

CONTROL Look for aphids on the stems from late spring onwards and if seen, spray immediately with thiacloprid, acetamiprid, bifenthrin, or pyrethrum. Once foliage has started to discolour, the process cannot be reversed, but in time plants usually recover. Damaged trees and hedges should be fed with a general fertilizer in spring and kept watered during dry spells to help them regain vigour.

DAEDALEOPSIS CONFRAGOSA

SYMPTOMS Otherwise known as the blushing bracket, this is a common bracket fungus. It attacks deciduous trees, in particular alders, birches and willow. The bracket-shaped, semi-circular fruiting bodies measure between 5 and 12cm in diameter (*see p.37*). The upper surface is smooth or slightly ridged, and reddish-brown in colour. The pores are circular but become elongate and red. The bracket has a corky, almost rubbery texture. It causes a white rot of the woody tissues and is generally associated with wounded or weakened trees.

CAUSE The fungus *Daedaleopsis confragosa*.

CONTROL Remove brackets to prevent spores spreading to other parts of the tree, or to other susceptible trees. Affected trees should be inspected by a professional arboriculturist to determine the extent of decay, and the overall health of the tree.

DAHLIA SMUT

SYMPTOMS Pale brown, elongate spots with yellow margins develop on leaves in mid-summer. The lower, older leaves are affected first, but younger leaves may then be affected too. The spots darken and increase in size and may join together, often killing the leaf.

CAUSE The fungus *Entyloma calendulae* f. *dahliae*, which may live in the soil around infected plants and on plant debris. It can persist in the soil in the absence of living dahlias for about five years.

CONTROL Dispose of severely infected dahlias, and plant replacements on a fresh site. The fungus does not seem to be present in the tubers, so provided any affected leaves are removed and the tubers appear clear of infected debris, they can be stored and should produce healthy plants the following year.

DAMPING OFF

SYMPTOMS Seedlings flop over, often showing discoloration and deterioration around the stem base (see *p.34*). Initially the basal lesions may appear water-soaked. The infection spreads rapidly, causing patches of seedlings to die off. An entire tray of seedlings may be killed within a few days. Fluffy white fungal growth may be seen on the surface of the dead and dying seedlings. If infection occurs before germination is complete, the seedlings only emerge patchily.

CAUSE Several soil- and water-borne fungi, in particular species of *Pythium*, *Phytophthora* and *Rhizoctonia*. Infection is encouraged by over-wet compost and prolonged raised temperatures. Seedlings which are not given adequate light, and which are from seed sown too densely, are particularly prone to infection. Damping off is most common when hygiene is poor – if pots, trays and implements are not properly scrubbed and cleaned, if unsterilized compost is used, or if non-mains water is used.

CONTROL Observe strict hygiene and only use clean trays, pots and implements and proprietary or sterilized compost. Do not use anything other than mains water, especially not water collected in a butt. Sow seedlings thinly and ensure they have adequate light and are not kept too warm for any longer than is needed to encourage germination. Drench compost with a suitable copper-based fungicide at sowing and treat seedlings with the same fungicide at intervals throughout their development.

DEER

Several species of deer cause problems in gardens. The most widespread is roe deer (*Capreolus capreolus*), but muntjac (*Muntiacus muntjak*), fallow deer (*Dama dama*) and red deer (*Cervus elephus*) also cause problems in some areas. A wide range of herbaceous and woody plants is eaten, with damage mainly occurring between dusk and dawn. Sometimes it is blooms rather than foliage that is eaten, as with tulips, but more usually whole shoots are lost. On woody plants, such as roses, the stems are partly bitten through cleanly but there remains a ragged edge to the stem (see *p.41*). This is because deer have no front teeth in their upper jaw and they cut off woody shoots by biting and tugging. The trunks of trees and side-branches can be damaged by male deer rubbing their antlers against the bark, a habit known as fraying.

CONTROL It can be difficult to exclude deer from gardens. They are agile animals good at jumping over obstacles, and a fence about 2m high is necessary to keep roe deer out. Proprietary animal repellents often give disappointing results; the effects tend to be short-lived and they may simply encourage deer to feed on something else in the garden. Deer are inquisitive feeders, attracted to new plantings, so give new plants some protection with netting. Although deer eat almost anything, it is a good idea to look at other gardens in the locality to see which plants tend to be left alone.

DELPHINIUM BLACK BLOTCH

SYMPTOMS Black blotches develop on leaves (see *p.12*), rapidly enlarging and spreading to the stems, petioles and flower buds. Entire leaves may turn black and die off.

CAUSE The bacterium *Pseudomonas delphinii*, spread from the soil or by rain or watering splash.

CONTROL Remove and dispose of infected plants promptly and avoid over-head watering. Plant new delphiniums on a fresh site.

DELPHINIUM LEAF MINER

SYMPTOMS The leaves of delphinium and monkshood develop large, brown, dried-up areas extending from the tips towards the leaf stalks (see *p.18*). White maggots up to 3mm in length may be found feeding within the mined areas.

CAUSE Larvae of a leaf-mining fly, *Phytomyza aconiti*, which has two generations during the summer months.

CONTROL For light infestations, the damaged parts of leaves can be picked off. There are currently no insecticides available for controlling leaf miners.

DELPHINIUM MOTH

SYMPTOMS During the spring, leaves at the shoot tips of delphinium and monkshood are spun together with silk threads by small green caterpillars heavily marked by black spots. Later the caterpillars feed openly on the expanded foliage. They are greenish-white by the time they pupate, in a pale yellow cocoon spun on the underside of leaves during late spring to early summer. The mature caterpillars are up to 40mm long, but often escape notice as they remain inactive on the underside of the foliage during the day.

CAUSE Caterpillars of a moth, *Polychrysia moneta*, which has one generation a year and overwinters as young larvae in the old plant stems.

CONTROL It is the young larvae feeding at the tips of new shoots in the spring that cause most harm. Look for bound-up leaves and remove the caterpillars by hand. For heavy infestations, spray with bifenthrin.

DELPHINIUM VIRUSES

SYMPTOMS Yellowing of the foliage, usually as mosaic patterns, mottling or vein-clearing (yellowing immediately adjacent to the leaf veins). Infected plants may be stunted and fail to flower properly, often dying within a year or so of the first sign of the symptoms.
CAUSE Various viruses may be involved, either singly or in combination. The most common is cucumber mosaic virus (*see p.121*). The viruses involved are not all spread in the same manner, but common means of spread include handling, and vectors include sap-feeding pests, such as aphids, and soil-inhabiting nematodes.
CONTROL Remove infected plants promptly and avoid handling other plants after handling one showing symptoms. Do not propagate from infected plants.

DIANTHUS ANTHER SMUT

SYMPTOMS Anthers of pinks and carnations (*Dianthus*) become swollen and darken as they fill with masses of black-purple spores. These are released in great quantity as the anthers rupture. Stems become stunted.
CAUSE The fungus *Ustilago violacea*, which appears to be present throughout the plant, although only producing symptoms on the anthers and flower stems.
CONTROL Dispose of infected plants promptly. Do not propagate from any part of an infected plant. Remove all plant debris and scrub down the greenhouse when plants are grown under glass. Do not grow *Dianthus* on the site for at least five years.

DIANTHUS FUSARIUM WILT

SYMPTOMS Carnations (*Dianthus*) start to fade, and the leaves wilt and develop a greyish or straw-coloured discoloration. The leaves may also develop a purple-red discoloration. The whole plant dies, but the symptoms rarely occur over the whole plant simultaneously.
CAUSE The fungus *Fusarium oxysporum* f. sp. *dianthi*, which has thick-walled resting spores that enable it to remain dormant in the soil for many years, even in the absence of carnations in the area.
CONTROL Remove infected plants promptly and grow new plants on a fresh site. Do not propagate from infected plants.

DIDYMELLA STEM ROT

SYMPTOMS Black-brown, sunken blotches develop on the stems of tomatoes (*see p.33*), and occasionally aubergines, at soil or ground level, often combined with the development of adventitious roots (those above soil level). The older leaves turn yellow. A blackish-brown rot, usually at the calyx end, develops on the fruit. Infected areas later develop numerous tiny, black, pinprick-sized fungal fruiting bodies on the surface.
CAUSE The fungus *Didymella lycopersici*, which can be seed-borne, but is usually spread from plant to plant by rain or watering splash, or by handling. The fungus can overwinter on debris from infected plants or in the soil.
CONTROL Remove and dispose of infected plants promptly; do not compost them. Clear up all crop debris at the end of the season and grow tomatoes on a fresh site. As aubergines may also occasionally be affected, choose a fresh site for these, too.

DIEBACK

SYMPTOMS The stems of plants start to die back. Dieback may start either at the tip, the base, or part-way up the stem; it rarely affects all stems simultaneously. Dark blotches or sunken patches may develop at the point where the deterioration starts. Leaves wilt, yellow and die off. The symptoms may spread down into the base of the plant, or its crown, so causing the whole plant to die.
CAUSE Various fungi, some of which can actively invade healthy stems, and others which only enter the plant through wounds or other points of damage. Dieback, especially from stem tips downwards, may also be caused by cultural problems, including poor establishment of young plants, drought and waterlogging.
CONTROL Prune out affected stems to a point well below any signs of damage. Improve cultural conditions to ensure increased plant vigour and plenty of replacement growth.

DOGS

Although not pests, dogs can cause problems in gardens. Their urine can scorch foliage, with bitches causing circular brown patches in lawns, and dogs damaging conifers and other plants that they cock their legs against.
CONTROL Spraying the foliage with water to dilute the urine will prevent damage, but this needs to be done soon after the dog has urinated. Proprietary animal repellent products are available from garden shops but these do not usually give reliable long-term protection.

DOGWOOD ANTHRACNOSE

SYMPTOMS Spots on the leaves and bracts of *Cornus* in late spring or early summer. These symptoms usually appear first towards the base of the tree, and then progress upwards. Cankers may develop on branches. As the disease develops, extensive dieback may occur and the tree may even be killed. The precise range of dogwoods affected is still under investigation, but *Cornus florida* and *C. nuttallii* are known to be particularly susceptible. *Cornus sanguinea* and *C. alba* are also affected, but on these, dieback is restricted to weak shoots.
CAUSE The fungus *Discula destructiva*, which is encouraged by cool, damp weather and which is most prevalent in areas of high rainfall.
CONTROL Prune out infected stems immediately to try to save the affected plant.

DOUGLAS FIR ADELGID

SYMPTOMS Small, fluffy white balls of waxy threads appear on the undersides of leaves on Douglas fir (*Pseudotsuga menziesii*) during the summer. Heavy infestations may cause some yellowing of the foliage, which can also be discoloured by the growth of sooty moulds on the pest's excrement.
CAUSE *Adelges cooleyi*, a blackish-brown sap-feeding insect, up to 2mm long, with

an aphid-like appearance. It can be found on Douglas fir throughout the year, and may also occur on its alternative host plant, Sitka spruce (*Picea sitchensis*), where it causes an elongate swollen gall at the shoot tips. The spruce generation is not normally a problem in gardens.

CONTROL Large trees cannot be sprayed adequately and the presence of adelgids has to be tolerated. On small trees, some protection can be given against over-wintering adelgid nymphs by spraying with bifenthrin during mild, dry weather in late February. At that time they are not protected by a waxy coating and are therefore more vulnerable to sprays.

DOWNY MILDEW

SYMPTOMS Yellow or otherwise discoloured areas develop on upper leaf surfaces. Each corresponds to a slightly fuzzy, greyish-white or, occasionally, purplish fungal growth beneath. The infection may spread, causing large areas or even the entire leaf to discolour and die. A wide range of plants may be affected, but downy mildew is especially common on young plants and those growing in moist environments. The symptoms may be confused with those of powdery mildew infection.

CAUSE Several different fungi, in particular species of *Peronospora*, *Bremia* and *Plasmopara*. These include *Peronospora parasitica* (downy mildew of cruciferae), *Peronospora destructor* (on onion), *Peronospora farinosa* f. sp. *spinaceae* (on spinach), *Peronospora viciae* (on pea), *Peronospora sparsa* (on

rose), *Bremia lactucae* (on lettuce) and *Plasmopara viticola* (on grape).

CONTROL Remove infected leaves promptly and improve air circulation around plants by increased spacing and good weed control. Increase ventilation in a greenhouse. Avoid overhead watering. Spray infected lettuce plants with mancozeb.

• *see also* BRASSICA DOWNY MILDEW *p.111*; GERANIUM DOWNY MILDEW *p.132*; HEBE DOWNY MILDEW *p.136*; LETTUCE DOWNY MILDEW *p.146*; ONION DOWNY MILDEW *p.154*; PANSY DOWNY MILDEW *p.155*.

DROUGHT

SYMPTOMS Symptoms vary depending on the plant affected and whether the drought has been recurrent or occasional. Common symptoms include poor growth and stunting, wilting of the foliage (*see p.28*) and, in extreme cases, other plant parts such as flower stems, stems and flowers (*see p.44*). Prolonged drought may also cause poor flowering, bud drop and the formation of small fruits. Drought followed by sudden watering or rainfall may cause splitting or cracking of fruits and stems.

CAUSE Most commonly, inadequate rainfall or watering; occasionally, an inability of the plant to take up adequate moisture because of a damaged root system. Very free-draining soils, such as thin soils over chalk or soils with a high sand content, are particularly prone to drying out. Plants with restricted root runs, such as those growing in containers, are also particularly prone.

CONTROL Take any measures possible to ensure that the soil or compost never dries

out completely. Where feasible, grow plants in open ground, or protect containers from the effects of direct sunlight. Improve soil moisture retention by incorporating bulky organic materials such as compost into light soils. Apply a mulch to improve soil moisture retention and reduce loss by evaporation.

DRY ROT OF BULBS AND CORMS

SYMPTOMS Foliage turns brown and dies, often with a clear dark discoloration just above soil level. Crocus, freesias, gladioli and montbretias (*Crocosmia*) are particularly affected. Whole plants may die back. The corms develop numerous small sunken lesions, each of which bears many pinprick-sized black dots or fungal fruiting bodies.

CAUSE The fungus *Stromatinia gladioli*, which can overwinter on stored corms and which may persist in the soil as small, hard, black resting bodies (sclerotia). These can remain viable in the soil for many years even in the absence of host plants.

CONTROL Always check bulbs are healthy and free from the small black sclerotia before planting. Remove any infected plants immediately. Plant replacement plants, and any others which are susceptible, on a fresh site.

DUTCH ELM DISEASE

SYMPTOMS Areas of the tree's crown show wilting and leaf yellowing. Affected leaves then turn brown and die. Dieback of twigs and branches usually follows and young shoots may curl over

to form "shepherds' crooks". The whole tree may be killed by the end of the first summer. If the bark on affected stems and branches is peeled away, longitudinal, discontinuous brown staining is apparent. If stems or branches are cut across, they may show rings of brown staining in the current year's growth. Feeding damage caused by the elm bark beetle (*see p.104*) may be seen in the crotches of young stems. The "galleries" of tunnels caused by the beetle which spreads the infection show beneath the bark on larger stems and trunks. Symptoms are much more obvious and the disease more damaging during hot, dry weather, as these conditions exacerbate the problems caused by the fungus blocking the tree's vascular system.

CAUSE The fungus *Ceratocystis ulmi*, which may cause this disease on elms and on *Zelkova*.

CONTROL There is currently no reliable chemical control method. Research is, however, being carried out into the breeding of resistant elms. Remove infected trees and dispose of by burning.

EARTHWORMS

Earthworms (*see p.61*) are generally beneficial animals that improve soil condition. Their tunnels provide aeration and drainage pathways through the soil, and worms help incorporate organic matter by pulling dead leaves into the soil. They can, however, be a nuisance when they excrete small piles of soil onto the surface of lawns (*see p.58*). These worm casts can spoil the appearance of a fine lawn, as well as making the surface muddy and slippery. When lawns are mowed, worm casts become smeared, and this creates patches of bare soil that lawn weeds can colonize. Not all earthworms produce worm casts. The culprits are usually the *Allolobophora* species of earthworms. They are mainly a problem during spring and autumn when warm, moist soil conditions encourage worm activity in the upper layers of the soil; drought or cold sends worms deeper and reduces surface casting.

CONTROL There are currently no chemicals approved for earthworm control available to amateur gardeners. Worm casts can be dispersed by brushing the lawn with a besom broom or using an up-turned wire rake. This is best done when worm casts are dry and crumbly. During wet periods, walking on the lawn should be kept to a minimum to avoid smearing the worm casts.

EARWIGS

Earwigs, *Forficula auricularia*, are up to 14mm long, tawny brown in colour and easily recognised by the pair of pincers on the rear end of the body. They eat ragged holes in the young foliage and flowers of many plants (*see p.46*), especially clematis, dahlias and chrysanthemums. During the day they hide in dark crevices or amongst the foliage and petals of host plants. They emerge and feed after sunset, so a torchlight inspection is often needed to find the culprits.

CONTROL Vulnerable plants

can be given some protection by placing earwig traps made of flower pots inverted on bamboo canes and placed level with the plant tops. The pots are loosely stuffed with hay or straw and will be used as daytime hiding places by earwigs. Inspect the pots daily and destroy any earwigs found inside. If earwigs are numerous and still cause damage, spray the plants with bifenthrin at dusk on a mild evening.

EELWORMS

Eelworms or nematodes are mostly microscopic worm-like animals, some of which are important plant pests. Chrysanthemum eelworm (*see p.118*), narcissus eelworm (*see p.151*), onion eelworm (*see p.154*), phlox eelworm (*see p.160*), potato cyst eelworms (*see p.164*) and root knot eelworms (*see p.171*) all feed within their host plants, but there are others, such as *Pratylenchus*, *Longidorus*, *Trichodorus* and *Xiphinema* species, that live in the soil and attack the root hairs of plants. The last three are able to transmit some viral diseases of ornamentals and soft fruits. Not all eelworms are pests however; the vast majority feed on bacteria, fungi and other micro-organisms, while some are predators. Slugs (*see p.178*), chafer grubs (*see p.116*), leatherjackets (*see p.145*) and vine weevils (*see p.188*) can be controlled by beneficial eelworms.

CONTROL There is no effective chemical control for soil-dwelling eelworms available to gardeners. Parts of gardens from which virus-infected plants have been removed should not be replanted with the same

types of plant, otherwise they may quickly be infected by the eelworms.

ELAEAGNUS SUCKER

SYMPTOMS The shoot tips are infested with small insects that excrete a sticky honeydew onto the foliage. The adults are about 2–3mm long and yellowish brown with transparent wings strongly patterned with black markings. The immature nymphs are creamy-orange with darker markings. Sucker nymphs have broad, flattened bodies with prominent wing pads on the older nymphs. Heavy infestations cause die back and leaf fall.

CAUSE A sap-feeding insect, *Cacopsylla fulguralis*, which originates from eastern Asia and became established in Britain in 2002. It has several generations between spring and early autumn but the worst damage seems to occur during the spring when heavy infestations build up on the new growth.

CONTROL Inspect the new growth on *Elaeagnus* plants and spray with bifenthrin, thiacloprid, or acetamiprid if suckers are seen. Many *Elaeagnus* spp. and hybrids are susceptible to this pest but *Elaeagnus angustifolia* and *E. multiflora* seem to be resistant.

ELEPHANT HAWK MOTH

SYMPTOMS Various garden plants, but particularly fuchsias, are defoliated (*see p.30*) by green or, more usually, brownish-black caterpillars. When fully grown these are up to 80mm long and as thick as a man's finger. They have two pairs of eye spot markings on the front end of the abdomen.

When the caterpillar is disturbed it retracts its head into its body, causing the front of the abdomen to swell into a false head and making the eye spots more prominent. In this state the caterpillar can be mistaken for a small snake.

CAUSE Caterpillars of a hawk moth, *Deilephila elpenor*, which are active during mid- to late summer. Feeding is generally done at night, and the caterpillars rest on the underside of leaves during the day. Despite their size, the caterpillars often go unnoticed until they have completed their feeding and are wandering about looking for somewhere to pupate.

CONTROL Damage to garden plants is infrequent. If the caterpillars are found they can be removed by hand and perhaps transferred to one of their wildflower hosts, such as rose bay willow herb or Himalayan balsam.

ELM GALL MITE

SYMPTOMS Elm leaves develop hard yellowish-green swellings on the upper surface (*see p.16*) during mid- to late spring that persist throughout summer.

CAUSE A microscopic gall mite, *Aceria ulmicola*, which begins feeding on the foliage in the spring. Chemicals injected into the leaves by the mites induce the galls to develop. The mites spend the summer months feeding and breeding within these hollow structures before emerging in late summer and overwintering under bud scales.

CONTROL Amateur garden pesticides will not control gall mites. Fortunately, apart from creating galls, this mite does little harm to elm trees so can be tolerated.

ENCHYTRAEID WORMS

These small animals are related to earthworms and are also sometimes known as "white" or "pot" worms. They have white bodies which are up to 20mm long and 1mm thick. They can be mistaken for eelworms or nematodes; but where these have completely smooth, unsegmented bodies, enchytraeid worms have segmented bodies, visible with the aid of a hand lens. Enchytraeids cause no harm as, like earthworms, they feed on decaying organic matter in the soil. They are sometimes abundant in soil that has been heavily dressed with farmyard manure or similar composts. They may also be found amongst the roots of a dead plant but would not be responsible for the plant's condition.

EUCALYPTUS GALL WASP

SYMPTOMS Eucalyptus leaves develop many slightly raised swellings about 1mm in diameter that are pinkish brown in spring. Similar markings can be caused by oedema (*see p.154*). The latter causes less regularly-shaped swellings that are solid. The insect-induced galls are hollow and contain tiny white grubs. Heavy infestations can result in excessive leaf fall in early summer.
CAUSE A tiny gall wasp, *Ophelimus* species, that originates from Australia but has become established in several European countries. It has been present in south east England since 2005 and will become more widespread. The tiny black adult wasps emerge from the galls in early summer and lay eggs in the new foliage. The galls gradually develop and mature the following spring.
CONTROL Some eucalyptus trees are too large to be sprayed effectively. On small trees, it is possible that spraying with thiacloprid or acetamiprid in early summer may control the young larvae.

EUCALYPTUS SUCKER

SYMPTOMS Grey or yellowish-orange insects, up to 2mm in length, cluster at the shoot tips of eucalyptus. The adults resemble winged aphids and are the overwintering stage. They lay yellow eggs at the shoot tips in early spring, and all stages in the life cycle can be found together during the summer. The nymphal stages are wingless, flattened insects that secrete wispy white wax filaments from their bodies. Heavy attacks cause stunted growth, and sooty moulds grow on the suckers' excrement.
CAUSE A sap-feeding insect, *Ctenarytaina eucalypti*, which originates from Australia but is now widespread in European gardens.
CONTROL Infestations on mature trees have to be tolerated because of the difficulty of spraying thoroughly. Small trees can be treated with pyrethrum, thiacloprid, acetamiprid, or bifenthrin.

EUONYMUS SCALE

SYMPTOMS This is a specific pest of evergreen euonymus, particularly *Euonymus japonicus*. The foliage becomes heavily coated with elongate, whitish scales about 2mm long (*see p.21*), which are the males. The females mainly cluster on the stems and are brownish-black, pear-shaped objects, up to 3mm in length, which are well camouflaged against the bark. Heavy infestations cause yellowish discoloration of the foliage, plants lack vigour and die back.
CAUSE A sap-feeding scale insect, *Unaspis euonymi*. In Britain it is mainly confined to the south coast of England, but can occur further inland.
CONTROL The newly-hatched scale nymphs are the most vulnerable stage and are present in early summer and early autumn. Spray at those times with thiacloprid or acetamiprid if the scale is present.

FAIRY RINGS

See TURF FAIRY RINGS *p.186.*

FASCIATION

SYMPTOMS Shoots, or less commonly flower stems, become enlarged and flattened, developing a ribbon-like, often slightly curled appearance. Leaves, buds and flowers develop normally on the distorted stem (*see p.43*). Sometimes many stems are affected simultaneously, but more commonly the majority of the plant remains unaffected. Where weather is involved, stems on the more exposed areas of the plant are more likely to be affected.
CAUSE Early injury to the growing point, such as that caused by frost or insect attack, slug damage or some form of mechanical injury. It has been suggested that viral or other infections may also be involved.

CONTROL Although rather peculiar in appearance, fasciation is harmless and does not affect the overall growth and development of the plant. Prune out affected stems, if desired.

FELT GALL MITES

SYMPTOMS Various trees and shrubs, particularly alder, beech, birch, crab apple, lime, maple and sycamore, develop a dense coating of hairs on the underside of the leaves. These hairs are usually creamy-white, but on purple-leaved trees such as copper beech, some maples and crab apples, the hairy patches will be reddish-pink (*see p.19*). Later in the summer the hairs dry up and become brownish. On some plants, such as walnut and grape vine, there is an up-ward bulging of leaves where felt galls have developed on the undersides.
CAUSE Microscopic mites of various species which are usually specific to a particular plant species. The mites feed by sucking sap from the foliage and while doing so they secrete chemicals which induce the growth of hairs that form felt galls.
CONTROL Light infestations on small trees can be checked by picking off affected leaves. No pesticides are available for controlling gall mites, so heavier infestations have to be tolerated. Fortunately, apart from creating the galls, these mites have little impact on the host plant's growth.
• *See also* ACER GALL MITES *p.98;* VINE ERINOSE MITE *p.188;* WALNUT BLISTER MITE *p.189.*

FELTED BEECH COCCUS

See BEECH BARK SCALE *p.106.*

FIGWORT WEEVILS

SYMPTOMS Small beetles, which are mainly black and mottled with brownish-white markings, infest the shoot tips of plants such as figworts (*Scrophularia*), phygelius, *Buddleja* and verbascum. The beetles have a snout-like projection on their heads and there is usually a black circular mark where the wing cases meet. Both the adults, which are about 4mm long, and their brownish-yellow slimy larvae feed on foliage and flowers.
CAUSE Several species of weevil, *Cionus* species, which have two generations between late spring and late summer. When fully fed, the larvae pupate inside brownish cocoons on the shoot tips and these can be mistaken for seed pods.
CONTROL Inspect plants for signs of infestation and if necessary spray with bifenthrin or pyrethrum.

FIREBLIGHT

SYMPTOMS Flowers wilt and wither, then die back, followed by adjacent leaves and stems shortly afterwards (*see* **p.35**). The plant may show extensive dieback and be killed within a few seasons. Infections usually occur scattered throughout the crown of the tree or shrub, healthy stems often growing side by side with infected ones. Infection commonly occurs at flowering time. The bark sinks inwards on young stems and, if pared back, a foxy-red discoloration is visible on the wood. Bacterial ooze may be produced from affected areas. The symptoms may be confused with those of blossom wilt (*see p.109*), but the infection only occurs in pome-fruit (apple-like) bearing members of the plant family Rosaceae.
CAUSE The bacterium *Erwinia amylovora*, which is encouraged by warm, wet weather. The bacteria usually enter the plant via the open flowers, but infection may also take place through stem injuries. Bacteria may be carried by rain splash and spread on pruning tools. In certain areas of the British Isles (Northern Ireland, Isle of Man, and the Channel Islands) fireblight is still a notifiable disease, so cases in gardens should be reported to the local Department for Environment, Food and Rural Affairs.
CONTROL Prune out affected stems to a point at least 60cm into healthy wood. Dispose of all prunings by burning, and sterilize all tools after use on an infected stem. If the infection is widespread, or the plant small, it may be better to remove it entirely.

FISTULINA HEPATICA (BEEFSTEAK FUNGUS)

SYMPTOMS Soft bracket fungi, usually growing singly and measuring 5–30cm in diameter, develop on branches, stems and tree stumps between late summer and late autumn. They are tongue-shaped or become semi-circular as they enlarge. Cream-coloured when young, they become reddish-brown and very juicy as they mature, the flesh often being marbled and resembling a slice of raw steak, giving rise to their common name. If cut or otherwise injured, blood-red juice is produced. The fruiting bodies produce spores, mainly from pores on the lower surface of the fruiting body. The wood on affected trees develops a streaky brown discoloration, called "brown oak" and later cubical cracking occurs. The fruiting bodies disintegrate with the onset of cold weather and frosts.
CAUSE The fungus *Fistulina hepatica*, which may enter trees via wounds and may cause the wood to become so brittle that it is weakened.
CONTROL Remove fruiting bodies in order to limit spore spread. The fungus itself will continue to remain active within the tree. An inspection of the tree to determine its health and safety is advisable.

FLATWORMS

There are many flatworms native to Britain; most live in water but some occur in soil. They are mostly small animals which cause no damage to plants. In recent years two species from Australasia have become established in Britain and are causing concern because they prey on earthworms and can reduce them to very low numbers. The loss of earthworms leads to impaired drainage and aeration of the soil, with consequent effects on plant growth. The New Zealand flatworm (*Arthurdendyus triangulatus*) is up to 17cm long when fully extended, and is purplish-brown on its upper surface, although the colour fades towards the sides of the body. It is widely distributed in Ulster and Scotland and is spreading southwards, particularly down the western side of Britain. The Australian flatworm (*Geoplana sanguinea*) mainly occurs in south-west England and is up to 10cm long. It is orange-brown in colour. Both species are mainly nocturnal and can be found during the day curled up under stones, logs or other objects which are in close contact with the soil. These animals are flat to the point of being ribbon-like and have entirely smooth, unsegmented bodies. They can be confused with soil-dwelling leeches, which are flattish but have fine lines running across their bodies to indicate their segmented body structure. Leeches also have a sucker underneath the rear end of their bodies; flatworms do not.
CONTROL There is currently no control for flatworms and it is likely that they will become more widespread. They do not move far by their own efforts but can easily be transported amongst the roots of container-grown plants. When receiving or purchasing plants, inspect the rootballs carefully. Immersing the pots in water for a couple of hours will encourage flatworms to emerge and be seen. Infested plants should be rejected or repotted in clean compost after all the old soil has been washed from the roots to remove flatworms and their shiny black egg capsules.

FLEA BEETLES

SYMPTOMS Small beetles, mostly 2mm long but sometimes 3–4mm, readily jump off foliage when disturbed. They are shiny black, sometimes with a yellow stripe running down each wing case, or can be metallic blue or yellowish-brown in colour. They eat small, rounded holes in the upper leaf surface (*see* **p.29**) which often do not go all the way through the leaf. The remaining damaged tissues dry up and turn brownish-white. Heavy attacks can kill seedlings and check the

growth of older plants.
CAUSE There are many different species of flea beetle. On brassicas, including radish, swede and turnip, it is mainly *Phyllotreta* species and these also attack wallflowers, alyssum, stocks, aubrieta and nasturtiums. Potatoes and ornamental *Solanum* species are sometimes damaged by *Psylliodes affinis*. The damage is caused by the adult beetles, with the larvae living in the soil and feeding on roots. On fuchsias, both adults and larvae of a metallic blue species, *Altica lythri*, feed on leaves at the shoot tips. Flea beetles are active between mid-spring and late summer.
CONTROL Protect seedlings by sowing when weather and soil conditions will allow rapid germination and growth through the vulnerable seedling stage. Keep young plants watered during dry spells. If necessary, control with thiacloprid, pyrethrum, or bifenthrin

FLIES

Flies are characterized by having only one pair of wings; the hind wings are reduced to small club-shaped structures called halteres. There are a number of fly pests of garden plants and in all cases damage is caused by the larval stage rather than the adults.
Root and seedling pests – *see* BEAN SEED FLY *p.105;* CABBAGE ROOT FLY *p.113;* CARROT FLY *p.115;* FUNGUS GNATS *p.131;* LEATHERJACKETS *p.145;* ONION FLY *p.154.*
Bulb pests – *see* NARCISSUS BULB FLY *p.151.*
Foliage pests – *see* ALLIUM LEAF MINER *p.98;* BEECH GALL MIDGES *p.106;* BEET LEAF MINER *p.106;* BLACKCURRANT GALL MIDGE *p.108;* CELERY LEAF MINER *p.116;* CHIRONOMID

MIDGES *p.117;* CHRYSANTHEMUM LEAF MINER *p.118;* DELPHINIUM LEAF MINER *p.123;* GLEDITSIA GALL MIDGE *p.134;* HOLLY LEAF MINER *p.137;* LABURNUM LEAF MINERS *p.142;* SEMPERVIVUM LEAF MINER *p.176;* VIOLET GALL MIDGE *p.188.*
Flower pests – *see* HEMEROCALLIS GALL MIDGE *p.136.*
Fruit pests – *see* PEAR MIDGE *p.158.*

FLUTED SCALE

SYMPTOMS Stems and foliage becomes encrusted with orange-brown sap-feeding insects that secrete white waxy filaments from their bodies. The mature females have a shell-like appearance and are about 4mm long. The females are perched on the edge of a mound of white waxy material in which the eggs are deposited. These egg masses have a distinctive grooved or fluted appearance. All stages in the life cycle can be found together throughout the year.
CAUSE A sap-feeding insect, *Icerya purchasi*, which is also known as the cottony cushion scale. This used to be an uncommon pest found only in heated glasshouses but it is now becoming established out of doors on patios and in sheltered inner city gardens, especially in London. A wide range of plants is attacked, especially *Acacia* spp, *Choisya*, and citrus plants. Heavy infestations soil plants with honeydew and sooty mould, with a severe loss of vigour in the plants.
CONTROL On small plants remove the scales and egg masses before the eggs hatch. Spray infested ornamental plants with thiacloprid or acetamiprid. On citrus plants with edible fruits, use fatty acids or

plant oil sprays. Several applications of these organic sprays will be required at one week intervals.

FOMES POMACEOUS

SYMPTOMS Hoof-shaped or rounded fungal fruiting bodies develop on the bark of dead or living stems or trunks of various deciduous trees (*see* **p.39**). They have a woody texture and measure about 5cm in diameter. The upper surface is pale grey-brown and may show concentric furrows. The margin of the developing fruiting body is velvety grey, later becoming cinnamon-coloured. The wood within the stem may decay, turning the centre of the stem or trunk crumbly and white with a dark brownish-purple zone around the edge.
CAUSE The fungus *Fomes pomaceous* (syn. *Phellinus pomaceous*) which produces spores from minute pores on its undersurface. The spores infect via wounds caused by pruning or other mechanical injury.
CONTROL Remove fruiting bodies to minimize spread of spores. However, the decay will continue within the branch or trunk. Have the tree inspected to determine its overall health and safety.

FOMES ROOT AND BUTT ROT

SYMPTOMS A dark reddish-brown, woody-textured bracket fungus (*see* **p.37**), usually measuring 5–10cm across but occasionally up to about 30cm in diameter, develops on the trunk or butt of trees, usually at ground level. Infection is usually seen on conifers but occasionally deciduous species such as birch, beech, alder and oak

are attacked. The lower surface of the bracket is creamy-white and it is from the minute pores on this undersurface that the spores are released. The fungus causes extensive deterioration of the woody tissues within the tree, causing rotting and deterioration of the wood. Trees are rendered unsafe, and may occasionally be killed. The fruiting bodies are fairly frost-resistant and usually perennial.
CAUSE The fungus *Heterobasidion annosum* (syn. *Fomes annosus*), which may infect stumps through cut surfaces, and then infects nearby trees by root contact.
CONTROL Trees are rendered unsafe, and because of the ease with which the infection spreads, should be felled promptly. Remove stumps, together with the soil in the immediate vicinity, to minimize the risk of spread.

FOOT AND ROOT ROTS

SYMPTOMS Deterioration of the tissues around the stem base causes the upper parts of the plant to wilt, discolour and die back. Soft-stemmed plants, in particular seedlings and cuttings, are commonly affected, but most annual, biennial and herbaceous perennial plants may show symptoms. Tomatoes, cucumbers and melons, especially those growing in greenhouses, are frequently affected. The tissues at the stem base may darken, atrophy, become discoloured (usually darkening) and, occasionally soften. Root rotting may occur simultaneously (*see* **p.55**).
CAUSE Various fungi, often similar to or the same as those responsible for damping-off disease of

seedlings (*see p.123*). Other fungi, including various species of *Fusarium* and *Verticillium*, may cause both wilt and foot rot symptoms.

CONTROL Once foot rot symptoms have developed, there is no cure available for infected plants. It may, however, be possible to prevent spread of the infection by very prompt removal of any infected plants, together with the compost or soil around their roots. Unhygienic cultivation techniques, including the use of unsterilized compost, dirty trays, pots and tools and the use of non-mains water, are often responsible for introducing or spreading this type of infection.

FORSYTHIA GALL

SYMPTOMS Numerous galls develop along the length of stems (*see p.40*). The whole or just certain areas of the plant may show the symptoms. Individual galls are usually up to 10–15mm in diameter, rough-surfaced and commonly near-spherical; occasionally several may be joined or fused together. They are the same colour as the forsythia stems. Overall growth and development of the plant is rarely affected but the plant's shape and outline can be spoiled.

CAUSE Unknown.

CONTROL If the forsythia becomes disfigured, prune out affected stems to a point several inches beneath the lowest gall.

FOXES

Foxes are becoming increasingly at home in urban gardens and often make their dens under garden sheds. They cause damage by digging up or trampling plants. Dog foxes use their faeces and urine to mark out their territories and this can create an unpleasant nuisance. Fox urine can scorch the foliage of plants, especially conifers, and has a distinctive pungent smell.

CONTROL It is difficult to prevent foxes from coming into a garden. They are agile animals that are as good as cats when it comes to scaling walls and fences, but have the additional ability to dig underneath obstacles. Proprietary animal repellents have little lasting effect. In gardens where foxes are a problem, the use of bonemeal or dried blood should be avoided. Foxes dig up plants given these fertilizers because the smell makes them think there is food beneath.

FROGHOPPERS

See CUCKOO SPIT *p.121.*

FROST DAMAGE, APPLE

See APPLE FROST DAMAGE *p.100.*

FROST DAMAGE (BUDS AND FLOWERS)

SYMPTOMS Buds, whether tightly closed or partially opened, discolour, usually turning brown and sometimes becoming soft and squashy. Petals of open blooms discolour, usually turning brown, and either retain their shape or wither (*see p.44*). The buds or flowers on the more exposed areas of the plant are usually the worst affected. Frost may only damage buds or flowers at certain developmental stages, killing some, damaging others and leaving some untouched. On fruiting plants, such as plums, there may be subsequent partial or total loss of crop.

CAUSE Frost, which is most likely to cause damage on slightly tender plants or those which have put on precocious growth. Late frosts, occurring once flower buds have opened, can also be very damaging. Plants which are subjected to the warming of early morning sun following overnight frosts are especially prone to injury.

CONTROL Choose planting positions carefully, especially for early flowering plants, or those known to be readily frost-damaged. Provide protection using horticultural fleece or similar material.

FROST DAMAGE (FOLIAGE AND STEMS)

SYMPTOMS Foliage, usually towards the shoot tips and particularly on the more exposed parts of the plant, turns brown or black or appears scorched (*see p.28*) and may wilt, wither and die back. Young stem growth adjacent to the frosted leaves may also die back. Isolated patches on otherwise sound stems may develop, but this is less common.

CAUSE Frost, which is most likely to cause damage on slightly tender plants or those which have put on precocious growth. Late frosts, occurring once flower buds have opened, can also be very damaging. Plants subjected to the warming of early morning sun following overnight frosts are especially prone to injury.

CONTROL Choose planting positions carefully, especially for plants known to be readily frost-damaged. Provide protection using horticultural fleece or similar material. An autumn dressing of sulphate of potash helps to harden stem growth and so may help to minimize frost damage.

FRUIT TREE RED SPIDER MITE

SYMPTOMS The foliage of apples and plums loses its glossy green colour and becomes increasingly dull and yellowish-green (*see p.20*). There is a fine pale mottling of the upper surface and large numbers of tiny mites and spherical eggs (*see p.40*) occur on the lower leaf surfaces. Heavy infestations can develop in hot, dry summers and result in early leaf fall.

CAUSE Sap-feeding mites, *Panonychus ulmi*, which are blackish-red in colour and less than 1mm long. This mite is generally not a problem on unsprayed trees where predatory mites and insects keep it at a low level. Spraying often kills beneficial mites and insects, allowing spider mites to breed unchecked. Overwintering eggs are laid in crevices on the underside of branches and can be so numerous that the bark has a reddish hue.

CONTROL Avoid unnecessary spraying against other pests as this may lead to spider mite problems. If spider mite needs controlling, spray thoroughly with bifenthrin. Three applications at weekly intervals may be needed.

FUCHSIA GALL MITE

SYMPTOMS Growth at the shoot tips becomes grossly distorted and often has a reddish yellow colour. The flowers are also deformed or fail to develop. New growth comes to a halt once a heavy infestation has developed.

CAUSE A microscopic gall mite, *Aculops fuchsiae*, which originates from South America but has spread to the US and Europe. It was first found in England in 2007. The mites can be spread on the bodies of bees when they visit the flowers.
CONTROL Cold winters will kill the mites on hardy fuchsias but the mites will survive in greenhouses and in mild districts. There are no effective pesticides available to amateur gardeners so affected plants should be destroyed to prevent the pest spreading.

FUCHSIA RUST

SYMPTOMS Numerous patches of pale orange or deep yellow fungal pustules develop on the lower leaf surface, older leaves towards the base of the plant usually showing symptoms first. The upper leaf surface shows corresponding areas of yellow or tan-brown discoloration (*see* **p.18**), followed by necrosis. Infected leaves fall prematurely, leaving leggy plants which may be killed.
CAUSE The fungus *Pucciniastrum epilobii*, which is encouraged to spread and infect by moist or overcrowded growing conditions and high humidity. Ornamental and weed species of *Epilobium* (willowherbs) may also be infected and so may harbour the fungus. In addition, the fungus may develop on the foliage of firs (*Abies*) but infection is much more common on fuchsia than on either of the other genera.
CONTROL Remove infected leaves promptly. Check willowherbs growing in the vicinity and remove them or control infections as appropriate.

FUNGUS GNATS

SYMPTOMS Greyish-brown flies with slender bodies, 3–4mm long, run over the soil surface of pot plants or fly slowly and alight on leaves. The larvae are whitish maggots, up to 5mm long, with black heads (*see* **p.53**). They live in the soil and feed mainly on decaying organic matter, such as dead roots, but may damage seedlings or the base of soft cuttings.
CAUSE Several species of fungus gnats occur in greenhouses and on house plants. Most are harmless but some *Bradysia* species damage fine roots on seedlings and tunnel in the stems of soft cuttings, such as pelargoniums.
CONTROL The adult flies can be trapped by hanging sticky yellow strips close to soil level. A predatory mite, *Hypoaspis miles*, which feeds on the maggots can be purchased by mail order from biological control firms. On established plants, fungus gnats are a nuisance rather than a pest, and so their presence can be tolerated. Maintain good hygiene in the greenhouse and remove dead leaves and flowers from the soil surface. Potting compost can be watered with thiamethoxam, which will give seedlings and cuttings protection from fungus gnat larvae.

FUSARIUM BULB ROT

See GLADIOLUS CORM ROT *p.133*; NARCISSUS BASAL ROT *p.150*.

FUSARIUM WILT

SYMPTOMS Soft-stemmed plants wilt, either in part or in their entirety. Woodier-stemmed plants may retain their overall shape, but foliage on affected stems wilts and withers. Internal staining of the vascular tissue (conducting elements in the stems and roots) may occur and is usually brown or black. This may be apparent just beneath the bark on the stems or within the central area of the roots. Death almost always follows, either rapidly or over several seasons, depending on the plant affected. Cool, damp conditions may encourage very pale pink or white, slightly fluffy fungal growth from the infected tissues.
CAUSE Various types of the fungus *Fusarium*, most commonly *F. oxysporum*. The fungi involved are usually fairly host-specific, attacking only one genus or a few very closely related genera of plants. The fungi are responsible for the whole or partial blocking (usually by gum-like formations) of the vascular tissues, so the symptoms are very similar to those caused by drought. Dry soil conditions may encourage rapid wilt symptoms, but unlike plants suffering from drought, those affected by fungal wilts do not recover permanently when watered. The fungus may persist in plant debris and may also be capable of remaining viable in the soil for several years, even in the absence of any suitable host plants. In *Fusarium oxysporum*, the specialized spores responsible are known as chlamydospores.
CONTROL There is no chemical control available. Remove infected plants promptly to minimize spread. Where feasible, remove soil or compost from the immediate vicinity of the roots. Do not grow the same or closely related plants on the site again, or at least for a minimum of five years.

GALL MIDGES

Gall midges are tiny flies, usually no more than 2–3mm long, and most of them have larvae that feed on plants. The maggots are orange-white and secrete chemicals that affect the development of plant parts and cause them to produce galls that enclose the larvae. These midges lay their eggs on a particular part of their host plant, such as developing leaves, shoot tips, flower buds or fruits, and the galls that subsequently develop have a distinctive and characteristic shape that enables the pest to be recognized. Not all members of the gall midge family are pests; one species, *Aphidoletes aphidomyza*, has larvae that feed on aphids and is used to give biological control in greenhouses.
• *See also* BEECH GALL MIDGES *p.106*; BLACKCURRANT GALL MIDGE *p.108*; GLEDITSIA GALL MIDGE *p.134*; HEMEROCALLIS GALL MIDGE *p.136*; PEAR MIDGE *p.158*; VIOLET GALL MIDGE *p.188*.

GALL MITES

These are microscopic animals that can be found on many plants, particularly trees and shrubs. The mites have tubular bodies with just two pairs of legs. Although too small to be seen without magnification, the effect that gall mites have on their host plants is often very characteristic. They feed mainly on foliage or in buds and they secrete chemicals that induce the plant to make abnormal growths. On foliage, gall mites cause raised structures, thickened and curled leaf margins, blisters, excessive growth of hairs or discoloured patches. Bud-feeding gall mites cause swollen buds which later dry up or proliferate into cauliflower-like structures, or shoots develop as witches' brooms. With a few exceptions, most gall mites do little harm to their host plants and the presence of galls can be tolerated.

• *See also* ACER GALL MITES *p.98*; BIG BUD MITES *p.107*; BROOM GALL MITE *p.110*; ELM GALL MITE *p.126*; FELT GALL MITES *p.127*; FUCHSIA GALL MITE *p.130*; LIME NAIL GALL MITE *p.147*; MOUNTAIN ASH BLISTER MITE *p.150*; PEAR LEAF BLISTER MITE *p.158*; PLUM LEAF GALL MITE *p.162*; RASPBERRY LEAF AND BUD MITE *p.169*; RED BERRY MITE *p.169*; VINE ERINOSE MITE *p.188*; WALNUT BLISTER MITE *p.189*.

GALL WASPS

Gall wasps are black or brown insects up to 5mm long, which are mainly associated with oaks and roses. The gall structures are produced by the plant in response to chemicals secreted by the larval stages, which are white, legless grubs. Although galls are sometimes numerous they have no harmful effect and so control measures are unnecessary.

• *See also* ACORN GALL WASP *p.98*; EUCALYPTUS GALL WASP *p.127*; FUCHSIA GALL MITE *p.110*; OAK GALL WASPS *p.153*; ROBIN'S PIN CUSHION *p.171*.

GANODERMA

SYMPTOMS Large, brown or tan-brown, tough fungal brackets develop, most frequently towards the base of the trunk. The lower surface of the bracket is usually a cream or pale brown colour and perforated by numerous closely packed, tiny pores, through which the spores are released. Initially, brackets may be rounded in outline, but they become increasingly bracket-shaped and flattened as they enlarge. They are perennial and woody in texture. Large quantities of dark brown spores are produced and may cover the vegetation beneath. Affected trees may initially show little sign of deterioration, but the woody tissues within the trunk may be seriously weakened, rendering the tree potentially dangerous. The tree may become thin in the crown, with a sparse covering of leaves, and show signs of dieback.

CAUSE Various species of the fungus *Ganoderma*.

CONTROL Remove the bracket to prevent further dissemination of spores. As the damage will continue within the tree, it must be inspected by a reputable tree surgeon or arborist to determine its overall safety. In most instances the tree has to be felled.

GERANIUM DOWNY MILDEW

SYMPTOMS Pale green or light brown, often angular areas of discoloration develop on the leaves (*see p.25*) of *Geranium* species and hybrids. Severely affected leaves wither and die. Off-white, slightly fuzzy fungal patches develop on the lower leaf surface, corresponding with the discoloured areas. Symptoms may be confused with the much whiter fungal growth which occurs on upper leaf surfaces and stems and is due to powdery mildew (*see p.167*).

CAUSE A species of the fungus *Peronospora*, which is encouraged to spread by moist, humid conditions and poor air circulation.

CONTROL Remove infected leaves promptly. Try to improve air circulation around the plants.

GERANIUM SAWFLY

SYMPTOMS This pest attacks *Geranium* species and hybrids, not pelargoniums. Rounded holes appear in the foliage and greyish-green caterpillar-like larvae, up to 12mm long, may be found on the underside of leaves, although the larvae often drop to the ground when disturbed.

CAUSE Larvae of a sawfly, *Protoemphytus carpini*, which has two generations between late spring and late summer.

CONTROL By late summer the foliage may be extensively holed but this has little effect on the plant's growth, so control may be unnecessary. If required, spray the undersides of the leaves with thiacloprid, pyrethrum or bifenthrin.

GEUM SAWFLIES

SYMPTOMS Three species of sawfly occur on geums. One has leaf-mining larvae up to 4mm long that feed within leaves, causing the damaged areas to dry up and form large brown blotch mines. The other two have pale green larvae up to 13mm long with short, forked hairs on their upper surface. They feed on the young leaves in early summer and can cause severe defoliation.

CAUSE The leaf-mining sawfly is *Metallus lanceolatus* and it has two generations between late spring and late summer. The other two sawflies are *Claremontia waldheimii* and *Monophadnoides rubi*, both of which have a single generation, with larvae active in late spring to early summer.

CONTROL Light infestations can be picked off, or spray infested plants with thiacloprid, pyrethrum or bifenthrin. Geum leaf-mining sawfly larvae is difficult to control with the pesticides available to amateur gardeners.

GHOST SPOT

See GREY MOULD *p.135*.

GLADIOLUS CORE ROT

SYMPTOMS A moist rot and blackening affects the central core of the corm. The base of the corm looks normal but is soft if squeezed. The rot spreads outwards until the whole corm becomes soft and brownish-black. Fluffy, white fungal growth develops beneath the outer scales. Black sclerotia (resting bodies) of the fungus develop on the surface in amongst the white fungal growth. Symptoms are usually noticed in storage. If a slightly diseased corm is planted, it may grow but produces only a few yellowed and stunted leaves, often with brown spots and red margins.

CAUSE The fungus *Sclerotinia draytonii*, which spreads rapidly during damp weather. The spores are spread by rain and water splash. The fungus may survive in the soil for many years, as the sclerotia are very resilient.

CONTROL Remove and dispose of plants showing foliage symptoms. Check all corms thoroughly before storing, and again at the end of storage before replanting. Allow plants to dry off naturally but quickly after lifting. After lifting, treat corms with sulphur dust. Store corms in a cool, dry place. Plant corms on a fresh site each year.

GLADIOLUS CORM ROT

SYMPTOMS Concentrically ridged brown markings develop on the corm, usually at its base and whilst in storage (see **p.55**). Crocuses and bulbous irises may also be affected. The corm dries out, becoming mummified but not developing fungal growth. If plants grow from infected corms, the foliage shows yellow flecking and later striping towards the tips of the outer leaves (see **p.55**). The discoloration spreads downwards and the leaves turn brown and die. Roots of affected corms blacken and may die back.
CAUSE The fungus *Fusarium oxysporum* f. sp. *gladioli*, which persists in the soil and usually enters the plant through the basal plate or roots of the corm.
CONTROL Dispose of plants showing foliage symptoms. Check corms before storing, and again before replanting. Dry plants off after lifting and treat corms with sulphur dust. Store corms in a cool, dry place. Plant on a fresh site each year.

GLADIOLUS DRY ROT

SYMPTOMS Corms develop numerous slightly sunken lesions which may join together. Rotting occurs and causes dark brownish-black areas of discoloration, often accompanied by black fungal resting bodies (sclerotia). If infected corms are planted, they may grow, but the foliage dies off and a dark

brown rot occurs at the base of the foliage, later becoming covered in tiny sclerotia.
CAUSE The fungus *Stromatinia gladioli*, which can persist as sclerotia both on the corms and in the soil.
CONTROL Remove and dispose of plants showing foliage symptoms. Check all corms thoroughly before storing, and again at the end of storage before replanting. Dry plants off naturally after lifting and treat corms with sulphur dust. Store corms in a cool, dry place. Plant on a fresh site each year.

GLADIOLUS HARD ROT

SYMPTOMS Small, purple-brown spots form on leaves and develop grey centres bearing numerous pycnidia. Dark, reddish-brown to black areas appear towards the base of the corms. Affected areas of the corm often sink in and enlarge. Gladioli, crocus, acidantheras and freesias may be attacked.
CAUSE The fungus *Septoria gladioli*, which is encouraged by wet weather and is most severe on gladioli growing in poor soils.
CONTROL Dispose of affected corms; plant only those which seem healthy.

GLADIOLUS SCAB AND NECK ROT

SYMPTOMS Small, red-brown spots develop on leaves and these darken as they enlarge. Leaf tips become shrivelled. Pale yellow to golden, sunken lesions develop on the corms and produce a resin-like, bacterial ooze. In addition to gladioli, freesias and crocuses are commonly affected by this disease.
CAUSE The bacterium *Pseudomonas gladioli*,

which is encouraged by wet weather.
CONTROL Do not plant affected corms. Dispose of any showing symptoms and once the disease has developed, do not grow susceptible plants on that site for at least five years.

GLADIOLUS THRIPS

SYMPTOMS The foliage of gladioli develops a fine, pale mottling with tiny black excrement spots during mid- to late summer. Flowers also have pale flecking (see **p.45**) and in severe attacks, flower buds fail to open.
CAUSE Sap-feeding insects, *Thrips simplex*, which tend to be more troublesome in hot, dry summers. Elongate, narrow-bodied insects, up to 2mm in length, they live in the flower buds and at the bases of the leaves. Adult thrips are brownish-black while the immature nymphs are pale yellow.
CONTROL Watch for signs of infestation and spray with acetamiprid, thiacloprid, or bifenthrin. Open flowers can be damaged by sprays, particularly if treated in sunny weather. In autumn burn infested foliage and seed heads. Some thrips overwinter on corms; store in a cool, frost-free place, otherwise thrips continue feeding over winter and weaken the corms.

GLASSHOUSE LEAFHOPPER

SYMPTOMS A coarse, pale mottling occurs on the upper leaf surface (see **p.20**) of many indoor plants. This insect also attacks garden plants including primulas, polyanthus, foxgloves (*Digitalis*) and nicotiana. The adults are pale yellow with

greyish markings, 3mm long and broadest at the head end, tapering to a point behind. Adults jump off plants when disturbed and so it is easier to find the less active, creamy-white, wingless immature nymphs. Both adults and nymphs live on the underside of leaves.
CAUSE Sap-feeding insects, *Hauptidia maroccana*, which breed continuously throughout the year in greenhouses and in sheltered places out of doors.
CONTROL Spray the underside of the leaves with bifenthrin, thiacloprid, or pyrethrum when leaf spotting is seen. Damaged leaves remain discoloured, but new growth remains green once the pest is controlled.

GLASSHOUSE RED SPIDER MITE

SYMPTOMS Leaves develop a fine pale mottling (see **p.22**) and the foliage becomes firstly dull green, but later increasingly yellowish-white. These symptoms can be confused with a mineral deficiency but examination of the lower leaf surface, preferably with a x10 hand lens, will reveal tiny spider mites and their spherical eggs. The mites are less than 1mm long and have four pairs of legs; they are yellowish-green with two large dark markings towards the head end of the body. They only become orange-red in the autumn and winter when the adult females are hibernating in sheltered places. When plants are heavily infested, mites will crawl over a fine silk webbing spun between the leaves and stems (see **p.43**). Leaves dry up (see **p.23**) and fall prematurely, so that only young leaves at the shoot tips remain.

CAUSE Sap-feeding mites, *Tetranychus urticae*, which attack a wide range of indoor and greenhouse plants. They also cause problems out of doors in warm dry summers, especially to beans, raspberries, strawberries, currants, roses and many other ornamental plants.

CONTROL Red spider mites breed rapidly under warm conditions, and some strains have gained resistance to some pesticides. Biological control with the predatory mite, *Phytoseiulus persimilis*, gives good results if introduced before a heavy infestation has developed. It also requires warm daytime temperatures and an absence of pesticides if it is to become established. *Phytoseiulus* can also be used on outdoor plants during the summer. Alternatively, plants can be sprayed with bifenthrin, thiamethoxam, abermectin, plant oils or fatty acids. Several applications will be necessary to break the mite's life cycle.

GLASSHOUSE THRIPS

SYMPTOMS Foliage develops a silvery white discoloration on the upper surface. The leaves are marked with brown excrement spots and narrow bodied insects up to 2mm long may be present. A wide range of indoor plants is attacked, especially palms and citrus plants. Climate change means that this greenhouse pest is now becoming a problem on outdoor plants in sheltered gardens, where it has a particular liking for *Viburnum tinus*.

CAUSE A small sap-feeding insect, *Heliothrips haemorrhoidalis*, which is active all year round on indoor plants. The adult thrips are blackish brown,

while the immature nymphs are creamy white.

CONTROL On ornamental plants, spray with bifenthrin, thiacloprid, or acetamiprid when signs of damage are seen. On fruiting citrus plants, use an organic spray such as pyrethrum, fatty acids, or plant oils.

GLASSHOUSE WHITEFLY

SYMPTOMS White-winged insects, 2mm long, readily fly up when plants are disturbed. The adults and the immobile, whitish-green, oval, scale-like nymphs live on the undersides of leaves of many houseplants and greenhouse vegetables and ornamentals. The upper leaf surfaces become sticky with honeydew excretion and a black, sooty mould often develops on this substance. Glasshouse whitefly can spread to garden plants during the summer, but will not survive a cold winter.

CAUSE A sap-feeding insect, *Trialeurodes vaporariorum*, which breeds continuously throughout the year on indoor plants.

CONTROL Pesticide resistance is a widespread problem with this pest. Biological control with a tiny parasitic wasp, *Encarsia formosa*, is often the best remedy on indoor plants between mid-spring and mid-autumn when temperatures are high enough to suit the parasite. *Encarsia* must be introduced before a heavy infestation has developed, as it needs time to breed before it can start reducing the whitefly population. Sticky yellow traps hung just above plants will catch adult whiteflies and indicate when they are starting to appear. Fatty acids or plant oils will reduce

whitefly numbers and have no lasting impact on *Encarsia*. Other insecticides, such as bifenthrin and pyrethrum, are dangerous to the parasite and may give poor control of whitefly if a resistant strain is present. Several applications at about five-day intervals will be required even if the whitefly is susceptible. Longer control on pot-grown ornamental plants can be achieved by spraying with or using thiacloprid, acetamiprid, or thiamethoxam as a compost drench. Thiacloprid can be sprayed on greenhouse tomatoes, cucumbers, peppers, and aubergines.

GLEDITSIA GALL MIDGE

SYMPTOMS Leaves at the shoot tips of honey locust (*Gleditsia triacanthos*) fail to expand into the normal pinnate form. The leaflets are folded over and make pod-like galls (*see p.27*) which contain up to seven whitish-orange maggots. These galls first appear in early summer; by late summer, infestations may be so severe that normal leaf production comes to a complete halt. The cultivar 'Sunburst' seems to be particularly susceptible and produces pale yellow galls, sometimes with a pink flush.

CAUSE A tiny fly, *Dasineura gleditchiae*, that originates from North America but arrived in Europe during the 1970s. Grey, and 2mm long, it has three generations, with female midges laying eggs in developing leaves at the shoot tips in late spring and mid- to late summer.

CONTROL Small trees can be sprayed with bifenthrin when the midges are seen on the shoot tips, particularly against the first generation in

late spring and early summer. Tall trees cannot be sprayed properly and this pest often has to be tolerated. Fortunately, the tree is able to make enough normal leaves to maintain growth before heavy infestations develop in mid-to late summer.

GLEOSPORIUM ROT

SYMPTOMS Also known as bitter rot, this is seen as circular brown, somewhat sunken lesions on the surface of apples and occasionally quinces and pears while in store or, occasionally, while still on the tree. On close inspection it may be possible to see many tiny fungal pustules arranged in concentric circles on the surface of the lesions. Occasionally a canker of old stems may also develop at the base of the spurs.

CAUSE The fungus *Glomerella cingulata*, which usually enters fruits through their lenticels (breathing pores in the skin).

CONTROL Remove infected apples. Check apples in store regularly and remove any showing symptoms.

GOOSEBERRY DIEBACK

SYMPTOMS Stems of gooseberries or sometimes blackcurrants die back. Occasionally the whole bush is affected, but more commonly only certain areas deteriorate while most of the bush appears to be perfectly healthy. Infection will spread, and if the main stem or base of the plant is affected, is it possible that the whole plant will die back. During damp weather, fuzzy, grey fungal growth may develop around the damaged areas. Foliage on infected stems yellows,

turns brown and then dies back, falling prematurely. Any developing fruits shrivel and die.
CAUSE The fungus *Botrytis cinerea*. The spores are spread on air currents and by rain or water splash.
CONTROL Prune out infected areas promptly, removing any wood that shows staining or discoloration and cutting back to a healthy-looking bud. The infection may also occur on the fruits, so these should also be removed immediately to help minimize the risk of spread of the infection.

GOOSEBERRY MILDEW

See AMERICAN GOOSEBERRY MILDEW *p.98*.

GOOSEBERRY RUST

SYMPTOMS In spring, reddish-orange pustules develop on the older, lower leaves and occasionally on the fruits and stems. In summer, the cluster cup stage of the fungus develops on the pustules, causing extensive distortion and curling of the leaves.
CAUSE The fungus *Puccinia caricina*, which may also attack sedges (its alternate host) and which is most damaging to gooseberries following dry weather in early spring.
CONTROL Remove infected areas promptly. Remove sedges growing in the vicinity, as these may harbour the infection.

GOOSEBERRY SAWFLY

SYMPTOMS Gooseberry bushes become rapidly and severely defoliated (*see p.31*), often being reduced to bare stems by the time the berries are ready for picking. Three

species of sawfly can attack gooseberry. They have pale green caterpillar-like larvae up to 20mm long, heavily marked with black spots in two of the species. Damage starts in mid- to late spring, but as there can be two or three generations, there may be further damage later in the season. Red and white currants are also attacked in a similar manner.
CAUSE Larvae of the common gooseberry sawfly (*Nematus ribesii*), the lesser gooseberry sawfly (*N. leucotrochus*) and the pale gooseberry sawfly (*Pristiphora pallipes*).
CONTROL Inspect gooseberry and currant bushes carefully for larvae from mid-spring. Eggs are often laid on the underside of leaves in the centre of bushes, so larvae may go unnoticed until they have eaten upwards and outwards and there are few leaves left. Spray affected plants with pyrethrum or thiacloprid. These can be used up to one or three days before harvesting respectively.

GREENFLY

See APHIDS *p.99*.

GREEN SPRUCE APHID

SYMPTOMS The foliage on spruce (*Picea*) develops a whitish-yellow mottling, then turns brown and drops off during the late winter and spring. The new growth in the spring is not affected. Dark green aphids with red eyes can be found amongst the foliage, which is often blackened by sooty mould growing on the aphids' sugary excrement.
CAUSE A sap-feeding aphid, *Elatobium abietinum*. Unlike most greenfly, it is active between autumn and spring

and can be particularly damaging in mild winters.
CONTROL Spray small trees with bifenthrin, thiacloprid, acetamiprid, or pyrethrum in the autumn, and again if symptoms are noticed in the winter or spring. Badly defoliated plants will take several years to regain their appearance; encourage better growth by feeding with a general fertilizer in the spring and keep trees watered in dry summers.

GREY MOULD

SYMPTOMS Grey or occasionally off-white or grey-brown fuzzy fungal growth develops on infected areas (*see p.34*). Grey mould is a very common fungus and, because it can live on most living and dead plant material, its spores are almost always present in the air. Grey mould may attack all above-ground parts of most plants, usually gaining entry via wounds or points of damage. Fruit infections may originate from infection via the open flowers, the fungus remaining dormant until the fruits start to ripen (*see p.50*). Prior to the development of the fungal growth, the infected plant tissue usually discolours, often turning brown and soft. Growth above points of infection may deteriorate, with leaves yellowing and wilting and flowers or fruits dying off (*see p.45*). Grey mould infection on petals or fruits may initially cause a hypersensitive reaction which results in colour changes but no rotting. On flowers, bleached white or pale brown spots (known as ghost spots) may form (*see p.42*). The fungus may also produce numerous small black sclerotia (resting

bodies) which are capable of withstanding a wide range of growing conditions and then infecting plants when conditions are at their most favourable for its growth.
CAUSE The fungus *Botrytis cinerea*. Its spores are readily spread by rain or water splash and on air currents. The fungus may also carry over from year to year as sclerotia in the soil or on infected plant debris.
CONTROL Control is difficult, because the fungus is so widespread. Remove all dead and injured parts before they have a chance to become infected. Remove infected areas promptly, cutting right back into healthy growth. Do not leave plant debris lying around.

GRIFOLA FRONDOSA

SYMPTOMS Overlapping, greyish-tan to honey-coloured fungal brackets develop around the base of a tree, on its roots or on tree stumps. The margin of each "frond" is white. The fruiting body may measure up to 30cm across, the individual "fronds" measuring 20–60mm across. If bruised, the fungus has a distinct mouse-like smell. The fungus causes a butt rot of susceptible trees, and infected wood becomes soft and pulpy. White mycelium (a fungal sheet) develops in decaying areas. The fruiting body usually appears in late summer or autumn and disintegrates with the first frosts.
CAUSE The fungus *Grifola frondosa*, easily confused with the much larger *Meripilus giganteus* (*p.148*).
CONTROL Affected trees may be rendered dangerous and so should be inspected by a reputable tree surgeon and felled if necessary.

GROUND BEETLES

Ground beetles or carabids are mostly predatory insects that feed on a variety of insects and other invertebrate animals. They vary in size from 2–30mm and are mostly black, brown or bronze-green. They have long, slender legs that enable them to run rapidly when disturbed. Adult ground beetles live and feed mainly in or on the soil surface but some species will climb up plants at night to prey on aphids, small caterpillars and other insects. The larvae live in the soil and are also predatory. Some adult ground beetles feed on plant seeds and while most are harmless, damage can occur on strawberry fruits.
• *see also* STRAWBERRY SEED BEETLES *p.181.*

GUMMOSIS

See CUCUMBER GUMMOSIS, *p.121.*

GYPSY MOTH

SYMPTOMS A wide range of trees and shrubs, including conifers, can be severely defoliated by caterpillars in early summer. The hairy, blackish-yellow caterpillars, up to 70mm long, are variable in colour but distinctive, due to a pair of blue warts on each of the front five seg-ments of their bodies. The other segments have reddish warts on the upper surface.
CAUSE Caterpillars of a moth, *Lymantria dispar.* An Asiatic form of this moth has moved into Europe in recent years and is proving to be a more formidable pest than its European counterpart. It has a wider range of hosts and the females fly more readily, thus dispersing the pest more widely. Neither form of gypsy moth is established in

Britain and suspected sightings should be reported to the Forestry Commission or the local office of DEFRA.
CONTROL Look for caterpillars, which hatch from overwintered eggs laid in batches on the bark, from mid-spring onwards. Small trees and shrubs can be treated with an insecticide such as bifenthrin. Small caterpillars are more susceptible than older larvae.

HALO BLIGHT

See BEAN HALO BLIGHT *p.105.*

HAWTHORN GALL MIDGE

SYMPTOMS In early summer, hawthorn shoot tips fail to extend and instead develop a rosette of small leaves. Several orange-white, legless larvae, up to 2mm in length, live within each rosette, which then dries up and turns brown later in the summer.
CAUSE A small fly, *Dasineura crataegi,* that has several generations during summer. Clipped hawthorn hedges are always more heavily infested than uncut hawthorns.
CONTROL This pest causes little real damage, although old, dried-up galls can be unsightly. There is no point in using insecticides; as only clipped hawthorns are particularly susceptible, the damaged shoots will be removed when the hedge or bush is trimmed.

HEBE DOWNY MILDEW

SYMPTOMS Yellow or buff-coloured discoloration

develops on the upper leaf surface, corresponding with off-white, fluffy fungal growth beneath. Infected leaves die and drop early, and as a result may leave the plant with sparse foliage. Lower leaves and those in congested parts of the plant are most likely to be attacked.
CAUSE The fungus *Peronospora grisea,* which is encouraged by moist, humid conditions and poor air circulation.
CONTROL Remove infected leaves promptly and destroy any severely infected plants.

HELLEBORE BLACK DEATH

SYMPTOMS Black streaks mostly along the leaf veins, sometimes in the form of ringspots. The stems and leaves become distorted and flowers may become infected. Cultivars and hybrids of *Helleborus hybridus* are particularly affected. Infected plants may show no symptoms for a year or more so can be a source of infection.
CAUSE A type of calavirus, most likely *Helleborus net necrosis* virus. The virus is believed to be spread by aphids feeding on hellebores and acting as vectors.
CONTROL There is no control available. Remove and burn infected plants promptly.

HELLEBORE LEAF BLOTCH

SYMPTOMS Slate-grey or greyish-brown lesions develop on the leaves (*see* **p.20**). They may join together, leaving large leaf areas discoloured and dead. Flower stems and occasionally the flowers themselves may also be attacked, causing flower stems to flop over.
CAUSE The fungus *Coniothyrium hellebori,*

which spreads especially rapidly in damp weather.
CONTROL Remove and dispose of affected leaves and other plant parts promptly. Even if the plant is left with only very sparse foliage, replacement leaves should soon be produced. In certain circumstances, *Helleborus orientalis* appears to show a good degree of resistance to this infection, and so in troublesome sites it may be preferable to concentrate on growing this rather than other species.

HEMEROCALLIS GALL MIDGE

SYMPTOMS The flower buds of daylilies (*Hemerocallis*) become abnormally swollen but fail to open before going brown and drying up (*see* **p.45**). Petals inside a galled bud are thickened, with a crinkled appearance, and there are many white maggots, 2–3mm long, at the base of the petals.
CAUSE A tiny fly, *Contarinia quinquenotata,* that deposits its eggs in the developing flower buds during late spring and early summer. There is only one generation a year, so buds developing after the egg-laying period and flowering from mid-summer onwards escape damage.
CONTROL Pick off and destroy galled buds before the maggots have a chance to escape into the soil to pupate. Chemical controls are likely to be ineffective since it is difficult to prevent the females from laying eggs, and the grubs are well protected inside the buds. By selecting *Hemerocallis* cultivars with main flowering periods from mid-summer onwards, the problem can be largely avoided.

HEMISPHERICAL SCALE

SYMPTOMS A wide range of indoor ornamental plants are susceptible to this pest. When mature, the scales (see **p.16**) have a dark brown, hemispherical, convex shell over their bodies. The shell is up to 4mm across and has an H-shaped ridge on it. The immature scales are flat, oval and yellowish-brown in colour. They occur on the underside of leaves and on stems. Their sugary excrement, called honeydew, falls down and makes upper leaf surfaces sticky. Sooty moulds often grow on this substance.

CAUSE A sap-feeding insect, *Saissetia coffeae*, which breeds continuously throughout the year on houseplants and in heated greenhouses.

CONTROL Heavily infested plants are best disposed of, especially ferns, which generally do not tolerate insecticides. Some scales and honeydew can be removed by wiping the foliage and stems with a damp cloth. Spray with plant oils, fatty acids, acetamiprid, thiamethoxam, or thiacloprid against the more vulnerable young scale nymphs. Several applications at seven-day intervals may be needed to break the insect's life cycle if using plant oils or fatty acids.

HIGH TEMPERATURE INJURY

SYMPTOMS Injury may be temporary or result in tissue death and permanent damage. Scorching or scalding of leaves, fruits and flowers is common, often being seen as dry, papery, buff or brown areas. Wilting may occur if the plant is losing more water than it can take up, due either to excessive transpiration or dry soil conditions. Trees or shrubs which have thin bark may show bark death down the exposed side of their stem or trunk.

CAUSE Excessively high temperatures, or the rapid onset of high temperatures immediately following cool weather.

CONTROL There is rarely anything that can be done once injury has occurred, other than to ensure that further injury is prevented by improving ventilation, provision of shading, damping-down of the greenhouse floor or other means appropriate to the situation in which the plants are growing.

HOLLY LEAF BLIGHT

SYMPTOMS Circular black spots develop on infected leaves. Premature leaf fall follows. Black lesions form on young shoots and may girdle them, so causing dieback. On holly hedges defoliation occurs in inverted "v" shapes.

CAUSE The fungus *Phytophthora ilicis*, which is spread from tree to tree by rain splash, and far greater distances by flooding and in areas of poor drainage. It has been recorded on *Ilex* x *altaclerensis*, *I. aquifolium* and *I. crenata*.

CONTROL The fallen leaves may harbour the infection so it is important that these are raked up and disposed of promptly. Avoid bringing infected soil or plants into the garden.

HOLLY LEAF MINER

SYMPTOMS Leaves develop yellowish-purple blotches (see **p.17**) where the internal tissues have been eaten out by a fly maggot. Leaf mines can be seen throughout the year and may be abundant on hollies trimmed to form a hedge.

CAUSE The larvae of a small fly, *Phytomyza ilicis*, which lays eggs on young foliage in early summer. There is one generation a year.

CONTROL Apart from creating discoloured patches on the foliage, this fly has little impact on the health and vigour of hollies. Mined leaves that turn yellow and drop in early summer are likely to be part of the normal loss of old foliage that occurs at that time. Holly leaf miner generally has to be tolerated, as insecticides do not penetrate the thick, waxy surface of holly leaves.

HOLLYHOCK RUST

SYMPTOMS Bright yellow or orange spots with orange-red centres develop on the upper leaf surface. Bracts may also be affected. The lower leaf surface develops numerous raised orange pustules (see **p.17**), each about the size of a pin-head and corresponding with the spots on the upper surface. The spots may join together, killing large areas of leaf tissue and causing it to fall away, leaving the leaves with a ragged and unsightly appearance. The pustules later turn buff-brown in colour.

CAUSE The fungus *Puccinia malvacearum*, which is very common, but most troublesome in damp seasons. The spores are ejected from the pustules and spread on air currents. The fungus overwinters on infected plant debris and on any infected leaves which may remain on the plant during mild winters.

CONTROL Inspect plants regularly and remove any infected leaves. Spray plants from an early stage at regular intervals with a fungicide such as penconazole or myclobutanil. Remove potential weed hosts such as wild mallow (*Malva* species) from the area. There are no resistant hollyhocks available, but the species *Althaea rosea* is said to be less susceptible.

HOLM OAK LEAF MINER

SYMPTOMS Deciduous oaks can be attacked by this leaf miner, but it is most commonly seen on evergreens, such as the holm oak, *Quercus ilex*. Affected leaves develop elongated, brownish-white, discoloured blotches on the upper leaf surface. The corresponding areas on the lower surface have a blistered appearance where the lower epidermis has become detached. Damage is particularly noticeable during winter.

CAUSE Caterpillars of a tiny moth, *Phyllonorycter messaniella*, which has three generations a year on holm oak.

CONTROL Treatment is difficult, as insecticides do not readily penetrate into the tough, shiny leaves of holm oak. Many trees may also be too large for effective spraying. Fortunately, oaks can support heavy infestations without adverse effects on their growth. Although the foliage may be unsightly in the winter when the year's accumulated damage is present, the tree's appearance will improve. In the spring the new growth will cover up the damaged foliage and some of the older leaves will be shed in early summer.

HONEY FUNGUS

SYMPTOMS Affected trees, shrubs, woody climbers, and sometimes woody herbaceous perennials start to die back. Occasionally other plants, like strawberries and potatoes, can become infected. Leaves may discolour and wilt, or fail to leaf up in spring. Death may appear to be quite rapid, or an infected plant may take several years to die. Flowering or fruiting plants may put on a particularly good display or set an unusually productive crop before dying. Resin may exude from around the base of the trunks of infected conifers. The roots and/or stem or trunk bases of infected plants develop a cream mycelium (fungal sheet) on them, sandwiched between the bark and the woody tissues beneath (*see p.34*). The mycelium has a characteristic mushroom-like smell. The fungus also produces tough blackish fungal strands called rhizomorphs. They may show branching and are hollow inside, the blackish outer "rind" being lined with white or faintly pink fungal growth. Rhizomorphs resemble old tree roots and are easily mistaken for them. They vary in length and may be plump or flattened, measuring from under a millimetre to several millimetres in diameter. The rhizomorphs feed on a woody host, or the remains of an old tree stump or dead plant. They may then grow through the soil and latch onto other food sources such as living trees or woody debris. The rhizomorphs may also grow beneath the bark of an infected plant, or be found growing over the surface of a root. Most are found within the top 20cm of the soil. Clumps of honey-coloured toadstools may appear around the base of infected plants in late summer or autumn. The toadstools may also appear around old tree stumps, following the root line of underground roots, or growing from underground rhizomorphs in the vicinity of an infected plant. Toadstools are very variable, but usually grow in clumps and have honey coloured caps which may be marked with dark brown flecks. The spores are white and may accumulate on the caps of other fungi in the clump. The toadstool's stipe (stem) usually bears a distinct creamy-white ruff or collar.

CAUSE Various species and races of the fungus *Armillaria*. These vary in their ability to infect potential hosts, some being far less pathogenic than others and living only on dead or very stressed plants. It is hard to differentiate between species, except on the basis of the damage they cause. Spread occurs either by rhizomorphs which grow through the soil at the rate of about 1m a year, or by root contact.

CONTROL Remove infected plants promptly, digging out stumps and as much of the root system as possible. Keep plants in good health by regular feeding, mulching, and other relevant care, as this makes them less prone to infection. The spread of disease from the toadstools is minimal, so it is not necessary to clear them away. Plants like birches (*Betula*), cedars (*Cedrus*), most common hedging conifers, cotoneaster, currants (*Ribes*), forsythias, hydrangeas, lilac (*Syringa*), *Malus*, peonies, privet, *Prunus*, rhododendrons, roses, willows (*Salix*) and wisterias, are known to be particularly susceptible and so should not be used to replace plants in gardens with honey fungus. Trees, shrubs, and climbers that show a good degree of resistance include beeches (*Fagus*), box (*Buxus*), mallows (*Malva*), oaks (*Quercus*), yews (*Taxus*), bamboos, abutilons, actinidias, carpenterias, celastrus, cercis, catalpas, ceratostigma, chaenomeles, choisyas, clematis, cotinus, elaeagnus, fothergillas, hebes, kerrias, passifloras, phlomis, photinias, pieris, pittosporum, rhus, romneyas, sarcococcas and tamarix.

HONEYDEW

This is the sugary excrement produced by some sap-feeding pests such as aphids, whiteflies, suckers, mealybugs and some scale insects. These pests suck sap from the phloem vessels in leaves and stems. The sap they imbibe is very rich in sugars, and is only partially digested by the insects, so their excrement is a sticky liquid due to its high sugar content. Although these pests usually live on lower leaf surfaces, it is upper leaf surfaces that becomes sticky. The insects flick honeydew droplets away from them and these then fall down onto leaves and other surfaces beneath. Under damp conditions non-parasitic fungi called sooty moulds grow, on honeydew and form a black or greyish-green deposit on the foliage.

HONEYSUCKLE APHID

SYMPTOMS Greyish-green aphids form dense colonies on the shoot tips and flower buds of honeysuckles in late spring and summer. This can result in the flower buds aborting and shoot tips drying up and browning.

CAUSE A sap-feeding insect, *Hyadaphis passerinii*.

CONTROL Check the new growth and flower buds for the presence of aphids from mid-spring onwards. If necessary spray with bifenthrin, thiacloprid, acetamiprid, pyrethrum, fatty acids or plant oils.

HORNBEAM WITCHES' BROOMS

SYMPTOMS Extensively branched twigs develop, closely clustered together, to form "witches' brooms" scattered throughout the crown of the tree. An off-white layer of fungal spores may develop on the surface of the leaves from late spring until mid-summer. During winter the growths may, at first, be mistaken for birds' nests and can spoil the outline of the tree or hedge, but during the summer they appear as dense, leafy areas and so are less obtrusive.

CAUSE The fungus *Taphrina carpini*, spread by water splash and on air currents.

CONTROL Prune out growths if desired. Spread seems to be very slow, but removal will help to limit the problem. No chemical control is available.

HORSE CHESTNUT BLEEDING CANKER

SYMPTOMS Drops of reddish-brown to black gummy ooze appear in scattered patches on the bark. Bleeding may then occur on the trunk, main stems and larger limbs (*see p.38*) and, as the weather warms up, crusty deposits develop. The surrounding bark is killed and the tree itself may die; young trees (10–30 years old) are more severely affected, as they have smaller trunks and are more easily girdled. The wood discolours beneath the bark in affected areas, and secondary wood-rotting fungi may appear from damaged areas. There may also be dieback, thinning of the tree's crown, yellowing of foliage (*see p.23*) and premature leaf drop. Both *Aesculus hippocastanum* and *A. carnea* are affected.

CAUSE The bacterium *Pseudomonas syringae* p.v. *aesculi* is the pathogen generally responsible. The fungi *Phytophthora citricola* and *P. cactorum* are now less commonly involved. Bleeding canker is worst during mild, moist weather in spring and autumn. It is most severe in southern parts of the UK, but does extend into Scotland.

CONTROL There are no chemical control measures available. Remove infected branches promptly. Burn or bury all infected wood; it must not be kept for later use as firewood, nor should it be chipped on site.

HORSE CHESTNUT LEAF BLOTCH

SYMPTOMS Chestnut to tan-brown, irregular leaf blotches (*see p.20*), each with a bright yellow margin, develop from mid-summer onwards. The damage is often most severe on the leaf margins and leaf tips. Numerous pinprick-sized, black fungal fruiting bodies may develop on the upper leaf surface over the spots. Severely infected leaves turn brown, curl and fall prematurely.

CAUSE The fungus *Guignardia aesculi*, which although causing early leaf fall has little effect on the tree's overall health, because most of the damage occurs quite late in the season.

CONTROL Collect up and dispose of fallen leaves in the autumn to minimize disease carry-over.

HORSE CHESTNUT LEAF-MINING MOTH

SYMPTOMS Whitish green blotch mines, which later become brown, develop between the leaf veins. These mines are very similar to the damage caused by the fungal disease horse chestnut leaf blotch. By holding an infested leaf up to the light it should be possible to see the caterpillars, or their circular silk cocoons in which they pupate, within the mines. The caterpillars are up to 5mm in length and the mines may be 40mm long by the time the larvae are fully fed. Heavy infestations can result in early leaf fall and loss of vigour.

CAUSE The larvae of a small leaf-mining moth, *Cameraria ohridella*. This first arrived in Europe in Macedonia in the 1970s and has since spread rapidly throughout the continent. It was first found in Britain in 2002 and is likely to become widespread. There are at least three generations between late spring and autumn.

CONTROL Because of the size of most horse chestnuts, there is no effective way of treating them with pesticides. There is also no insecticide available to gardeners at the present time which will give good control of leaf miners. A biological control in the form of a parasite or predator is needed but as yet this pest seems to have no effective natural enemies in Europe. *Aesculus indica* is resistant and could be a replacement for *A. hippocastanum*.

HORSE CHESTNUT MARGINAL LEAF SCORCH

SYMPTOMS Marginal leaf browning or scorching occurs, usually in early to mid-summer. Leaves curl slightly and scorched areas turn tan-brown. All or part of the tree may be affected. If the damage is only limited, the rest of the tree usually appears perfectly healthy.

CAUSE The precise cause is unknown, but it is believed that this disease may be due to a xylem-limited bacterial infection (a bacterial infection restricted to the xylem, or water-conducting elements, within the tree).

CONTROL None known, but symptoms may be less apparent in some years than others.

HORSE CHESTNUT SCALE

SYMPTOMS Many trees and shrubs, including horse chestnuts (*Aesculus*), limes (*Tilia*), sycamore (*Acer pseudoplatanus*), maples (*Acer*), dogwoods (*Cornus*), magnolias, bay trees (*Laurus nobilis*), elms (*Ulmus*) and *Skimmia japonica*, are attacked. Infestations are most recognisable in early to mid-summer when the females deposit their eggs under white, waxy threads; these are secreted from the rear edge of the shell-like covering over the insects' bodies. Mature scales are brown, up to 5mm long and are found on the trunks and larger branches. After egg-laying, the scales die and fall away, leaving circular white egg patches on the bark (*see p.40*). During the summer young scales – flat, oval, yellowish objects – live on the undersides of leaves. In the autumn they move onto the bark and overwinter as blackish-yellow immature nymphs. Little problem with honeydew and sooty mould occurs with this scale, except on *Skimmia japonica*.

CAUSE A sap-feeding insect, *Pulvinaria regalis*, which has one generation a year. The heaviest infestations tend to occur on trees growing in streets and car parks where there is a warmer microclimate due to heat being reflected off concrete, tarmac and brickwork.

CONTROL On large trees, horse chestnut scale has to be tolerated. Fortunately, it causes little real harm apart from the unsightliness of the egg masses. On small trees and shrubs, scrape off the scales and egg masses before the eggs have a chance to start hatching in mid-summer. If necessary, spray the undersides of leaves with thiacloprid or acetamiprid in mid-summer.

HOVERFLIES

Many of these flies have black-and-yellow striped abdomens and can be seen sitting on flowers or hovering near them. Some have larvae that feed on greenfly and other aphids. They may consume as many as 400 before they are ready to pupate and are helpful in controlling aphids. Other types of hoverfly have larvae that feed in rotting vegetation, in wasp nests, in the mud in pools or in rot holes in trees. The only hoverflies to cause problems in gardens are the narcissus bulb fly (*see p.151*) and sempervivum leaf miner (*see p.176*).

HYDRANGEA LEAF SPOT

SYMPTOMS Grey, brown or purple-brown spots of varying sizes are usually most apparent on the older foliage. Pinprick-sized fungal fruiting bodies may develop within the damaged areas. A yellow halo appears around the edge of the leaf spot, often indicating that a bacterial infection is responsible. Although spoiling the appearance of the foliage, none of these leaf spots causes long-term damage.

CAUSE Various fungi and bacteria, mostly weak pathogens, only able to attack hydrangeas which are suffering from other problems.

CONTROL Remove severely infected leaves. Improve the vigour of the plant by regular watering, feeding and other appropriate measures.

HYDRANGEA POWDERY MILDEW

SYMPTOMS Dense, white powdery deposits form on the upper leaf surfaces, and occasionally on petioles and new stem growth. The affected leaves may become distorted, particularly if infection occurs when they are still young. The leaves usually remain on the plant but may show patchy necrosis. They are rarely killed outright.

CAUSE The fungus *Microsphaera polonica*, which is encouraged by moist or humid air around the foliage and dry soil conditions around roots.

CONTROL Remove any severely affected leaves. Keep the plant well watered at the base, but avoid overhead irrigation. If the disease appears early in the year and threatens to become extensive.

HYDRANGEA SCALE

SYMPTOMS This pest is most visible during late spring and early summer when the females deposit eggs amongst a mass of white, waxy fibres on the stems and foliage of host plants (*see p.15*). Hydrangea is the most frequently attacked plant but this scale also occurs on *Prunus* and *Acer*. The egg masses are oval in shape and up to 5–6mm long. The mature scale is brownish-yellow but drops off when the insect dies, leaving the egg mass as the most obvious sign of infestation. Eggs hatch in mid-summer and the flat, oval, pale yellow scale nymphs live on the underside of leaves next to leaf veins during the rest of the summer. They move onto the bark in the autumn and overwinter as immature nymphs.

CAUSE A sap-feeding insect, *Pulvinaria hydrangeae*, which has one generation a year. Heavy infestations reduce the plant's vigour and result in dieback.

CONTROL Spray with acetamiprid or thiacloprid in mid-summer against the more vulnerable newly hatched scales. Hydrangeas can be damaged by some pesticides, so avoid treatment when plants are exposed to bright sunlight or high temperatures. Water the plants first if they are dry at the roots.

INONOTUS HISPIDUS

SYMPTOMS Annual, bracket-shaped, semi-circular fungal fruiting bodies develop on large limbs or the main trunk of trees. Each measures between 10 and 35cm across and is 2–10cm deep. They are covered in dense, hair-like growth and are an orangey-rust colour initially, turning dark brownish-black as they age. Their presence causes a yellowish-orange discoloration of the wood.

CAUSE The fungus *Inonotus hispidus*, which usually enters through wounds or stubs left by poor pruning.

CONTROL Infected trees may be rendered dangerous, their safety depending largely on where the decay has occurred. Employ a reputable tree surgeon to advise as to the best course of action.

IRIS INK DISEASE

SYMPTOMS Blue-black streaking appears on the outer surface of the bulb. In extreme cases, or if left unchecked, the whole bulb turns black and rots. If infected bulbs produce foliage, it is streaked yellow initially and later develops a red discoloration, particularly towards the tips. Black necrotic spots may develop on the yellowed areas.

CAUSE The fungus *Drechslera iridis*, which produces spores that are spread on air currents. The fungus can overwinter on infected bulbs, bulb debris, and possibly in soil around infected bulbs.

CONTROL Dispose of any bulbs showing symptoms. Remove any plants which show foliar symptoms, and grow irises on a fresh site.

IRIS LEAF SPOT

SYMPTOMS Water-soaked spots develop on the foliage, later turning brown with grey centres. The spots join together as they enlarge, so destroying the foliage. Affected irises are weakened by the loss of functional foliage and so are less vigorous and flower poorly.

CAUSE The fungus *Mycospharella macrospora*, which is encouraged by mechanical injury and wet weather. It produces spores which are spread on air currents and by rain and water splash. It overwinters on infected leaves and plant debris.

CONTROL If infection occurs after flowering, cut off badly affected areas. Keep plants well fed and ensure they are planted in a suitable, sunny site; they are less susceptible to infection if growing vigorously.

IRIS RHIZOME ROT

SYMPTOMS Foliage yellows and withers, starting at the leaf tips or leaf margins. A soft, often foul-smelling rot develops at the base of the affected leaves and rapidly spreads down the rhizome (*see p.55*). As the infection spreads, the foliage symptoms become more extreme, and leaves may yellow completely and topple over.

CAUSE The bacterium *Erwinia carotovora* subsp. *carotovora*, which is soil-borne and is particularly prevalent in heavy, wet soils and in wet seasons. The bacteria enter the iris via wounds.

CONTROL Plant irises on a well-drained soil. Choose only healthy rhizomes, and plant them shallowly. Avoid injuring the rhizomes and control pests such as slugs which may damage them. Remove and dispose of badly affected irises. It may be possible to save rhizomes which are only slightly affected by cutting off all damaged areas right back to sound, healthy tissue.

IRIS RUST

SYMPTOMS Numerous small, raised, lens-shaped orange to yellow pustules, measuring about 1mm in length, develop on the leaves (*see* **p.15**). These enlarge slightly and erupt to produce masses of yellow-orange spores. The infected leaves discolour, turning yellow at first and later, if the infection is severe, browning and dying off. Older leaves are generally the most badly affected.
CAUSE The fungus *Puccinia iridis*, which is encouraged by wet weather and excessively soft leaf growth.
CONTROL Cut off severely affected leaves. Infection usually occurs towards the end of the summer, so it does not cause too much loss of vigour. However, in severe cases or if the infection develops earlier in the year, spray with a suitable fungicide such as penconazole.

IRIS SAWFLY

SYMPTOMS V-shaped notches are eaten from the leaf margins initially (*see* **p.29**), but later there is extensive defoliation, particularly on the upper parts of the leaves. Only waterside irises are attacked, especially yellow flag iris, *Iris pseudacorus*.
CAUSE The caterpillar-like larvae of a sawfly, *Rhadinoceraea micans*. It is of local occurrence in Britain, but is numerous where it is found. The greyish-brown larvae, up to 25mm long, are present during early to mid-summer. The adult insects are black, with smokey-grey wings, and can be seen resting on waterside irises in late spring and early summer. There is one generation a year.

CONTROL The larvae can be controlled by spraying with pyrethrum or bifenthrin, but insecticides should not be used where there is a risk of pesticides getting into the water. Fish, frogs and other pond wildlife are very vulnerable to insecticides. Removal of the larvae by hand would be the best option in most gardens.

IRIS SCORCH

SYMPTOMS The upper parts of affected leaves wilt and bend over and become reddish-brown. The symptoms may initially be confused with those of iris rhizome rot, but the leaf discoloration is different and there is no deterioration of the rhizome. The roots die back; some rot and others disintegrate internally, leaving just a hollow outer "tube" with the cotton-threadlike remains of the central cylinder inside.
CAUSE The precise cause is not known, but the problem seems to be caused by root injury in winter or early spring. Winter waterlogging may be responsible.
CONTROL Ensure good drainage. It can sometimes be possible to save rhizomes that are only slightly affected by lifting, removing all deteriorating roots and leaves, washing thoroughly and then replanting on a new, very well-drained site. Foliar feeding, in an attempt to stimulate the production of replacement roots, can also be beneficial.

IRON DEFICIENCY AND LIME-INDUCED CHLOROSIS

SYMPTOMS Usually seen in combination with manganese deficiency. Leaf yellowing, or chlorosis (*see* **p.24**), is often

combined with the development of brown areas of discoloration which start at the leaf margins and then spread between the veins. Younger growth is usually affected earlier and more severely than older growth.
CAUSE Acid-loving plants and lime-hating plants have roots which are poorly adapted for the absorption of necessary trace elements from an alkaline soil. If conditions are too alkaline, they readily develop deficiency symptoms, in particular of iron and manganese.
CONTROL Avoid growing lime-hating plants on a soil which is not sufficiently acid. Treat affected shrubs with a chelated compound containing iron, manganese and other trace elements which will be available to the plant because they are in a form which does not become "locked up" in the alkaline soil. Use acidic mulches such as chopped, composted bracken or conifer bark. Incorporate acidic materials such as these into the planting hole at planting. Feed plants with a fertilizer formulated for use on acid-loving (lime-hating) plants. Before planting, consider acidifying soil using sulphur or aluminium sulphate and ferrous sulphate. Sulphur treatments can also be used around existing plants. Follow instructions carefully.

IRREGULAR WATERING

SYMPTOMS Fruits are unusually small and may have a particularly thick skin. The surface may be ruptured, with splits caused when the fruit swells in response to the sudden availability of water (*see*

p.57). Once the fruit's surface is broken it is open to infection by secondary organisms, such as grey mould, which may then cause further, rapid deterioration. Other fruit symptoms may include blossom end rot (*see p.108*). Leaves fail to grow at a regular rate and so develop an uneven, almost puckered surface (*see* **p.26**). They may also appear unusually small. Buds may fail to form, or form and later drop, and bark may split (*see* **p.36**).
CAUSE An irregular or erratic supply of water at the plant roots. Plants growing on a light, sandy soil with a low organic matter content are particularly prone, as are any plants growing in containers.
CONTROL Keep plants adequately watered at all times. Incorporate plenty of bulky organic matter into light soils before planting. Apply a bulky mulch around the root areas to encourage soil moisture retention.
• *see also* BUD DROP *p.112*; BUD, DRY *p.112*.

IVY LEAF SPOT

SYMPTOMS Brown or grey, near-circular spots develop on ivy foliage. On pale-leaved or variegated forms, the spots may be surrounded by a narrow band of reddish-purple discoloration. On variegated leaves, leaf-spotting is usually most severe on the pale or white areas of the leaf. Numerous, pinprick-sized, raised fungal fruiting bodies may develop on the spots, usually in concentric rings.
CAUSE Various fungi, including *Mycosphaerella hedericola*, *Colletotrichum trichellum* and *Phyllosticta hedericola*.
CONTROL Remove severely

infected leaves to limit disease spread. Although unsightly, ivy leaf spot infections will not lower the overall vigour of the plant.

IVY ON TREE TRUNKS

SYMPTOMS Ivy growing up the trunk and, if left unchecked, into the crown of trees. On young trees, growth may be slightly constricted, but this is unusual on larger trees.
CAUSE Ivy, with its climbing, self-clinging habit, moves onto and up suitable surfaces, including tree trunks. It is, however, rare for it to cause damage, but young or poorly growing trees may have their stem or trunk growth constricted. Dense ivy growth on trunks may also mask the development of other problems, such as decay-causing bracket fungi, but it does provide a useful wildlife habitat.
CONTROL Rarely justifiable. If desired, the ivy should be removed regularly, before it develops a lot of growth. Established plants can be cut off at ground level and, once dead, their top growth pulled away from the trunk and branches of the tree.

JUNIPER APHIDS

SYMPTOMS Leaves and stems become coated with black, sooty moulds that grow on the sugary excrement of aphids. Heavy attacks cause yellowing and dieback in mid- to late summer.
CAUSE Two species of aphid occur on junipers. *Cinara fresai* is 2–4mm long and

brownish-black. Colonies on stems close to the soil often have a protective covering of soil provided by ants, which collect honeydew from the aphids. The other species, *C. juniperi*, measures 2–3mm in length, is pale brown, and occurs mainly on younger growth. Both feed by sucking sap.
CONTROL Look for aphid infestations from mid-spring onwards and, if seen, spray with bifenthrin, acetamiprid, thiacloprid, or pyrethrum.

JUNIPER SCALE

SYMPTOMS Flat, rounded, or elongate and rectangular white objects with a small yellowish patch develop on the foliage of junipers, cypresses, Leyland cypress and also thuyas. The rounded scales are protective covers for the females and measure up to 1.5mm long. The smaller rectangular scales cover the males. Heavy infestations can cause yellowing (chlorosis) and dieback of green shoots (*see p.37*).
CAUSE A sap-feeding insect, *Carulaspis juniperi*, which has one generation a year.
CONTROL On plants small enough to be sprayed, apply thiacloprid or acetamiprid in early summer when the more vulnerable newly-hatched scale nymphs are present.

JUNIPER WEBBER MOTH

SYMPTOMS Patches of brown foliage bound together with silk threads develop. Dark brown caterpillars with paler stripes running along their bodies, up to 11mm long, live within the webbing (*see p.32*). The leaves damaged by the caterpillars dry up and die in early

summer. Compact, upright junipers, such as 'Skyrocket', can be badly damaged.
CAUSE Caterpillars of a moth, *Dichomeris marginella*, which has one generation a year.
CONTROL Infestations can sometimes be dealt with by cutting out webbed shoots in late spring or early summer. Alternatively, spray forcibly with bifenthrin so as to penetrate the webbing.

KABATINA SHOOT BLIGHT

SYMPTOMS Cankers develop on the shoots of Leyland cypress, certain thuyas, junipers, and cypresses. The cankers usually develop at the base of small branches and cause dieback and, occasionally, death.
CAUSE Fungi of *Kabatina* species. The fungal spores are carried on air currents and by rain splash.
CONTROL Remove affected shoots promptly to minimize spread of the disease.

KEITHIA THUJINA NEEDLE BLIGHT

SYMPTOMS Small, black fungal fruiting bodies, measuring about 1mm across, are embedded in the foliage of *Thuja plicata*. These later split open as the spores are released, leaving small, brown sunken pits, or holes, in the leaves. Foliage, particularly that towards the base of the plant, may turn brown (*see p.28*) and die back if the infection is extensive.
CAUSE The fungus *Didymascella thujina*. The

spores are air-borne and most infections take place in late summer or early autumn, especially in humid or damp weather. They may over-winter in mucilage attached to the damaged areas.
CONTROL Prune out infected areas promptly, before the spores are released. Remove severely infected stems.

KNOPPER GALLS

See ACORN GALL WASP *p.98*.

LABURNUM LEAF MINERS

SYMPTOMS Whitish-brown, dried-up patches develop on laburnum leaves (*see p.18*) between late spring and early autumn. If the leaf is held up to the light it is often possible to see leaf-mining larvae feeding within the leaf.
CAUSE Two leaf miners commonly occur on laburnum. The larvae of a leaf-mining fly, *Agromyza demeijerei*, cause narrow linear mines along the leaf margin which later enlarge into irregular blotch mines extending from the margin. Caterpillars of a leaf-mining moth, *Leucoptera laburnella*, cause roughly circular mines in the centre of leaves, but in heavy attacks the mines join together and lose their distinctive shape.
CONTROL Attacks in late summer may be extensive and unsightly, but they have little effect on the tree's health and vigour and so can be tolerated. None of the insecticides available to amateur gardeners is likely to give control.

LACEBUG

See PIERIS LACEBUG *p.161;*
RHODODENDRON LACEBUG *p.170.*

LACEWINGS

The larvae of lacewings are beneficial insects that feed on aphids and other small insects. Some lacewing larvae carry the sucked-out remains of earlier meals amongst clumps of bristles on their upper surface. A well-fed larva is up to 10mm long and, apart from a pair of sharp jaws protruding at the front, is hidden under a tangled mass of dead aphids. The harmless adult lacewings have slender, elongate bodies which are pale green, black or brown. They have two pairs of transparent wings, sometimes with darker markings, which have many veins – hence the name lacewing. The adults have long, thin antennae and fccd on pollen and nectar or honeydew. One species of green lacewing, *Chrysoperla carnea*, overwinters as adults and is often found indoors.

LACKEY MOTH

SYMPTOMS Gregarious, hairy caterpillars occur on a wide range of trees and shrubs from mid-spring to early summer. They are up to 48mm in length and have slate-grey heads with red, white, black and blue stripes on their bodies. The caterpillars cover their feeding area with silk webbing and can cause extensive defoliation (*see* **p.30**).
CAUSE Caterpillars of *Malacosoma neustria*, which hatch in the spring from eggs laid in late summer. These form a solid band of several hundred eggs glued

together around a twig.
CONTROL Prune out egg bands or the webbed shoots when they are seen. For extensive infestations, use an insecticide such as bifenthrin or pyrethrum against the young larvae.

LADYBIRDS

Most of these beneficial insects, including the harlequin ladybird, feed on greenfly and other aphids, both as adult beetles and larvae; some ladybirds also prey on scale insects or red spider mites, while others feed on fungal spores. Adult ladybirds are familiar insects, especially those with wing cases that are red with black spots (*see* **p.60**), but some species are yellow and black, or brown with white spots. The larval stages are generally slate-grey in colour with orange or white spots on the upper surface. One ladybird, *Cryptolaemus montrouzieri*, is commercially available for controlling mealybugs in greenhouses.

LAETIPORUS SULPHUREUS (CHICKEN OF THE WOODS)

SYMPTOMS Overlapping groups of annual, bright yellowish-orange fungal brackets, measuring between 10–40cm in total diameter, develop on the trunk or large limbs of host trees (*see* **p.38**). Each bracket is thin and has a wavy margin, with soft flesh resembling cooked chicken, and a yellow undersurface. Hardwoods, including oak, sweet chestnut, cherry (*Prunus*) and occasionally beech, and conifers, including larch, yew and

pine, may become infected. The infected wood discolours, turning yellowish-orange at first and later breaking up into brown, cubical blocks. Off-white to yellow mycelium (fungal sheets) develop between the blocks of decaying wood.
CAUSE The fungus *Laetiporus sulphureus*, which produces numerous spores from its lower surface.
CONTROL Infected trees may become dangerous as the strength of wood is reduced. An arboriculturist should examine the tree and advise on its health and safety.

LARCH ADELGID

SYMPTOMS Small patches of fluffy white wax appear on the foliage in early summer. Heavy infestations can cause leaf yellowing, and sooty moulds develop on the pests' sugary excrement (*see* **p.32**).
CAUSE Small, black sap-feeding insects, *Adelges laricis*, that resemble aphids and secrete waxy fibres.
CONTROL Large trees cannot be sprayed effectively but are usually able to tolerate infestations. Small trees, however, can be sprayed with bifenthrin, fatty acids, or plant oils.

LAVATERA LEAF AND STEM ROT

SYMPTOMS Red-brown necrotic spots develop on stems and occasionally on leaves and petioles of lavateras and hollyhocks. The spots enlarge and join together; they may girdle stems and so cause dieback or even death. Tiny, black, bristle-like spore-bearing structures may develop on the surface of the lesion.
CAUSE The fungus *Colletotrichum malvarum*,

which can be seed-borne.
CONTROL Remove and dispose of infected plants promptly. Grow replacement plants on a fresh site and do not save seed from those showing infection.

LAVENDER SHAB

SYMPTOMS The plant wilts, discolours and dies back, usually stem by stem. The whole plant may be killed. Numerous tiny, blackish-brown, raised fungal fruiting bodies the size of pinpricks develop on the dead foliage and stems. The symptoms resemble those caused by wet soil or winter cold injury, but are unique in having the fungal fruiting bodies.
CAUSE The fungus *Phomopsis lavandulae*, which enters the plant through pruning wounds or other points of damage on the bark.
CONTROL Prune out infected areas immediately, cutting well back into healthy wood. It may not be possible, however, to save infected plants, as symptoms develop rapidly.

LEAF AND BUD EELWORMS

SYMPTOMS Brown islands or wedges develop in leaves (*see* **p.25**) where the internal spread of the eelworms is restricted by the larger leaf veins. Eventually the entire leaf will turn brown and dry up. Typically, the youngest leaves at the shoot tips are free of infestation, but the older leaves lower down the stem show progressively more severe damage.
CAUSE Two microscopic species, *Aphelenchoides ritzemabosi* and *A. fragariae*, attack a wide range of plants in gardens and greenhouses.

CONTROL These eelworms cannot be controlled with the pesticides available to amateur gardeners. Badly infested plants should be burnt. With indoor plants, avoid wetting the foliage, as eelworms can emerge from leaves and spread rapidly in a surface film of water.

• *see also* CHRYSANTHEMUM EELWORM *p.118.*

LEAF CURL

See PEACH LEAF CURL *p.157.*

LEAF-CUTTING BEES

SYMPTOMS Circular or elliptical pieces are cut from the leaf margins (*see p.30*) of roses, wisterias, epimediums and other plants. The smooth outline and regular shapes of the missing pieces readily distinguish leaf-cutting bee damage from caterpillar or other damage. The female bee uses the leaf pieces to make thimble-shaped cells in her nest. This may be in a hollow plant stem or in a tunnel dug in rotten wood or soil. Quite often a bee makes use of soil in flower pots and the cells, resembling old cigar stubs, come to light when the plant is repotted. When a cell has been built, the bee stocks it with nectar and pollen, lays a single egg, and completes the cell by capping it with circular leaf pieces. A nest may contain up to twenty cells.

CAUSE Leaf-cutting bees, *Megachile* species, which resemble honeybees but have ginger hairs beneath their abdomens. They are solitary bees, each female constructing her own nest during the summer. They frequently visit a plant repeatedly to collect material and by the time their nest is finished they may have removed several hundred pieces.

CONTROL Like other bees, these are useful pollinators and should be tolerated unless small plants are being badly defoliated. If necessary, watch the plant until the bee returns, then swat it.

LEAFHOPPERS

Leafhoppers are sap-feeding insects, mostly 2–3mm long, which live on the undersides of leaves. The adults readily jump off leaves when disturbed, but their wingless nymphs are less active. Their feeding usually causes a coarse, pale mottling of the upper leaf surface. Garden plants frequently attacked are blackberries, plums, *Phlomis*, salvias, mints, beeches and many others. General control measures for leafhoppers are given in the entry for rose leafhopper (*p.173*).

• *see also* GLASSHOUSE LEAFHOPPER *p.133;* RHODODENDRON LEAFHOPPER *p.170.*

LEAF MINERS

Leaf-mining is a habit adopted by many insects, mainly the larvae of some flies, moths, sawflies and beetles. The damaged parts of the leaf dry up and become white or brown. The mines may be straight or meandering lines, circular or irregular blotches, and the form of the mine is usually distinctive for any particular leaf-mining insect. If a mined leaf is held up to the light, the larva or pupa can often be seen at the end or edge of the mined area. Leaf miners are difficult to control due to the lack of suitable insecticides.

• *see also* ALLIUM LEAF MINER *p.98;* APPLE LEAF MINER *p.101;* BEECH LEAF-MINING WEEVIL *p.106;* BEET LEAF MINER *p.106;* CELERY LEAF MINER *p.116;* CHRYSANTHEMUM LEAF MINER *p.118;* DELPHINIUM LEAF MINER *p.123;* GEUM SAWFLIES *p.132;* HOLLY LEAF MINER *p.137;* HORSE CHESTNUT LEAF-MINING MOTH *p.139;* LABURNUM LEAF MINERS *p.142;* LEEK MOTH *p.145;* LILAC LEAF MINER *p.147;* PYRACANTHA LEAF MINER *p.167;* SEMPERVIVUM LEAF MINER *p.176.*

LEAF SCORCH

SYMPTOMS The plant grows well, but its foliage is spoiled by dry, brown scorching, which is usually first seen around the leaf edges. The whole leaf may ultimately be affected, or the central area may retain its normal colour until the end of the season. If the problem is severe, or occurs in several years in succession, plant growth is affected and dieback may occur. On maples and sycamore (*Acer*), stems which deteriorate as a result of scorch often develop a white surface discoloration. In extreme cases extensive dieback, and occasionally even death, may occur.

CAUSE Various factors may be involved. The most common are an over-wet or excessively dry soil, cold winds or hot, bright sun. Plants lacking in vigour due to some other cause, or those which are not yet properly established, are particularly prone to scorch.

CONTROL Attempt to determine which adverse condition or conditions are involved, and improve growing conditions accordingly. If poor drainage is responsible, improve this and apply a foliar feed throughout the growing period to encourage replacement root growth. If the soil tends to become too dry, water well and apply a bulky mulch around the root area. Consider erecting windbreaks or planting to provide wind protection. If necessary, and feasible, consider moving the plant to a more suitable site in the autumn.

LEAF SPOTS (BACTERIAL)

SYMPTOMS Necrotic patches, usually circular or angular in outline, develop on foliage. Many bacterial leaf spots are surrounded by a bright yellow "halo" (*see pp.13, 29*). There are no minute, raised fungal fruiting bodies (pycnidia) as seen on fungal leaf spots.

CAUSE Various bacteria may be involved, and these are usually spread by rain or watering splash from leaf to leaf or, occasionally, from stem lesions to leaves (for example, bacterial canker may cause leaf spotting on *Prunus*). In most cases the leaf spotting itself is not very damaging, but may indicate that the plant is suffering from another, more serious problem. Examples include *Pseudomonas maculicola* on brassicas, *Pseudomonas lachrymans* on cucumber, black blotch on delphiniums and *Pseudomonas mori* on mulberry.

CONTROL Remove infected leaves promptly and avoid overhead irrigation.

LEAF SPOTS (FUNGAL)

SYMPTOMS Necrotic spots, usually circular and either grey or brown (*see p.13*), develop on the leaves and may join together, causing large areas of dead leaf tissue to develop (*see p.12*).

Occasionally leaves may be killed, but in most cases leaf spots are not very damaging in their own right, though their presence may indicate that the host plant is suffering from some other, rather more fundamental problem. The spots often show concentric zones of discoloration, and may bear numerous, raised, pinprick-sized, black or brown fungal fruiting bodies. Many common examples have entries elsewhere under their host name; others include *Phyllosticta antirrhini* on antirrhinums; *Discula quercina* on oaks and beeches; *Ramularia beticola* and *Cercospora beticola* on beet crops; *Septoria* on chrysanthemums, pinks and carnations (*Dianthus*), foxgloves (*Digitalis*), hebes, hawthorn and cyclamens; *Ascochyta cinerariae* on cinerarias; *Colletotrichum clivae* on clivias; *Marsonina daphnes* on daphnes; *Phyllosticta magnoliae* on magnolias; *Ramularia* on primroses and polyanthus, and *Phyllosticta wistariae* on wisterias.

CAUSE A wide range of fungi (see above, and also references to specific entries below).

CONTROL Remove affected leaves if necessary. Rake up and dispose of fallen leaves at the end of the season in order to minimize infection carry-over. A fungicide such as mancozeb can be used to control certain leaf spots, but check the product label for details.

• see also ALTERNARIA LEAF SPOT *p.98;* ARBUTUS LEAF SPOT *p.102;* BRASSICA LIGHT LEAF SPOT *p.111;* CELERY LEAF SPOT *p.116;* HYDRANGEA LEAF SPOT *p.139;* IVY LEAF SPOT *p.141;* PANSY LEAF SPOT *p.155;* PEA LEAF AND POD SPOT *p.156;* RHODODENDRON LEAF SPOT *p.170;* STRAWBERRY LEAF SPOT *p.181;* WATER LILY LEAF SPOT *p.190.*

LEAF WEEVILS

SYMPTOMS Ragged holes are eaten in the foliage of many deciduous trees, especially *Prunus, Malus, Sorbus,* birches (*Betula*), oaks (*Quercus*) and beech (*Fagus*), in late spring and early summer. Turquoise-green or golden-brown beetles, 3–7mm in length, may be present in large numbers.

CAUSE Several *Phyllobius* species occur in gardens and may cause extensive damage to young foliage in mid-spring to early summer.

CONTROL Small trees can be sprayed with thiacloprid or bifenthrin if leaf weevils are numerous enough to cause problems.

LEAFY GALL

SYMPTOMS Numerous small, bunched, often somewhat distorted, thickened or fasciated leaves develop around stem bases (*see p.39*). The plant may also produce normal growth.

CAUSE The bacterium *Corynebacterium fascians* (syn. *Rhodococcus fascians*), which is soil-borne and enters wounded plant tissue.

CONTROL Remove and dispose of affected plants, together with the soil or compost in which they are growing. Avoid spreading the bacteria by washing hands and tools after handling infected plants. Thoroughly scrub pots, trays and greenhouse staging.

LEATHERJACKETS

SYMPTOMS Legless, greyish-brown tubular larvae, up to 45mm long and without obvious heads, which occur in the soil (*see p.53*). They are mainly a lawn pest, causing yellowish-brown patches from late winter to mid-summer where the grubs have severed the roots (*see p.58*). Leatherjackets can also cause problems in vegetable gardens and flower borders where they kill seedlings and young plants.

CAUSE Leatherjackets are the larvae of crane flies (daddy-long-legs). Many species occur in gardens, mainly *Tipula* and *Nephrotoma* species. Eggs are laid in the soil in late summer.

CONTROL Leatherjackets in lawns can be controlled in early autumn with thiacloprid. Biological control with the pathogenic nematode *Steinernema feltiae* can be used against the larvae when the soil is moist and warm (at least 14°C/57°F). Leatherjackets can be removed from lawns by covering small areas with black polythene overnight after heavy rain or irrigation. The grubs will come up onto the lawn surface and can be removed when the polythene is lifted.

LEEK MOTH

SYMPTOMS Onion and leek foliage develops whitish-brown patches (*see p.23*) where caterpillars up to 12mm long have been feeding as leaf miners. The older caterpillars also bore into the stems of leeks and bulbs of onions. There are two generations during the summer, with leaf mining occurring in late spring to mid-summer and late summer to mid-autumn. When fully fed, the cater-pillars emerge and pupate in silk cocoons on the foliage.

CAUSE Caterpillars of a small moth, *Acrolepiopsis assectella*, which in Britain is mainly a problem in southern England.

CONTROL There are no insecticides approved for use against leek moth. Where leek moth is a regular problem, grow leeks under horticultural fleece to exclude the egg-laying moths. Search for pupae in their net-like cocoons on the leaves, and crush them.

• see also ALLIUM LEAF MINER *p.98.*

LEEK RUST

SYMPTOMS The outer leaves develop numerous bright orange, lens-shaped, raised fungal pustules, each 1–2mm in length (*see p.17*). These erupt producing conspicuous masses of bright orange spores. Affected leaves yellow and die back. The inner growth is rarely, or only mildly, affected.

CAUSE The fungus *Puccinia allii*, which may also affect related plants such as chives, onions, shallots, garlic and, occasionally, ornamental alliums. The disease is encouraged by damp conditions and wet weather.

CONTROL Remove and dispose of affected leaves at harvest. Dispose of all debris at the end of the season and grow leeks and other susceptible crops on a fresh site. Avoid excessive use of high-nitrogen fertilizers, as these encourage the production of soft growth which is then more readily infected. At planting, apply sulphate of potash at 15–20g per m² to harden growth slightly. Space plants widely to improve air circulation. Where available, choose leek cultivars resistant to this infection, such as 'Walton

Mammoth', 'Neptune', 'Apollo', 'Titan', 'Poristo', 'Poribleu' and 'Splendid'.

LEEK WHITE TIP

SYMPTOMS Leaf tips turn yellow at the end of summer. As the season progresses, the damage spreads down the leaves; they then either dry out and become pale brown or become bleached and papery in texture, or they rot.
CAUSE The fungus *Phytophthora porri*, which can persist in plant debris in the soil for several years after infection, even in the absence of any new leek crops during this time.
CONTROL Remove affected leeks promptly. Do not grow either leeks or other susceptible plants, such as onions or ornamental alliums, in the contaminated soil for at least five years.

LEOPARD MOTH

SYMPTOMS Apple trees (*Malus*) are the principal host plant, but *Pyrus*, *Sorbus*, sycamores and maples (*Acer*), birches (*Betula*), hawthorn (*Crataegus*), oaks (*Quercus*) and many others are also attacked. The creamy-yellow or white caterpillar, which measures up to 55mm in length, has many black spots on its body. It tunnels in branches or trunks which have a diameter of up to 10cm, and is often detected when the weakened trunk or branch snaps on a windy day. Another sign of infestation is pellets of compacted sawdust (the caterpillar's excrement) coming from a hole in the trunk or branch.
CAUSE Caterpillars of a moth, *Zeuzera pyrina*, which take two years to complete their feeding. Often the infestation

consists of a sole caterpillar, but this can have disastrous consequences if it is present in the trunk of a young tree.
CONTROL Preventive measures are not possible since attacks are sporadic, and spraying against the adult moths during summer is unlikely to be effective. If caterpillars are detected, damaged branches can be pruned out, or a piece of wire inserted into the tunnel to skewer the caterpillar.

LETTUCE DOWNY MILDEW

SYMPTOMS Angular yellow areas of discoloration develop on the leaves (*see p.22*), the older, outer leaves being affected first. The infected areas turn pale brown and either papery or soft. Off-white, fluffy fungal growth, occasionally with a slightly purple tinge, develops on the lower leaf surface, corresponding with the patches of discoloration. Severely affected leaves die off. Seedlings and mature plants may be affected.
CAUSE The fungus *Bremia lactucae*, which is encouraged by humid or damp growing conditions.
CONTROL Remove infected leaves promptly, or the entire plant if the infection is particularly severe. Spray with a fungicide containing mancozeb. Improve ventilation and/or air circulation. Grow resistant lettuces such as 'Avoncrisp', 'Avondefiance', 'Beatrice', 'Court', 'Debby', 'Lakeland', 'Musette', 'Plenty' or 'Saladin'.

LETTUCE GREY MOULD

SYMPTOMS Fluffy grey fungal growth develops on lettuce leaves (*see p.28*). If the grey

mould infection takes place towards the base of the plant, the stem rots, often together with the internal tissues there, causing a yellowish-orange or brown slimy rot. The whole plant may then die off. Seedlings are also susceptible.
CAUSE The fungus *Botrytis cinerea*, which is encouraged by damp growing conditions and often enters the plant through wounds. It may also be seed-borne in lettuce. The spores are carried on air currents. Black fungal resting bodies (sclerotia) may form on the infected tissues, but will not yet have developed if the plants are removed promptly.
CONTROL Remove all plant debris which may harbour the fungus. Avoid injury to the plant. Control likely sources of injury, such as downy mildew or slugs. Improve ventilation, and water early in the evening to allow time for the foliage to dry off before nightfall. Remove leaves infected with the fungus.

LETTUCE ROOT APHID

SYMPTOMS Lettuce plants tend to wilt and make slow growth in sunny weather. When the plants are dug up, a white, powdery wax can be seen on the roots and soil particles. Creamy-yellow aphids, measuring up to 2mm long, will be found on the roots.
CAUSE Sap-feeding insects, *Pemphigus bursarius*, which feed on the roots of outdoor lettuce during mid- to late summer.
CONTROL Some lettuce cultivars, such as 'Avoncrisp', 'Avondefiance', 'Debby', 'Lakeland' and 'Sigmaball',

are resistant to lettuce root aphid. Root aphids are more difficult to control than leaf-feeding types. Infestations are unlikely to be controlled by the insecticides currently available to amateur gardeners. Keep the plants well watered and grow lettuce crops on a new site each year.

LETTUCE VIRUSES

SYMPTOMS Leaves show yellow mottling, mosaic patterning or vein clearing (yellowing immediately adjacent to the veins). Plants are stunted and the foliage may be distorted.
CAUSE Various viruses, including lettuce big vein, lettuce mosaic, cucumber mosaic and beet western yellows virus.
CONTROL Remove plants showing symptoms promptly. Control pests such as aphids, which may spread the viruses. Wash hands thoroughly after handling infected plants. Grow lettuces on a fresh site. Choose resistant cultivars, where available. Those showing resistance to lettuce mosaic virus include 'Action', 'Corsair', 'Debby', and 'Musette'.
• *see also* CUCUMBER MOSAIC VIRUS *p.121*.

LILAC BLIGHT

SYMPTOMS Angular brown spots develop on lilac leaves. Affected leaves die back. Buds blacken and die, and young shoots may be girdled and wilt and die (*see p.33*). Large, elongated cankers may form on older stems. Secondary infection by grey mould is common and may mask the symptoms.
CAUSE The bacterium

Pseudomonas syringae, which is readily spread on air currents and by rain and water splash.
CONTROL Prune out affected areas of the plant promptly, cutting well back into perfectly healthy tissue.

LILAC LEAF MINER

SYMPTOMS The leaves of lilac, privet and ash develop brown shrivelled areas (*see* **p.19**) where the internal tissues have been eaten by leaf-mining caterpillars. When half-grown, the whitish-green caterpillars emerge from their mine and roll up the leaf tip to form a tube held in place by silk threads. There are two generations, with damaged leaves being seen from late spring to early summer and from late summer to early autumn.
CAUSE Caterpillars of a small moth, *Caloptilia syringella*.
CONTROL Pick off affected leaves; clipping privet hedges will remove many of the mined or rolled leaves.

LILY BEETLE

SYMPTOMS Bright red beetles, measuring up to 6–7mm long, and with black heads, appear on lilies (*Lilium*) and fritillaries from early spring onwards. In mid-summer the larval stage also occurs on the foliage. The grubs have rotund bodies, up to 8–9mm long, and are orange-red with black heads. They cover themselves with their own wet, black excrement. Both the grubs and the adults eat leaves, flowers and seed pods, and can ruin plants.
CAUSE A leaf beetle, *Lilioceris lilii*, active as an adult from spring to autumn. This pest occurs throughout England and in parts of Scotland,

Wales, and Ireland.
CONTROL The long period during which the adults and larvae are active makes this a difficult pest to control. With small numbers of lilies and fritillaries it is possible to protect them by picking the adults and grubs off by hand. Spray with thiacloprid or acetamiprid; the grubs are more vulnerable than adult beetles.

LILY DISEASE

SYMPTOMS Dark green, water-soaked spots develop on the foliage. Affected leaves soon turn brown or bleached and die off. Leaves wither from the base of the stem upwards, wilt and remain hanging on the stem. Flower buds are killed or damaged, so that they become very distorted, and plants fail to flower. If infection spreads into the stem, the plant topples over. The bulb usually remains healthy but may start to rot.
CAUSE The fungus *Botrytis elliptica*, which produces numerous spores that are spread both on air currents and by rain and water splash. The fungus may persist from year to year as black sclerotia (fungal resting bodies) which develop on infected plant debris. Wet weather encourages both the development and spread of the disease.
CONTROL Remove and dispose of infected leaves, or the whole of the top growth if necessary. Bulbs from infected plants can be kept, provided they show no rotting. Remove all debris at the end of the season. Some species, notably *Lilium hansonii*, *L. martagon*, *L. pyrenaicum* and *L. davidii* var. *wilmottiae*, seem to show a degree of resistance.

LILY VIRUSES

SYMPTOMS Foliage shows yellow mottling, streaking and mosaicing. Plants are usually stunted, and leaves show twisting or other deformity. Flower buds may not develop, or turn dry and brown and die off.
CAUSE Various viruses, including tulip-breaking virus, lily mosaic and lily mottle. Tulip-breaking virus also causes flower breaking symptoms on tulips and so lilies and tulips should not be grown in close proximity.
CONTROL None available. Remove infected plants, and wash hands thoroughly after handling them. Spray against aphids, as these may be vectors of the viruses.

LIME-INDUCED CHLOROSIS

See IRON DEFICIENCY AND LIME-INDUCED CHLOROSIS *p.141.*

LIME NAIL GALL MITE

SYMPTOMS From late spring onwards, slender, tubular, red or yellowish-green projections, up to 5mm long, develop on the surface of lime tree leaves (*see* **p.14**).
CAUSE Microscopic gall mites, *Eriophyes tiliae*, which induce the galls' growth by secreting chemicals as they feed on the underside of the leaves in early summer. The galls are hollow and the mites live within them during the summer months.
CONTROL Apart from creating the galls, the mites have no harmful effects on the tree and can be tolerated.

LUPIN APHID

SYMPTOMS Dense colonies of whitish-grey aphids, which can be up to 4.5mm long,

occur on herbaceous and tree lupins during late spring and summer (*see* **p.41**). They live on the underside of leaves and on flower spikes which become sticky with honeydew excreted by the pests. Infestations can be so heavy that plants wilt and die.
CAUSE A sap-feeding insect, *Macrosiphum albifrons*, which is a North American species that arrived in Britain in 1981 and is now widespread in Europe.
CONTROL Look for aphids on leaves from mid-spring; spray with thiacloprid, acetamiprid, bifenthrin, plant oils, pyrethrum or fatty acids before a damaging infestation has developed. Use one of the last three if it is necessary to spray during flowering.

MAGNESIUM DEFICIENCY

SYMPTOMS Yellowing develops between leaf veins (interveinal chlorosis) and also around the leaf margins, leaving clear green bands immediately adjacent to the veins (*see* **p.22**). As the green pigment deteriorates, other pigments may become visible, so instead of yellowing, red, purple or brown discoloration develops between the veins. Camellias affected by this deficiency often develop distinct, raised, brown interveinal patches on the lower leaf surface. When magnesium is in short supply, it is transported out of the older leaves into new, developing leaves; older, lower leaves are therefore the ones most readily affected.

CAUSE A deficiency of magnesium, which often occurs in plants growing in very acid soils or composts, or after periods of heavy rain or watering. Magnesium is readily leached through the soil, particularly when soil water levels are high. The use of high potash fertilizers or sulphate of potash may also increase the occurrence of this deficiency, as the magnesium may be rendered unavailable because of the high levels of potassium in the soil.

CONTROL For a rapid response, apply magnesium as a foliar spray. Use Epsom salts at the rate of 200g in 10 litres (8oz in 2½ gallons) of water to which is added a wetting agent such as a few drops of mild liquid detergent or soft soap. Several applications are usually needed, at weekly to fortnightly intervals. Alternatively, apply Epsom salts to the soil at 40g per m² or Kieserite at 70–140g per m².

MAHONIA RUST

SYMPTOMS Bright orange spots develop on upper leaf surfaces (*see p.24*), with corresponding dark brown spore-filled fungal pustules beneath (*see p.12*). The leaf develops bright, often red or orange discoloration, and falls prematurely.

CAUSE The fungus *Cumminsiella mirabilissima*, which is encouraged by moist or humid air.

CONTROL Clear up and dispose of fallen leaves to minimize overwintering of the fungus. If feasible, improve air circulation around the plant. Infection is rarely damaging enough to merit treatment, but if necessary, spray with a suitable fungicide such as penconazole.

MANGANESE DEFICIENCY

See IRON DEFICIENCY AND LIME-INDUCED CHLOROSIS *p.141.*

MEALY CABBAGE APHID

SYMPTOMS All brassicas are susceptible to this pest between mid-spring and mid-autumn. Yellow patches appear on the foliage, and the underside of these areas will have dense colonies of whitish-grey aphids (*see p.32*). During early summer the aphids infest leaves developing at the shoot tips of young plants. These leaves become distorted and have a pale, mottled appearance. Particularly heavy attacks can kill shoot tips, causing secondary buds to develop side-shoots, resulting in multi-headed plants.

CAUSE Sap-feeding insects, *Brevicoryne brassicae*, which sometimes also attack swede and radish.

CONTROL Spray with bifenthrin, thiacloprid, fatty acids or plant oils, especially if young plants are under attack.

MEALY PLUM APHID

SYMPTOMS The shoot tips and undersides of leaves on plums have dense colonies of whitish-green aphids on them during late spring to mid- or late summer. The foliage and fruits become sticky with honeydew excreted by the aphids. A black, sooty mould often grows on this substance.

CAUSE A sap-feeding insect, *Hyalopterus pruni*.

CONTROL Look for the aphid from late spring onwards, and if necessary spray with pyrethrum, thiacloprid, fatty acids or plant oils. Sooty mould can be wiped off fruits with a damp cloth.

MEALYBUGS

SYMPTOMS Cacti and succulents are the main host plants, but many other house plants and those grown in greenhouses are attacked. Mealybugs are soft bodied, greyish-white or pink insects up to 4mm long. They frequently infest leaf axils (*see p.32*) and other inaccessible places, such as between twining stems (*see p.37*). A white, fluffy wax is secreted from the insects' bodies and this also conceals the eggs. Infested plants become sticky with honeydew excreted by the pest. Sooty moulds may grow on the honeydew.

CAUSE Several sap-feeding mealybugs (*Pseudococcus* and *Planococcus* species) occur in greenhouses.

CONTROL During mid-summer, when temperatures are high, biological control with a ladybird predator, *Cryptolaemus montrouzieri*, can reduce infestations. At other times spray thoroughly with thiacloprid, acetamiprid, thiamethoxam, fatty acids or plant oils.

• *see also* PHORMIUM MEALYBUG *p.161;* ROOT MEALYBUGS *p.171.*

MELON COTTON APHID

SYMPTOMS Small, dark green or black aphids form dense colonies on the foliage and flowers of chrysanthemums (*see p.43*). Other plants likely to be attacked include cucumbers, melons, begonias, dahlias and cinerarias. Heavy infestations cause stunted growth and the early death of blooms. Plants are soiled, with cast aphid skins, a sticky excrement called honeydew and sooty moulds.

CAUSE A sap-feeding insect, *Aphis gossypii*, which can also damage plants by transmitting tomato spotted wilt virus. It is mainly a greenhouse problem, where it can remain active throughout the year.

CONTROL This aphid has a high degree of pesticide resistance, so most of the usual chemicals available to amateur gardeners fail to work. Biological control with parasitic wasps, *Aphidius colemani*, or predatory midge larvae, *Aphidoletes aphidomyza*, can help during the warmer summer months. Thiacloprid, acetamiprid, and thiamethoxam sprays are effective.

MERIPILUS GIGANTEUS

SYMPTOMS Large, fleshy, golden brown fungal brackets develop around the base of affected trees or in close proximity to the trunk or roots (*see p.39*). The lower surface is creamy-white (but bruises dark when handled). The fruiting bodies are conspicuous for their large size (up to 30cm across) and the number that are produced (they may total up to about 12kg in weight). They usually appear in late summer or autumn, and deteriorate into a smelly, black mass when frosted. The fungus causes root decay, often attacking the lower roots first, so the damage may not be apparent on initial investigation. The roots become hollow and so make the tree unsafe. Infected trees often show dieback of branches; they produce small, sparse leaves, and may be blown over in heavy winds.

CAUSE The fungus *Meripilus giganteus*, occasionally known as the giant polypore. Mature or over-mature trees of beech (*Fagus*), oak (*Quercus*) and occasionally robinia may be attacked.

CONTROL There is no control available, and prompt inspection by a professional arboriculturist is advisable, as trees may be rendered dangerous. Removal of the fungal growth will not prevent further damage to the tree but will prevent spores being spread to other potential hosts.

• see also GRIFOLA FRONDOSA *p.135*.

MICE

Several types of mice, such as *Apodemus* and *Mus* species, cause problems in gardens. They dig holes where peas, beans and sweetcorn have been sown, or where crocus corms have been planted. Green shoots are left lying on the soil surface but the corms and seeds are eaten (*see p.51*). Mice also cause damage in the autumn to spring period, when they come indoors and feed on stored fruits, vegetables and seeds.

CONTROL Set mouse traps in places where mice are causing damage. In the garden, put traps under the cover of logs or bricks, away from birds and pets. After planting crocuses, press the ground down firmly to make it harder for mice to locate the corms.

MICHAELMAS DAISY MITE

SYMPTOMS Infested plants are stunted and have a brownish scarring on the stems. The flowers are reduced in size and number, with many of them being converted into rosettes of small green leaves (*see p.45*). *Aster novi-belgii* cultivars are particularly susceptible.

CAUSE Microscopic sap-feeding mites, *Phytonemus pallidus*, which live at the tips of shoots and in the flower buds.

CONTROL None of the pesticides available to amateur gardeners is effective. Infested plants should be scrapped, as their condition will not improve. Other forms of Michaelmas daisy, such as *Aster novae-angliae* and *A. amellus*, are less susceptible and should flower satisfactorily.

MICHAELMAS DAISY WILT

SYMPTOMS The foliage wilts and later withers. The plant is usually affected in sections. Dark brown discoloration of the vascular tissue (conducting elements) within the stem may be apparent if the bark is peeled back. The whole plant is usually killed.

CAUSE The fungus *Phialophora asteris*, which is soil-borne. It may enter via damaged roots, or via pruning tools which have been used on an infected plant and then on a healthy plant without disinfecting them.

CONTROL No control is available. Remove infected plants and the soil in the vicinity of their roots. Do not plant Michaelmas daisies on the same site.

MIDGES

See CHIRONOMID MIDGES *p.117*; GALL MIDGES *p.131*.

MILLIPEDES

SYMPTOMS Millipedes feed mainly on rotting organic matter, but will sometimes damage seedlings and other soft plant tissues, such as strawberry fruits. They can also extend damage initiated by other pests, such as slugs on potato tubers. Millipedes have long, segmented bodies, up to 40mm long, which may be cylindrical or flat. Each body segment has two pairs of legs, unlike the superficially similar predatory animals called centipedes (*see p.116*), which have one pair of legs per segment.

CAUSE Many species can be found in garden soils. The one most often associated with plant damage is the spotted snake millipede, *Blaniulus guttulatus* (*see p.54*). It is cream-coloured, with a row of red dots on either side of a cylindrical body up to 25mm long.

CONTROL Protect strawberry fruits by using straw to lift the berries clear of the soil. Damage to potato tubers is generally initiated by slugs, so it is necessary to control them first (*see p.178*). Millipedes are encouraged by organic fertilizers and these should be avoided if millipedes are numerous enough to cause problems. They are not controlled by any of the insecticides available to amateur gardeners.

MINT BEETLE

SYMPTOMS Emerald-green beetles, up to 10mm long, occur in mid-summer on wild and cultivated mints (*see p.31*). In late summer, rotund, matt-black grubs also appear. Both adults and larvae eat holes in the foliage.

CAUSE A leaf-feeding beetle, *Chrysolina menthrastri*, which has one generation a year.

CONTROL Hand-removal is often sufficient, as the adult beetles are easy to see.

MINT RUST

SYMPTOMS Leaves and stems become flecked with yellow and distorted. The stem may be conspicuously curled, often shortly after growth has begun in spring. Numerous orange-grey, cup-like fungal bodies develop on the lower leaf surface and the affected areas of stem. These become yellowish-orange and later blackish-brown. Mint is the most common host, but several other herbs, including marjoram and savory, may also be affected.

CAUSE The fungus *Puccinia menthae*, which survives the winter on the remains of the plants or as resistant, resting spores in the soil.

CONTROL It may be possible to kill the fungus by burning off all of the top growth of mint plants and scorching the soil surface at the end of the season. This is, however, difficult to do safely and reliably; it may be more advisable to dispose of all infected plants, replanting fresh stock on a new site.

MITES

Mites are allied to spiders and most have four pairs of legs, although gall mites have only two pairs. All the plant pest species of mites are less than 1mm in length. They feed by sucking sap from their hosts. Gall mites secrete chemicals as they feed that cause the plant to produce abnormal growths that enclose the mites. Most mites capable of damaging plants belong to the red spider mite group. These cause fine, pale mottling of foliage and this can lead to defoliation. Heavily infested plants may become festooned with silk webbing. Tarsonemid mites infest the shoot tips and flower buds, causing stunted and distorted growth. Many mites are harmless, since they feed on fungal growth on rotting plants or on algae (see oribatid mites, p.154). Some mites are predators of pest insects or of other mites, and can be used to give biological control in greenhouses. Examples are *Phytoseiulus persimilis* for the control of glasshouse red spider mite, *Hypoaspis miles* for fungus gnats, and *Amblyseius* species for western flower thrips.

- *see also* **Gall mites** – ACER GALL MITES *p.98;* BIG BUD MITES *p.107;* BROOM GALL MITE *p.111;* ELM GALL MITE *p.126;* FELT GALL MITES *p.127;* FUCHSIA GALL MITE *p.130;* LIME NAIL GALL MITE *p.147;* MOUNTAIN ASH BLISTER MITE *below;* PEAR LEAF BLISTER MITE *p.158;* PLUM LEAF GALL MITE *p.162;* RASPBERRY LEAF AND BUD MITE *p.169;* RED BERRY MITE *p.169,* VINE ERINOSE MITE *p.188;* WALNUT BLISTER MITE *p.189.*
- **Red spider mites** – BOX RED SPIDER MITE *p.110* BRYOBIA MITES *p.112;* CONIFER RED SPIDER MITE *p.120;* FRUIT TREE RED SPIDER MITE *p.130;* GLASSHOUSE RED SPIDER MITE *p.133.*
- **Tarsonemid mites** – BULB SCALE MITE *p.112;* MICHAELMAS DAISY MITE *p.149;* STRAWBERRY MITE *p.181;* TARSONEMID MITES *p.183.*

MOLES

Moles, *Talpa europaea*, are infrequently seen, as they spend most of their lives below soil level. They create a system of underground tunnels and chambers in which they live and feed on earthworms and other soil-dwelling invertebrates. The soil excavated during tunnel construction is deposited in heaps on the surface of lawns and flower-beds (*see p.59*). All of the mole activity in a small garden could be the work of a single animal. Moles also make shallow tunnels above which the soil is pushed up as a ridge.

CONTROL A variety of controls and deterrents is available. The cheapest and often most effective method is to use a mole trap. The position of a mole's tunnel is located by probing the soil around a recent molehill with a stick. When found, open up the tunnel with a trowel at least 30cm away from the molehill. Insert the trap, taking care to align it with the direction and depth of the tunnel. Cover the trap with turf or an inverted bucket to stop light entering the tunnel. If the mole keeps pushing soil into the trap, reset it elsewhere in the tunnel system. An alternative to trapping is an electronic device. These emit noises said to drive moles away, but are expensive and give variable results.

MOTHS

Moths, like butterflies, are characterized by having wings that are covered by many small, overlapping scales. These scales are often coloured, and their arrangement on the wings gives the distinctive colour patterns by which the different moth species can be recognized. Moths vary considerably in size; some of the smallest, like some of those with leaf-mining larvae, have a wing-span of about 5mm, while the biggest hawk moths have a wing-span of up to 80mm. In all cases it is the caterpillars which cause the damage to plants, rather than the adults, which take nectar from flowers. Most moths are not pests, as their caterpillars do not feed on cultivated plants, but those listed below can cause damage to garden plants.

- *see also* APPLE LEAF MINER *p.101;* BROWN-TAIL MOTH *p.112;* BUFF-TIP MOTH *p.112;* CABBAGE MOTH *p.113;* CARNATION TORTRIX MOTH *p.115;* CHINA MARK MOTH *p.117;* CODLING MOTH *p.119;* COTONEASTER WEBBER MOTH *p.120;* CURRANT CLEARWING MOTH *p.122;* DELPHINIUM MOTH *p.123;* ELEPHANT HAWK MOTH *p.126;* GYPSY MOTH *p.136;* HORSE CHESTNUT LEAF-MINING MOTH *p.139;* JUNIPER WEBBER MOTH *p.142;* LACKEY MOTH *p.143;* LEEK MOTH *p.145;* LEOPARD MOTH *p.146;* LILAC LEAF MINER *p.147;* MULLEIN MOTH *p.151;* OAK PROCESSIONARY MOTH *p.153;* PEA MOTH *p.157;* PLUM MOTH *p.162;* PYRACANTHA LEAF MINER *p.167;* SMALL ERMINE MOTHS *p.178;* SWIFT MOTHS *p.182;* TOMATO MOTH *p.184;* TORTRIX MOTHS *p.185,* VAPOURER MOTH *p.187,* WINTER MOTHS *p.193;* YELLOW-TAIL MOTH *p.194.*

MOUNTAIN ASH BLISTER MITE

SYMPTOMS The foliage of rowan or mountain ash and other *Sorbus* species develop a whitish-green blistered appearance during late spring or early summer. Later in the summer, the pale areas darken and become brown. Sometimes nearly every leaf is affected, and there may be some early leaf fall.
CAUSE A microscopic gall mite, *Eriophyes sorbi*, which lives inside the blistered areas of the leaves.
CONTROL None of the pesticides available to amateur gardeners gives good control. Light infestations on small trees can be dealt with by removing affected leaves. Heavier infestations should be tolerated, since despite the unhealthy appearance of the foliage, the mites have little effect on the tree's vigour. The degree of infestation can vary greatly from year to year.

MULBERRY BACTERIAL BLIGHT

SYMPTOMS Small, black, angular spots, each with a yellow "halo", develop all over the leaf surface. As the infection progresses, the entire leaf may become yellowed and, if infection is severe, the tree's overall vigour may be reduced. Roughened areas may develop on the bark of shoots and can result in dieback.
CAUSE The bacterium *Pseudomonas mori*, which overwinters in the stem lesions.
CONTROL No control measures are available. Remove affected leaves and prune out any stems showing lesions.

MULBERRY CANKER

SYMPTOMS Shoots die back due to the presence of small cankers which girdle the affected stems (*see p.38*). Tiny, reddish-brown pustules develop around the cankered area. Symptoms are usually first noticed in the summer.
CAUSE The fungus *Gibberella baccata*.
CONTROL Prune out cankered shoots and areas of dieback promptly in an attempt to limit spread.

- *see also* NECTRIA CANKER *p.152.*

MULLEIN MOTH

SYMPTOMS Caterpillars up to 50mm long, greyish-white with black and yellow markings, are found during summer on verbascums, figwort (*Scrophularia*) and *Buddleja*. They feed on the foliage and flower spikes.
CAUSE Caterpillars of a moth, *Cucullia verbasci*.
CONTROL The caterpillars feed in exposed positions and can be removed by hand. Where this isn't feasible, spray with pyrethrum or bifenthrin before extensive damage is caused.

MUSSEL SCALE

SYMPTOMS This pest lives on the older stems of woody trees and shrubs, including apple (*Malus*), cotoneaster, box (*Buxus*), dogwoods (*Cornus*), heathers and ceanothus. The scales are greyish-brown, 2–3mm long, and in shape and appearance resemble the shellfish of the same name. Heavy infestations can occur to the extent that the bark is entirely covered (*see p.36*), and scales may spread onto fruits and berries. Such plants lack vigour and shoots may die back.
CAUSE Sap-feeding insects, *Lepidosaphes ulmi*, which have one generation a year.
CONTROL On ornamental trees and shrubs spray with thiacloprid or acetamiprid in early summer to kill the more vulnerable, newly-hatched scale nymphs. If the plant is in flower, spray at dusk to avoid harming bees. On fruiting apples, bifenthrin may give some control.

MYCOPLASMA INFECTION

SYMPTOMS Symptoms vary from host to host and with the precise identity of the organism. They may resemble those produced by viruses, and may include greening of flowers and yellow markings on leaves, often combined with distortion and stunting.
CAUSE Various organisms, usually described as "virus-like", thought to be similar in structure and activity to both bacteria and viruses.
CONTROL None is available. Remove any affected plants promptly.

NARCISSUS BASAL ROT

SYMPTOMS Infection occurs at the basal plate (the flattened area at the base of the bulb), causing it to discolour and rot. Discoloration and rotting spread upwards through the inner scales, turning them chocolate-brown. A pale pink fungal growth may develop on the basal plate and amongst the affected scales. Bulbs in store turn brown all over, dry out and become mummified. Those in the ground usually rot off completely, and by the end of the season there may be no trace of them. Leaf symptoms on growing bulbs include yellowing, stunting and sparse foliage.
CAUSE The fungus *Fusarium oxysporum* f. sp. *narcissi*, which is soil-borne. It persists in soil on infected bulbs that are not sufficiently affected for the symptoms to have been noticed. The disease is most damaging during or following very hot summer weather.
CONTROL None available. Dusting with sulphur, and early planting in autumn, may reduce the risk of infection developing.

NARCISSUS BULB FLY

SYMPTOMS Bulbs fail to grow or produce only a few thin leaves in the spring. If the bulb is cut in half vertically, it can be seen that the centre has been eaten away and filled with a fly larva's muddy excrement (*see p.54*). The plump maggot is a dirty cream colour and is up to 16mm long. There is usually only a single maggot in each bulb. Several maggots, each up to 8mm long, indicate that the bulb has been colonized by small bulb fly larvae (*Eumerus* species). These flies lay eggs on bulbs that are already damaged by other pests or diseases, whereas narcissus bulb fly attacks sound bulbs.
CAUSE Narcissus bulb fly, a hoverfly (*Merodon equestris*), resembles a small bumblebee and lays eggs on the necks of bulbs in early summer, when the foliage is dying down. It also attacks other bulbs, including hippeastrum and snowdrops.
CONTROL Chemical controls against this pest are not effective, but cultivation techniques can reduce losses. The adult flies prefer warm, sunny places, so plant narcissus bulbs in shady or exposed sites. Mound up soil around the stem bases when plants start to die down, to make them less attractive as egg-laying sites. Valuable bulbs can be protected by growing them under horticultural fleece. This needs to be in position from late spring to mid-summer. When small bulb fly larvae are found, investigate whether the bulbs are suffering from other pests such as eelworm or narcissus fly, or from fungal diseases.

NARCISSUS EELWORM

SYMPTOMS Plants produce stunted and distorted foliage and flower stems. Some viruses cause similar symptoms, so the presence of eelworms needs to be confirmed by cutting the bulb in half transversely. Eelworm activity in the bulb causes concentric brown rings or arcs where the pest has damaged the tissues (*see p.54*). In large plantings of daffodils, the pest gradually spreads, so that each year the area of killed and damaged plants gets larger. New bulbs planted as replacements in infested soil quickly succumb to eelworm.
CAUSE Microscopic nematodes (*Ditylenchus dipsaci*) that live within the bulb and foliage.
CONTROL There are no effective pesticides available to amateur gardeners. Dig up and burn daffodils showing symptoms; others growing within a radius of 1m may be infested and should also be removed. Avoid introducing narcissus eelworm into a garden by planting sound bulbs purchased from a reputable supplier. Hot water treatment of bulbs can be attempted, although it is difficult to do this properly without access to equipment that will keep the temperature constant. In late summer, dormant bulbs must be immersed for three hours in water heated to 44.5°C (112°F), after which they are cooled and planted in fresh soil. Too much heat will damage the bulbs, while too little allows eelworms to survive.

NARCISSUS LEAF SCORCH

SYMPTOMS As the leaves emerge, they become red-brown and scorched, and the symptoms then spread down the leaves (see **p.29**). Brown spots appear on the foliage and numerous tiny, pinprick-sized, raised fungal bodies may develop on damaged areas. Occasionally flower stems and spathes may also be attacked. The bulbs do not show rotting or discoloration. In addition to daffodils and narcissi, amaryllis, crinums, hippeastrums, nerines, snowdrops and sprekelias may be infected.
CAUSE The fungus *Stagonospora curtisii*, which may survive in the dry, papery outer scales of bulbs.
CONTROL Remove and dispose of infected leaf tips.

NARCISSUS SMOULDER

SYMPTOMS Infected bulbs produce deformed, chlorotic shoots with blackish-brown, withered leaf tips. Only a few leaves may be affected. In wet weather they may become covered with masses of grey spores. Infected leaves may rot off, and the deterioration then spreads down into the centre of the bulb. Numerous tiny, black sclerotia (fungal resting bodies) develop on the outer scales of the bulb and on infected leaves and debris.
CAUSE The fungus *Sclerotinia narcissicola*, which may persist in the soil as sclerotia on infected bulbs and debris.
CONTROL As soon as symptoms appear on leaves, cut back affected areas. Inspect bulbs carefully and plant only those which appear perfectly healthy and free from sclerotia.

NARCISSUS VIRUSES

SYMPTOMS Many different viruses may infect daffodils and narcissi, alone or in combination. The symptoms depend on the cultivar, the virus and growing conditions. Faint yellow mottling may occur in mild cases, particularly towards the leaf base. Flowers may show bleached streaks (*see* **p.23**), and occasionally yellow streaking on the flower stem. Certain viruses, in particular narcissus yellow stripe virus, cause bright yellow leaf stripes, often combined with distortion and stunting, and the plant may fail to flower.
CAUSE Various viruses, transmitted by aphids, soil-living eelworms or handling. The vector varies depending on the virus.
CONTROL Mild symptoms are quite common and may never develop seriously. All infected plants are, however, a source of infection, so disposal is usually the best option, particularly for those showing symptoms of narcissus yellow stripe virus.

NECTAR ROBBING

Many flowers produce nectar in order to attract bees and other pollinating insects. As these insects push their mouthparts into the nectaries at the base of a flower, their bodies become dusted with pollen from the anthers. When they visit other flowers of the same type, some of this pollen rubs off onto the stigma and pollination is achieved. Sometimes, however, bees cheat by biting a hole in the base of the flower (*see* **p.46**), enabling them to reach the nectar without touching the anthers and stigma. It is mostly short-tongued bumblebees that do this when they cannot reach the nectaries from the front of the flower. Once a short-cut to the nectar exists, other insects will take advantage. Nectar robbing is most frequently encountered on runner beans and may contribute to poor flower set. It is not possible to prevent nectar robbing, but flower set is usually adequate if beans are kept well-watered and do not suffer from high temperatures.

NECTRIA CANKER

SYMPTOMS Woody plants show roughened, flaked areas of bark on their branches, twigs and stems. The cankers enlarge and may girdle the stem, causing dieback. During the summer, white fungal pustules may develop on cankered areas; then in winter, these are replaced by numerous small, red fruiting bodies.
CAUSE Various species of the fungus *Nectria*.
CONTROL Prune out any affected areas immediately after symptoms develop. Spray with a fungicide containing copper.

NEEDLE BLIGHTS AND CASTS OF PINES

SYMPTOMS On young pines, the needles discolour, often showing diffuse brown or black lines. Infected needles may be shed prematurely. Damage to mature trees is insignificant, but young pines may be weakened. Another blight may cause the development of distinct bands of discoloration on the needles.
CAUSE *Lophodermium* species and *Dorthistroma pini*.
CONTROL None available. Rake up and dispose of infected needles as they fall.

NEEDLE RUSTS OF CONIFERS

SYMPTOMS On spruce (*Picea*), needles become discoloured. If the fungus *Chrysomyxa abietis* has attacked them, they may develop orange fungal pustules but will remain attached to the tree through winter; or, if attacked by *Chrysomyxa rhododendri*, they may become covered with orange pustules, followed by white spore tendrils, and fall from the tree in autumn. This latter fungus is less common and spends part of its life cycle on rhododendrons. Norway and Sitka spruce are most frequently affected, but most spruces can be attacked. On pines another fungus causes small yellowish spots on the needles, and resin-like exudations may occur. Later, off-white, column-like growths develop on the needles and erupt to produce bright orange spores. This fungus may also attack various herbaceous hosts, including campanulas and some weeds, in particular sow thistle and groundsel.
CAUSE The fungi *Chrysomyxa abietis* and *C. rhododendri* on spruce; the fungus *Coleosporium tussilaginis* on pine and other plants.
CONTROL None available. Dispose of infected needles.

NEMATODES

See EELWORMS *p.126*.

NITROGEN DEFICIENCY

SYMPTOMS Leaves are small and chlorotic, and general growth is reduced. On certain plants, red or purple leaf discoloration may develop as the green pigment levels drop. The lower, older leaves are affected first, but if the deficiency progresses, all parts of the plant may be similarly affected. Flowering, fruiting and tuber or root formation are also reduced.

CAUSE A deficiency of nitrogen, which can develop in any soil, particularly if it is light and has a low organic matter content, or if it has been heavily cropped. The use of large quantities of wood chippings may also result in nitrogen deficiency, as nitrogen is removed from the soil as the lignin in the wood is broken down.

CONTROL Apply nitrogen-rich fertilizers, composts and mulches to the soil regularly. Grow legumes, which are capable of fixing nitrogen (*see p.68*) using bacteria in their root nodules.

NUT WEEVIL

SYMPTOMS In late summer, hazel and cob nuts have a round hole, 1–2mm in diameter, in their shells (*see p.52*) where beetle grubs have bored their way out before pupating in the soil. The kernel inside the nut shell is eaten by a white maggot with a pale brown head.

CAUSE A brown weevil, *Curculio nucum*, which is about 10mm long. Almost half the body length is due to a slender snout or rostrum on the front of the head.

CONTROL Usually only a small proportion of the crop is attacked, so routine control measures are not required. This is fortunate as there are no pesticides approved for use against the nut weevil.

OAK GALL WASPS

SYMPTOMS English oaks (*Quercus robur* and *Q. petraea*) are the host plant of more than 30 species of gall wasps. Many of these have complex life cycles, involving alternating sexual and asexual (female only) generations which create different forms of gall (*see p.17*) on different parts of the tree. Some examples are the oak apple gall wasp and the common spangle gall wasp. The former lays eggs in buds in the winter and this causes them to develop into pithy, creamy-pink, bun-shaped objects in the spring. The adults that emerge in mid-summer go into the soil and lay eggs on the roots, giving rise to spherical swellings which contain the wasp's grubs. When adults emerge they crawl up the tree and complete the two-generation cycle by laying eggs in the buds. The common spangle gall wasp creates flat, yellowish-brown, disc-like galls (*see p.17*) on the underside of oak leaves in late summer. The next generation causes spherical yellowish-red galls on the male catkins in the spring, with adults emerging to lay eggs in the foliage in mid-summer. With all gall wasps it is the larval stage that secretes chemicals that induce the plant to produce gall tissues.

CAUSE Oak apple gall wasp, *Biorhiza pallida*, and common spangle gall wasp, *Neuroterus quercusbaccarum*. Other gall wasps frequently encountered on oaks are the marble gall wasp, *Andricus kollari*, which makes woody

spherical galls that measure up to 25mm across on the stems, and the artichoke gall wasp, *Andricus fecundator*, which makes greatly enlarged buds at the shoot tips. The silk button gall wasp, *Neuroterus numismalis*, has galls shaped like ring doughnuts, 3mm across, on the underside of leaves in late summer (*see p.17*). The cherry gall wasp, *Cynips quercusfolii*, produces reddish-yellow spherical galls measuring up to 18mm across under leaves in late summer.

CONTROL Control measures are unnecessary. None of the gall wasps has any serious effect on the growth of oak trees, even when numerous.
• *see also* ACORN GALL WASP *p.98*.

OAK PHYLLOXERA

SYMPTOMS In late spring or early summer, yellow spots develop on the upper surface of oak (*Quercus*) leaves. Underneath these spots on the lower leaf surface are orange aphid-like insects which measure up to 1mm in length. The females are often surrounded by a ring of yellow eggs. Later in the summer, the spots on the leaves tend to join together and damaged parts of the leaf dry up and become whitish-brown.

CAUSE A sap-feeding insect, *Phylloxera glabra*, which is a specific oak pest and over-winters on the tree as eggs. The heaviest infestations tend to be on saplings rather than mature trees.

CONTROL Encourage good growth by watering young trees in dry summers. On small trees, phylloxera can be controlled by spraying with bifenthrin or pyrethrum in early summer if damage is seen.

OAK POWDERY MILDEW

SYMPTOMS Tiny, almost imperceptible cinnamon-coloured spots develop on young leaves (*see p.24*) from late spring onwards. These are soon followed by a layer of white fungal spores and mycelium. Severely affected leaves may be killed, but the overall vigour of the tree is rarely affected.

CAUSE The fungus *Microsphaera alphitoides*, which overwinters in buds.

CONTROL Only really necessary on very young trees which are more likely to suffer serious effects. Keep the plants well watered and mulched, avoiding overhead irrigation.

OAK PROCESSIONARY MOTH

SYMPTOMS Oaks, *Quercus* species, can be defoliated during early summer by large numbers of gregarious caterpillars. Young caterpillars are brown but they become grey as they mature. Their bodies are covered in long white hairs. When not feeding, the caterpillars rest in dense clusters under silk webbing on the trunk or branches.

CAUSE Caterpillars of a moth, *Thaumetopoea processionea*, which is a native of central and southern Europe but has spread more widely in recent years. It was found in the London area in 2006 and so far attempts at eradicating it have been unsuccessful. In addition to devouring the leaves of oaks, the caterpillars are a health risk. Hairs on the caterpillars' bodies readily break off and have an irritating effect on skin, eyes, and mouths. The adult moths lay their eggs in

late summer to early autumn and they hatch the following spring. The caterpillars are present during late spring to early summer, when they pupate under silk webbing on the bark.
CONTROL At the present time in Britain, infestations of this pest should be reported to Forest Research (*see p.224*) which will advise on the necessary control measures. Avoid close contact with the caterpillars because of their irritating hairs.

OEDEMA

SYMPTOMS Raised, wart-like outgrowths develop, most frequently on the lower leaf surfaces (*see p.19*). These are the same colour as the leaf initially but may later become corky and brown. Affected leaves may sometimes be distorted but do not always die.
CAUSE The plant takes up more water than it can lose through the leaves by the process of transpiration. As a result, small groups of leaf cells swell up as they become over-filled with water, producing pale green warty growths. If the growing conditions do not improve, cells rupture and die off and so turn brown and corky. Oedema usually develops when humidity is high and water levels around the roots are excessive.
CONTROL Do not remove leaves showing the symptoms. This disorder is not infectious and removal of leaves further reduces the rate of moisture loss and so worsens the problem. Reduce watering and/or improve drainage. Increase air circulation around the plants by spacing them further apart and improving ventilation, as appropriate.

OLEANDER SCALE

SYMPTOMS Flat, round, whitish-brown objects up to 2mm across develop on the stems and foliage of many greenhouse plants. These include oleander, jasmine, asparagus ferns, azalea, palms and ivies; it also occasionally occurs on spotted laurel (*Aucuba japonica*) growing in sheltered places in gardens. Heavy attacks cause loss of vigour and dieback.
CAUSE A sap-feeding insect, *Aspidiotus nerii*, which breeds continuously throughout the year in greenhouses. Eggs are laid under the protection of the females' bodies and shell-like wax scales. No honeydew is excreted, so there is no sooty mould on the foliage.
CONTROL Prune out heavily infested shoots or leaves. Spray with thiacloprid, acetamiprid, or fatty acids or plant oils, particularly on the lower leaf surface. Two or three applications at fortnightly intervals may be needed to control the young nymphs as they hatch from the eggs.

ONION BOLTING

SYMPTOMS Onions produce flower stems before reaching the normal stage of maturity, so the bulb size is smaller than it should be.
CAUSE Exposure to cold temperatures earlier in the year promotes early flowering. This is usually brought about by a late, cold spring and is common on early cultivars. Very hot, dry conditions and day length may also induce bolting.
CONTROL Little can be done, but the incidence will be reduced by use of late cultivars which are less

likely to be subjected to early cold weather.

ONION DOWNY MILDEW

SYMPTOMS Affected leaves turn grey, wither and collapse, often with a "boiled" appearance. Close inspection may reveal dense, slightly fluffy, off-white fungal growth. However, secondary infections, particularly of the fungus *Alternaria*, may rapidly mask this. If left on the plant, the rotting of the leaves may spread down into the bulb and so cause further losses. The earlier in the season that the disease appears, the greater the reduction in bulb size.
CAUSE The fungus *Peronospora destructor*, which is particularly common in mild, wet weather and is encouraged by humid growing conditions.
CONTROL Remove infected leaves promptly. Improve air circulation around the plants by increased spacing and regular weeding. Clear up and dispose of all debris at the end of the season.

ONION EELWORM

SYMPTOMS Young onion plants are stunted and abnormally swollen. The tissues have a soft mealy texture and readily succumb to secondary rots (*see p.52*), so the plants usually die before reaching maturity. Onions attacked later in the summer do produce bulbs, but these will also be infested and will rot in store.
CAUSE Microscopic worm-like eelworms or nematodes, *Ditylenchus dipsaci*, that infest the foliage, bulbs and seeds of onion and

allied plants. It can also attack some other vegetables and common weeds, without necessarily causing obvious symptoms, but these alternative host plants allow onion eelworm to survive when onions are not being grown.
CONTROL There are no effective insecticides available to amateur gardeners, so remove infested plants as soon as they are spotted. Buy onion sets and seeds from reputable suppliers in order to reduce the risk of introducing the pest into a garden. When onion eelworm has occurred, crop rotation can be effective in eliminating it. Good weed control and growing vegetables not affected by onion eelworm, such as lettuce, turnip, swede and any of the brassicas, for one or two years will clear the soil.

ONION FLY

SYMPTOMS Onion is the most susceptible plant but leeks, shallots and garlic can also be attacked. In early summer young plants may collapse as the roots are eaten by white maggots up to 8mm long. Larvae of the second generation in late summer feed on roots and also bore into onion bulbs, allowing secondary rots to gain entry.
CAUSE Maggots of a fly, *Delia antiqua*, which is similar to a housefly in appearance.
CONTROL Onions grown from sets are less susceptible to first generation maggots than seedlings. There are no insecticides approved for use against onion fly. Infested plants should be lifted carefully and destroyed before the maggots go into the soil to pupate. Growing

onions and other host plants under a horticultural fleece will stop female flies laying eggs near the plants.

ONION NECK ROT

SYMPTOMS Scales of infected onions become soft, pale brown and semi-transparent (*see p.52*). A dense, grey fungal growth develops over the affected areas and the tissues start to dry out, so that the onion becomes mummified. Sclerotia (resilient black fungal resting bodies), which may be several millimetres in length, develop on the affected areas, particularly around the neck end of the bulb.
CAUSE The fungus *Botrytis allii*, which can persist in the form of sclerotia on onion debris or, having fallen from the bulb, loose in the soil. Infections usually spread by spores or by sclerotia. The symptoms are usually first noticed in store.
CONTROL Buy seed and sets only from reputable sources, as the fungus can easily be brought in on infected material. Do not grow onions on the same site for more than two years in succession. Improve cultural conditions so that hard, well-ripened bulbs are produced. Do not apply fertilizers after mid-summer and avoid excessive use of high nitrogen fertilizers. Keep crops evenly and adequately watered. Grow red or yellow bulbs instead of white, as these are generally less susceptible. Keep onions dry as they are drying off.

ONION THRIPS

SYMPTOMS The foliage of onions and leeks develops a fine, white mottling during the summer. Narrow-bodied

insects, up to 2mm long, and black or pale yellow in colour, occur on the leaves. Thrips are more troublesome in hot, dry summers and heavy infestations will check the plant's growth.
CAUSE The adults and nymphs of a sap-sucking insect, *Thrips tabaci*, which can also feed on many ornamental plants.
CONTROL Light infestations can be tolerated, but if much of the green colour is being lost before late summer, spray with pyrethrum, but this pest is not easily controlled.

ONION WHITE ROT

SYMPTOMS Dense, white, slightly fluffy fungal growth which develops around the base of the bulb and the roots (*see p.52*). Numerous small, black sclerotia subsequently develop in amongst this growth. The sclerotia may fall off and into the soil where they may remain viable for more than seven years. The foliage on affected bulbs becomes chlorotic and wilts.
CAUSE The fungus *Sclerotium cepivorum*.
CONTROL There is no chemical control available. Remove and burn infected plants as soon as they are noticed and do not grow onions or related plants on the same site for at least eight years.

ORCHID VIRUSES

SYMPTOMS Various symptoms, depending on the orchid cultivar and the virus or viruses involved. Yellow or very dark brownish-black streaking of the foliage may appear (*see p.12*). Yellow or brownish-black spots or

rings may also develop. Flowering may be affected and the plant's overall vigour declines.
CAUSE Various viruses, which may infect alone or in combination. The mode of transmission varies but handling, pruning tools and sap-feeding insect pests may all be involved.
CONTROL None is available. Remove and dispose of infected plants promptly and control potential means of spread.

ORIBATID MITES

Oribatid mites, such as *Humerobates* species, also known as beetle mites, commonly occur on the bark of a variety of trees and shrubs, and also sometimes on fences. They have shiny black or reddish-brown bodies, about 1mm long, and during the day the mites tend to cluster together in crevices where they can be mistaken for eggs. They are often noticed on dead or dying plants, but are not the cause of the plant's decline. Oribatid mites feed on algae on the bark and are mainly active after dark.
CONTROL No control measures are required, as these mites do not harm plants.

PANSY DOWNY MILDEW

SYMPTOMS Purple-brown spots appear on the upper surface of the leaves, often combined with a yellow "halo". These may be confused with the symptoms caused by pansy leaf spot

(*see below*). The lower leaf surface develops small, grey or off-white patches of fuzzy fungal growth which correspond with the spots. Affected leaves become chlorotic and may wither and die off. In severe cases the whole plant may become so weakened that it dies.
CAUSE The fungus *Peronospora violae*, which is encouraged by humidity and moist growing conditions.
CONTROL Remove affected leaves.

PANSY LEAF SPOT

SYMPTOMS Brown, purple-brown or very pale buff-coloured spots develop on the foliage (*see p.13*). The older leaves are affected first and most severely. If the affected leaf tissue is killed, the whole area may fall away, producing shothole symptoms.
CAUSE Various fungi, including species of *Ramularia*, *Cercospora* and *Phyllosticta violae*.
CONTROL Rarely necessary; remove any affected leaves.

PANSY POWDERY MILDEW

SYMPTOMS Powdery, white fungal growth develops on the upper leaf surface. This may cause distortion and "leaf yellowing" (chlorosis), occasionally followed by leaf death.
CAUSE *Oidium* species, which may be more troublesome if soil conditions are dry and the plants are growing in overcrowded conditions.
CONTROL Ensure that the soil or compost is adequately watered during dry weather, but avoid overhead watering. Pick off severely affected leaves.

PANSY SICKNESS

SYMPTOMS The whole plant wilts and does not recover, even if watered. The foliage yellows and dies off and the plant is killed. The symptoms may take several weeks to develop fully.

CAUSE The soil-borne fungus *Pythium violae*, and, occasionally, other fungi associated with root and foot rots. The fungus builds up in the soil, particularly when pansies or violas are grown in the same site for several years in succession. The fungus seems to be able to survive in soil for several years, even in the absence of suitable host plants.

CONTROL Remove and dispose of affected plants. Do not incorporate them into the compost heap. Plant new stocks of pansies or violas on a fresh site.

PAPERY BARK

SYMPTOMS The bark peels off as a thin, papery, brown sheet (*see p.40*). All woody parts of the tree may be affected but small- to medium-sized shoots more frequently show these symptoms than do large branches or tree trunks. Shoots may also die back.

CAUSE Problems associated with inadequate functioning of the root system of the tree, most commonly waterlogging.

CONTROL Improve growing conditions, in particular soil drainage. Prune off all dead stems and those showing the symptoms. Prune back to a healthy bud growing from perfectly healthy-looking wood. The removal of the dead wood is necessary not because the problem is infectious, but because

once it is dead or injured, the wood is prone to invasion by other organisms which may subsequently cause further dieback.

PARSNIP CANKER

SYMPTOMS Roughened cankered areas develop on the root, usually reddish-brown, orange-brown or black in colour (*see p.57*). The shoulder of the root is the area most commonly affected, but lesions may occur anywhere.

CAUSE The fungus *Itersonilia pastinacae* or, less commonly, *Mycocentrospora aceria*. The fungus may be spread into the soil from leaf spots and usually infects via injured fine roots. Other sources of injury, such as that caused by carrot fly, may also provide the initial point of entry.

CONTROL No control is available. Reduce the risk of infection by growing resistant cultivars such as 'Archer', 'Avonresister', and 'Cobham Improved Marrow'. Improve soil drainage and avoid or control any source of injury to the roots. Sowing late (during late spring), in closely spaced rows produces parsnips with smaller roots which seem to be less susceptible than larger-rooted ones.

PARSNIP VIRUSES

SYMPTOMS Yellow markings on the leaves, usually as bands adjacent to the leaves or mosaic-like patterns or flecking. The foliage may be slightly stunted and distorted and overall growth retarded.

CAUSE Various viruses, in particular parsnip yellow fleck virus which is spread by aphids.

CONTROL Remove infected plants promptly to prevent further spread of the infection. Try to control aphids to decrease the risk of infection.

PEA AND BEAN WEEVIL

SYMPTOMS Peas and broad beans have U-shaped notches eaten in their leaf margins (*see p.31*) by greyish-brown beetles which are 3–4mm long.

CAUSE Adult beetles, *Sitona lineatus*; the larval stage lives in the soil and feeds on the nitrogen-fixing nodules on the roots of peas and beans.

CONTROL Normally only a small proportion of the foliage is eaten, and plants can usually tolerate the damage. Control measures are only required where small plants are being heavily attacked. In such cases, spray plants with bifenthrin, but protect bees by not spraying while broad beans are in flower.

PEA DOWNY MILDEW

SYMPTOMS The foliage develops yellow blotches of discoloration, each of which corresponds with patches of fluffy, off-white fungal growth on the lower surface. Occasionally, yellowish patches of discoloration may also develop on the pods, but this symptom is less common. Infection early in the season may cause serious stunting and little, if any, pod set.

CAUSE The fungus *Peronospora viciae*, which is only likely to cause serious damage in wet seasons.

CONTROL Remove any signs of early infection. The fungus forms spore stages which can persist in the soil, so it is important to remove all debris at the end of the season and grow peas on a fresh site. Grow resistant cultivars such as 'Hurst Green Shaft' and 'Kelvedon Wonder'.

PEA LEAF AND POD SPOT

SYMPTOMS Brown or yellow, often sunken spots develop on the leaves (*see p.14*), stems, pods (*see p.51*) and flower stems. Numerous tiny, pinprick-sized, raised fungal fruiting bodies may develop on the lesions.

CAUSE Various fungi, including *Ascochyta pisi* and *A. pinodes*, which most frequently attack fully grown peas. They may also occasionally attack seedlings, which may be killed. The fungus is persistent and carries over from year to year in infected plant debris and, if seeds are collected from infected pods, the resultant seedlings will succumb, too.

CONTROL Remove diseased plant material at the end of the season. Plant fresh seed on a new site. Do not save seed from infected plants.

PEA MARSH SPOT

SYMPTOMS Circular brown areas of discoloration develop within the seed. A brown cavity later develops. This spoils the eating quality of the crop and also renders them useless for saving as seed.

CAUSE A deficiency of manganese. This is most likely to develop on poorly drained soils, particularly those which are alkaline and have a high organic matter content.

CONTROL Improve soil conditions and apply sulphur to lower the pH.

PEA MOTH

SYMPTOMS Caterpillars, measuring up to 6mm in length, live inside pea pods and feed on the developing seeds (*see p.51*). The caterpillars have black heads and small dark spots on creamy-white bodies.
CAUSE Caterpillars of a moth, *Cydia nigricana*. Eggs are laid on peas that are in flower, so early or late sown peas that flower outside the moth's flight period from early to late summer will escape damage.
CONTROL Spray peas that flower during the moth's flight period with bifenthrin about seven days after the onset of flowering.

PEA POWDERY MILDEW

SYMPTOMS Powdery white fungal growth develops on the leaves and occasionally the leaf stalks and the stems. Leaves may become slightly distorted and yellowed.
CAUSE The fungus *Erysiphe polygoni*, which is encouraged by dry conditions at the plant roots, and by poor air circulation around the foliage.
CONTROL Control is only necessary if the infection occurs early in the season. More usually it develops towards the end of summer and so has little if any effect on productivity. Grow resistant cultivars including 'Kelvedon Wonder', 'Oregon Sugar Pod II', 'Sugar Gem', and 'Top Pod'.

PEA THRIPS

SYMPTOMS Pea pods are discoloured with a silvery or brownish scarring and the pods may contain only a few seeds at the stalk end. The foliage is also discoloured. Adult thrips are black and have narrow elongate bodies 2mm long. The immature nymphs have a similar appearance but are pale yellow.
CAUSE Sap-feeding insects, *Kakothrips robustus*, which can be troublesome in hot, dry summers.
CONTROL Watch for signs of thrips' activity as the pods develop and, if necessary, spray with bifenthrin or pyrethrum.

PEA VIRUSES

SYMPTOMS Yellow mosaicing, streaking, spotting or flecking occurs, mainly on the leaves; occasionally symptoms may also be seen on the stems. Plants are stunted and distorted and may fail to flower or may flower but fail to set much of a crop. Pods formed may be discoloured and distorted.
CAUSE Various viruses, including pea enation virus, pea mosaic virus and bean leaf roll virus, the first being spread by several different species of aphid.
CONTROL Remove and burn affected plants promptly in order to minimize the risk of the infection spreading. Keep pests such as aphids under control. Avoid handling healthy-looking plants after handling those which are showing symptoms. If necessary, scrub your hands thoroughly. Do not save seed. Grow the cultivars 'Ambassador' and 'Sugar Pod II', which show resistance to pea enation virus.

PEA WILT

SYMPTOMS Rapid wilting of the plant, usually in association with yellowing and dieback of the leaves (*see p.28*). Infection can occur on plants of any age, but is most common towards the middle of the season. Dark brown discoloration of the vascular tissue (the conducting elements) in the stem develops but is only visible if the outer parts of the stem are gently stripped away.
CAUSE The fungus *Fusarium oxysporum* f. sp. *pisi*, which is soil-borne and so particularly troublesome if peas or related plants are grown on the same site for several years in succession.
CONTROL No control is available. Remove infected plants and grow any replacements or future crops on a fresh site. Grow cultivars showing a degree of resistance, such as 'Greenshaft', 'Kelvedon Wonder', 'Sugar Snap', 'Sugarbon', and 'Onward'.

PEACH APHIDS

SYMPTOMS Peach, nectarine and almond foliage becomes crinkled and curled, with a yellowish-green colour, during late spring to early summer. Green or black aphids suck sap from the underside of the leaves, although they may die out later in the summer, leaving only whitish cast skins.
CAUSE Several aphids occur on peach and related plants. The more important ones are the peach potato aphid, *Myzus persicae*, which is yellowish-green in colour, and a black species, the peach aphid, *Brachycaudus schwartzi*.
CONTROL Look for aphids on the foliage from mid-spring onwards and spray with pyrethrum, fatty acids or plant oils if they are seen.

PEACH LEAF CURL

SYMPTOMS Leaves may become infected before or as soon as they unfurl in the spring. They develop symptoms of puckering and blistering (*see p.26*), being pale green at first and soon turning bright red or purple. A white, powdery spore layer then develops over the leaf surface. Affected leaves drop prematurely. A second flush of foliage is usually produced later in the season and this almost invariably remains healthy. The overall vigour and cropping of the tree is only likely to be affected if the infection occurs several years in succession. On nectarines, but not on peaches, the smooth skin on the fruit may develop slightly raised, rough patches as a result of infection spreading to the fruits. A very similar disease may affect poplars.
CAUSE The fungus *Taphrina deformans*, which produces spores that are carried by wind and rain, and lodge in cracks and crevices both in the bark and in bud scales. This overwintering allows them to cause such early infections.
CONTROL Remove infected leaves promptly, preferably before the spore layer develops. Keep the tree well watered and fed to encourage the development of plenty of new replacement growth. Peaches and nectarines which grow under glass are rarely attacked because the glass acts as a shelter, preventing the spores from landing. Temporary protection may work well — this can be provided by the

erection of an open-sided polythene shelter over fan-trained trees. To be effective, the shelter must be in position from mid-winter until late spring. Spray with a copper fungicide or mancozeb several times between mid- and late winter to try to protect the developing foliage from attack. The spraying must be completed before the flower buds start to open. Repeat the spray in the autumn, just before leaf fall.

PEAR AND APPLE CANKER

SYMPTOMS Areas of bark sink inwards and become pale in colour. As the canker enlarges, it becomes elliptical and the bark flakes in concentric rings. The branch may swell around the cankered area. The growth above the canker is poor and the foliage may be sparse and discoloured. Cankers often enlarge sufficiently to girdle the stem or branch, and the growth above this point then proceeds to die completely. The canker generally starts near a bud, or in a wound such as a leaf scar or insect injury. During the summer months numerous small, raised, white, fungal spots develop on the cankered area, and then in the winter these are replaced by similar but red fruiting bodies.
CAUSE The fungus *Nectria galligena*, which produces most of its spores in the spring. The spores are wind-borne. Poor growing conditions, such as a very heavy soil or inadequate feeding may increase the likelihood of canker being troublesome.
CONTROL Remove all of the cankered stems and dispose

of them as far away as possible from any nearby apple, beech, ash, willow, *Sorbus* and pear trees. On large branches or trunks, carefully cut out the cankered area using a sharp knife, so that every bit of the diseased bark and wood is removed. Treat wounds with a canker paint that contains cresylic acid. Some apple cultivars have a tendency to show a degree of resistance under certain growing conditions: these include 'Bramley's Seedling', 'Newton Wonder', 'Laxton's Superb' and 'Lane's Prince Albert'. Avoid cultivars which are notably susceptible, such as 'Cox's Orange Pippin', 'James Grieve', 'Worcester Pearmain', and 'Spartan'.

PEAR AND CHERRY SLUGWORM

SYMPTOMS Pale yellow, caterpillar-like larvae, which secrete a black, slimy mucilage over their bodies, giving them a slug-like appearance. They feed on pear and cherry foliage between late spring and mid-autumn. Other plants attacked include hawthorn, Japanese quince, mountain ash and plums. The larvae can be up to 10mm long and are swollen at the front end, making them club-shaped. They feed by grazing away the upper leaf surface, causing the remaining damaged tissues to dry up and turn brown (*see p.19*).
CAUSE Larvae of a sawfly, *Caliroa cerasi*, which has two or three generations during the summer.
CONTROL On trees small enough to be sprayed, slugworms are easy to control with most contact

insecticides, such as pyrethrum, thiacloprid, and bifenthrin.

PEAR BEDSTRAW APHID

SYMPTOMS Leaves at the shoot tips on pear become yellowish-green and curled during the late spring. Whitish-grey aphids infest the undersides of the leaves and are sometimes present in such large numbers in late spring to early summer that they swarm over the branches and trunk. The aphids excrete a sugary honeydew that can allow sooty moulds to grow on the foliage.
CAUSE A sap-feeding insect, *Dysaphis pyri*, which in mid-summer migrates to wild flowers known as bedstraws (*Galium* species).
CONTROL The aphid overwinters on pears as eggs laid near the buds. In the spring, inspect the foliage and spray where necessary with thiacloprid, bifenthrin, or pyrethrum if aphids are seen. Poor control is likely once the foliage has become curled.

PEAR BORON DEFICIENCY

SYMPTOMS Distorted fruits are marked internally with numerous small, brown spots. Most fruits are affected. The fruit's texture is unpleasant because of the patches of dead, brown cells. Some stem dieback may occur and overall vigour may be reduced. The symptoms may be confused with those of the far more serious pear stony pit virus (*see p.159*).
CAUSE A deficiency of boron.
CONTROL *See* BORON DEFICIENCY, *p.109*.

PEAR LEAF BLISTER MITE

SYMPTOMS In mid- to late spring, the foliage of pears develops yellowish-green or pink blisters or raised blotches. On expanded leaves these form broad bands on either side of the midrib (*see p.16*). By mid-summer the blisters have darkened and become black. Less than 3mm in diameter, they should not be confused with the larger black lesions caused by pear scab disease (*see p.159*).
CAUSE Microscopic gall mites, *Eriophyes pyri*, which live within the leaves. The blistered appearance of the foliage is induced by chemicals secreted by the mites as they feed.
CONTROL None of the pesticides that are currently available to amateur gardeners gives good control. Fortunately, however, although blister mite gives a tree an unhealthy and unsightly appearance, it has little impact on the tree's ability to produce fruit. On small trees, affected leaves can be picked off if the infestation is light, otherwise this pest has to be tolerated.

PEAR MIDGE

SYMPTOMS Pear fruitlets make rapid initial growth but, a few weeks after petal fall, start to turn black at the eye end of the fruits (*see p.48*). In the centre of the fruitlets are whitish-orange maggots which are up to 2mm long. Affected fruitlets become extensively blackened and fall from the tree in late spring and early summer. Sometimes nearly all the potential fruit is lost.
CAUSE Larvae of a small fly or gall midge, *Contarinia*

pyrivora, which go into the soil when the fruitlets drop. They pupate inside silk cocoons and emerge as adults the following spring.
CONTROL On small trees, pick off and destroy infested fruitlets before the maggots complete their feeding. Prevent the adult flies from laying eggs by spraying with bifenthrin when the blossom is at the white bud stage – when the petal colour can be seen, but before the flowers open.

PEAR RUST

SYMPTOMS In the spring the leaves develop red spots on the upper surface. Each spot is 1–3cm in diameter and may have a bright margin, with the spot darkening with age. Cankers may develop on branches and so spoil the shape of the tree.
CAUSE *Gymnosporangium fuscum* (= *sabinae*), a fungus whose alternate host is *Juniperus sabina*.
CONTROL The only control is to remove any juniper bushes in the vicinity.

PEAR SCAB

SYMPTOMS Blackish-brown scabby patches develop on the fruits (*see* **p.48**), and similar but more greenish-grey spots develop on the leaves. In the most severe cases, fruit may be almost entirely covered with these scabby patches and is small, and misshapen. Infected fruits may also crack or split and so become infected with secondary organisms such as brown rot. Affected leaves yellow and fall early.
CAUSE The fungus *Venturia pirina*, which overwinters on scabby patches on the young stems and also on fallen infected leaves. The disease

tends to be most prevalent on trees that have crowded branches, and during damp seasons.
CONTROL Rake up and dispose of affected leaves as they fall. Prune out cracked and scabby shoots to limit the overwintering of the fungus. Spray with a fungicide such as mancozeb. Grow cultivars showing some resistance, including 'Jargonelle', and 'Catillac'.

PEAR STONY PIT VIRUS

SYMPTOMS The fruits become knobbly and pitted (*see* **p.48**). Within the flesh are found many dead sclerenchyma (stone cells), which make the fruits unpleasant to eat. The symptoms of this virus may be confused with boron deficiency. Affected fruits rarely appear all over the tree, but are frequently borne on just one branch.
CAUSE Stony pit virus, which is most common on old trees. The virus particles are only known to be transmitted by grafting.
CONTROL There is no known cure. Avoid cultivars known to be susceptible, including 'Doyenné du Comice', 'Laxton's Superb', 'Winter Nelis', and 'Anjou'.

PEAR SUCKER

SYMPTOMS Pear leaves become very sticky with honeydew excreted by blackish-green insects up to 2mm long. The immature nymphs have flattened bodies, while the adults resemble winged aphids. Sooty moulds frequently develop where honeydew has accumulated on the upper surface of leaves.

CAUSE A sap-feeding sucker or psyllid, *Psylla pyricola*, which has several generations during summer and overwinters as an adult.
CONTROL Pear sucker is less frequent on pears in gardens than in commercial orchards. Routine control measures are not needed, but look for the pest during late spring to late summer and, if necessary, spray with bifenthrin.

PEDICEL NECROSIS

SYMPTOMS Flower buds develop normally, but before or just after opening, the pedicel (area of flower stem immediately behind the bud) discolours, shrinks inwards and flops over. The flower bud then turns brown and fails to open, or opens and dies off.
CAUSE Extremely rapid or soft growth, such as that produced by excessive use of high-nitrogen fertilizers. A sudden check in growth caused, for example, by drought, inadequate fertilizer use, or the wrong amount of potassium, may also be involved. Plants with very large or double flowerheads may be more prone to this than single varieties.
CONTROL Remove damaged flowerheads. Feed the plant with sulphate of potash at about 20g per m² in spring and autumn. Avoid excessive use of high nitrogen fertilizers. Mulch the soil if necessary to improve its ability to retain moisture.

PELARGONIUM RUST

SYMPTOMS Rings of dark brown fungal pustules develop on the lower leaf surface, often in concentric rings (*see* **p.15**). Corresponding yellow blotches develop on the

upper surface. Occasionally, the pustules appear on the upper surface, too. The infection spreads rapidly and infected leaves become extensively discoloured, wither and die. If the disease spreads unchecked, the plant may be killed.
CAUSE The fungus *Puccinia pelargonii-zonalis*, which produces spores that are spread on air currents and by water splash. Spread of the fungus is encouraged by humid atmospheres, and so is usually most severe during the winter months and on plants growing in crowded conditions under glass.
CONTROL Remove infected leaves. Spray with a fungicide containing mancozeb, penconazole or myclobutanil. Improve air circulation to reduce humidity. Space plants further apart, or if under cover, open vents, windows and doors, as appropriate.

PELARGONIUM VIRUSES

SYMPTOMS Yellow markings, such as flecks, ring spots and streaks (*see* **p.21**), on the foliage. Leaves may, occasionally, be distorted and flowers may show colour-breaking symptoms (pale streaks on the petals). The overall vigour of the plant is rarely affected, but the appearance of foliage and flowers may be spoiled.
CAUSE Various viruses, either alone or in combination, some of which may be transmitted by aphids, others by handling, or by eelworms in the soil.
CONTROL There is no control available, but it is always advisable to inspect plants carefully before purchase and to dispose of any showing severe symptoms.

PEONY WILT

SYMPTOMS Shoots of herbaceous peonies wilt and turn brown at the stem base. In damp conditions, fuzzy grey fungal growth may develop at the base of the stem. The affected stem withers and dies, shortly followed by other stems on the plant. Sclerotia (shiny black fungal resting bodies) form on and within the infected stem. Occasionally the fungus attacks higher up the stem, causing similar symptoms (*see p.43*). Wilt is most common on herbaceous peonies, but tree species may also be affected.
CAUSE The fungus *Botrytis paeoniae*, which produces numerous spores that are carried on air currents and by water splash. The sclerotia are usually responsible for wilt infections, which appear early in the season; subsequent infections may be due to spore spread.
CONTROL Cut out infected stems promptly, cutting right back into healthy growth and below soil level if necessary. If the stems are not removed, the sclerotia fall into the soil where they can persist for several years. Remove all debris at the end of the season, as this may harbour sclerotia. If a plant has been severely damaged by the infections, scrape off the uppermost few millimetres of soil in the autumn in case it contains sclerotia – this must be done with care so as to avoid injuring the roots.

PERIWINKLE RUST

SYMPTOMS Dark brown spore pustules develop beneath the leaves (*see p.15*). The foliage becomes pitted and distorted and the whole plant also appears distorted

and may fail to flower.
CAUSE The fungus *Puccinia vincae*, which produces a fungal mycelium that invades all parts of the plant, including the roots.
CONTROL As the fungus is present throughout the plant, there is little point in attempting control, and all infected or obviously distorted plants should be removed and disposed of promptly.

PESTALOTIOPSIS

SYMPTOMS Grey-brown necrotic patches develop on the foliage and may also appear on shoots. The tree or shrub appears scorched. Leaves drop prematurely and stem infection may cause extensive dieback.
CAUSE The fungus *Pestalotiopsis*; most frequently *P. guepini*, which is encouraged by high humidity.
CONTROL Remove affected leaves and prune out infected stems.

PETUNIA FOOT ROT

SYMPTOMS The plant wilts and dies, often quite rapidly. The stem base may also become blackened and atrophied.
CAUSE Various species of the soil-borne fungus *Phytophthora*, which produces resilient resting bodies that enable it to persist in the soil for several years.
CONTROL No control is available. Remove infected plants promptly, together with the soil in the immediate vicinity of the roots. Do not grow petunias on that site for several years; instead choose other bedding plants such as zonal pelargoniums, French and

African marigolds, begonias and heliotrope, since these are not susceptible.

PETUNIA VIRUSES

SYMPTOMS Yellow flecking, spotting, mosaicing and streaking on the foliage. The leaves may also be stunted and generally chlorotic and may be deformed. The whole plant appears stunted and the flowers may show colour breaking (pale streaking) in the petals.
CAUSE Various viruses, some of which may be transmitted by aphids and other sap-feeding pests; others may be spread by handling or when propagating from cuttings.
CONTROL Remove and dispose of affected plants promptly. Always wash hands thoroughly after handling infected plants. Control pests which are potential vectors.

PHELLINUS IGNARIUS

SYMPTOMS Pale brown to buff-coloured fungal bracket, with a rounded outline, develops on the branches or trunk of the tree. There may be no apparent loss of vigour of the tree but internally the wood is weakened.
CAUSE The polypore fungus *Phellinus ignarius*.
CONTROL No control is currently available. Infected trees may stand safely for many years, but they may be rendered dangerous, so it is better to employ a reputable tree surgeon to examine the tree and advise as to the best course of action.

PHELLINUS POMACEOUS

See FOMES POMACEOUS p.129.

PHLOX EELWORM

SYMPTOMS In early to mid-summer, phlox stems become abnormally swollen and are liable to split. Growth is stunted and leaves at the shoot tips are greatly reduced in width, sometimes being little wider than the midrib (*see p.33*). Badly infested plants rot and die.
CAUSE Microscopic eelworms or nematodes, *Ditylenchus dipsaci*, that feed within the stems and foliage. They overwinter in dormant buds at the base of the plant.
CONTROL Phlox eelworm does not attack the roots, so clean plants can be propagated by taking root cuttings. There are no effective chemicals currently available to gardeners, so affected plants should be removed before the pest spreads.

PHLOX POWDERY MILDEW

SYMPTOMS White, powdery fungal growth develops on the upper and occasionally lower leaf surfaces. Infected leaves may be distorted and puckered and turn yellow and die off. The stems may show similar symptoms.
CAUSE The fungus *Sphaerotheca fuliginea*; this produces numerous spores which are spread by water splash or on air currents; it is encouraged by dry soil conditions and humid air.
CONTROL Treat affected plants with a fungicide such as penconazole or sulphur. Water more regularly, but avoid splashing water over the foliage.

PHOMOPSIS CANKER

SYMPTOMS Dieback of shoots, often accompanied by girdling of the stem by cankers. Girdled areas become constricted, leaving wider areas of stem above and below. On some species the cankers are accompanied by resin production. Many conifers may be attacked – in particular, Douglas fir, and occasionally various species of cedar, larch and other firs.

CAUSE The fungus *Phacidiopycnis pseudotsugae*, which enters stems and branches through wounds.

CONTROL Prune out and burn affected areas where feasible. Prune carefully, avoiding leaving stumps, and carry out any extensive pruning during the active growing season.

PHORMIUM MEALYBUG

SYMPTOMS A white, waxy substance accumulates at the ensheathing leaf bases on New Zealand flax plants. This is secreted by greyish-white, soft-bodied insects that measure up to 4mm in length. Particularly heavy infestations cause a loss of vigour and plants may eventually be killed.

CAUSE A sap-feeding insect, *Trionymus diminutus*. Most mealybugs are greenhouse pests, but this species is not harmed by winter frosts. It breeds continuously throughout the warmer months of the year.

CONTROL Applying acetamiprid or thiacloprid may reduce but not eliminate infestations. When purchasing a phormium, inspect the leaf bases carefully to ensure that it is free of this pest.

PHOSPHORUS DEFICIENCY

SYMPTOMS Growth is weak and reduced and leaves are small and may fall early. Leaves develop a blue-green or purplish discoloration, or

scorching around the edges. Plants may show delayed or poor flowering and fruiting.

CAUSE A deficiency of phosphorus which is most likely to occur on acid soils or after periods of heavy rain or watering.

CONTROL Apply fertilizers containing a high proportion of phosphorus, or apply super-phosphate at the rate and in the manner specified by the manufacturer.

PHYTOPHTHORA

SYMPTOMS Rotting at the roots and base (collar) of woody plants causes poor growth. Foliage is sparse and may become discoloured. Stems show signs of dieback and the whole plant may be killed. Inspection of the roots reveals that the finer ones are killed, and the larger roots show a blackish-brown discoloration. If the bark is removed at the base of the main stem or trunk, a reddish- or blackish-brown discoloration is visible.

CAUSE Various species of the soil- or water-borne fungus *Phytophthora* – *P. cinnamomi*, *P. citricola,* and *P. cryptogea* in particular. *P. ramorum* (the cause of sudden oak death) is becoming increasingly common, but at present it is rarely seen in domestic gardens.

CONTROL There is no cure available to gardeners. Remove infected plants together with the soil in the immediate vicinity of the roots. Improve drainage, as a poorly drained soil may encourage the development and spread of the disease.

PHYTOTOXICITY

Plants can be damaged by insecticides or fungicides; this is known as

phytotoxicity. Follow the manufacturer's instructions carefully to ensure that the correct dilution is being applied. Plants known to be harmed by a chemical will be listed by the manufacturer, but this information is not known for all potential plant/chemical interactions. The risk of spray damage occurring can be minimized by not treating plants that are already under stress as a result of exposure to bright sunlight, extremes of temperature or dryness at the roots. Seedlings will be more susceptible than mature plants, and petals are more likely to be scorched than foliage. If in any doubt, and there are many plants to be sprayed, it is sensible to treat a few and watch for adverse reactions over the next few days. Damaged plants may suffer a significant check in growth, but usually make a recovery.

PIERIS LACEBUG

SYMPTOMS The main host plant is *Pieris* spp. but *Rhododendron* can also be attacked. The insect lives on the underside of the foliage and causes a coarse pale mottling on the upper surface. In heavy infestations the foliage is extensively bleached and the underside of the leaves is covered in sticky tar-like spots of excrement. The adult bugs measure 4mm in length and have wings with many lace-like veins. The wings are held flat over the insect's body and black markings on the wings form a distinctive "H" shape pattern when the bug is at right angles to the observer. The nymphs are wingless dark brown insects with spine-covered bodies.

CAUSE A sap-feeding insect (*Stephanitis takeyai*) that originates from Japan but became established in Europe during the 1990s. This insect can be confused with the closely related rhododendron lacebug (*see p.170*), which only attacks rhododendrons.

CONTROL Check regularly, and spray with bifenthrin, thiacloprid, acetamiprid, or pyrethrum whenever the pest is seen. It is active in most months of the year except late winter and early spring.

PIGEONS

The wood pigeon, *Colomba palumbus*, can cause damage at any time of year, but they are particularly troublesome in the winter when snow or frost encourages them to feed in gardens. Brassicas, such as cabbages, Brussels sprouts and cauliflowers, can have their leaves stripped to the midribs (*see p.30*).

CONTROL Shooting is usually not a feasible proposition in gardens, so other means are needed to protect brassicas. Netting is the only certain way of keeping birds away. Various scaring devices, such as scarecrows, humming tapes, glitter strips, and models of cats or birds of prey, are generally not particularly successful. Birds quickly become used to scarers, particularly if they are hungry and other food supplies are scarce.

PINE ADELGID

SYMPTOMS The new shoots of pines, especially Scots pine, develop a fluffy, white, waxy coating in mid- to late spring (*see p.39*). Black aphid-like insects can be found underneath the wax. The insects and wax persist

PINE, NEEDLE BLIGHTS AND CASTS

through the summer but gradually become less obvious.

CAUSE Sap-feeding insects, *Pineus pini*, which overwinter on the tree as immature nymphs and have several generations during the summer.

CONTROL Spraying is usually unnecessary, and although the waxy secretions may be unsightly, the adelgids have little obvious effect on the tree's vigour once it has become established. Young trees could be sprayed with bifenthrin, fatty acids or plant oils on a dry, mild day in late winter against the overwintering nymphs before they produce their waxy covering. Treatment later in the spring is less effective, since the wax acts as a protective barrier that is difficult to penetrate with pesticides.

PINE, NEEDLE BLIGHTS AND CASTS

See NEEDLE BLIGHTS AND CASTS OF PINES *p.152.*

PINEAPPLE GALL ADELGID

SYMPTOMS In early summer, spruce trees (*Picea* species), especially the Norway spruce or traditional Christmas tree, develop swellings at the shoot tips that resemble green pineapples (*see* **p.38**). These galls contain many small aphid-like insects. In late summer the galls begin to dry up and turn brown. Slits open up in the gall as it dries, allowing the winged adult insects to emerge. Old galls can persist on trees for many years and spoil their appearance, particularly if they are to be used as Christmas trees.

CAUSE Sap-feeding insects,

Adelges abietis, which initiate the galls by secreting chemicals as they feed on the buds in the spring. These chemicals prevent normal shoot development and, instead, the bases of the leaves become swollen and fused together, while leaving cavities within which the adelgid nymphs continue their feeding and growth

CONTROL Pineapple gall adelgids have to be tolerated on trees that are too tall to be sprayed. Apart from the visual effect of the old galls, little real damage is done. On small trees, galls can be picked off or, as a preventive measure, trees can be sprayed on a mild, dry day in late winter with bifenthrin, fatty acids or plant oils. This treatment is aimed at overwintering nymphs before they mature and lay eggs on the buds.

PLANE ANTHRACNOSE

SYMPTOMS A brown, discoloured patch develops at the base of the leaf bud in winter. This may not be easy to see, but in spring affected buds die without producing a leaf. Adjacent areas of stem turn brown and die back and a small cankered area may appear on the stem. Buds which do open into leaves may show more obvious symptoms as the newly-produced growth is killed rapidly. Very dark brown, necrotic areas develop on leaves, primarily adjacent to the main veins. Throughout the summer, leaves which started off perfectly healthy may develop symptoms due to the leaf blight stage of the disease and this is the stage most frequently noticed.

CAUSE The fungus *Apiognomonia veneta* syn.

Gnomonia platani, which is encouraged by cool weather conditions in the spring and which may overwinter on fallen infected leaves.

CONTROL None available. If the infection is mild and the tree well-sized, it may be possible to prune out all infected areas. Rake up and dispose of fallen leaves.

PLUM BLOSSOM WILT

SYMPTOMS Blossom withers and dies shortly after appearing. Trusses of dead blossom remain hanging on the tree. Infection may then spread to adjacent leaves, causing them to wilt, turn brown, die and remain hanging on the branch. Tiny, pinprick-sized, raised, buff-coloured fungal pustules develop on infected areas. Localized dieback may occur.

CAUSE The fungus *Sclerotinia laxa*, which may overwinter as cankers on infected stems, or pustules on infected flowers and foliage. The spores are wind-borne and spread is most rapid during damp weather.

CONTROL Prune out infected flower trusses, preferably before the infection spreads to leaves or into the spur.

PLUM LEAF GALL MITE

SYMPTOMS Whitish-green, pouch-like swellings or galls develop on the leaves of plums, damsons and gages during the early summer, mainly on the upper leaf surface, especially around leaf edges (*see* **p.16**).

CAUSE Microscopic gall mites, *Eriophyes similis*, which suck sap from the underside of the leaves during the spring. They also secrete chemicals into the leaf,

causing the growth of the galls, which are hollow and enclose the mites.

CONTROL Nothing currently available to gardeners is effective against gall mites. Fortunately, the mites have little harmful effect, even when galls are numerous, so their presence can be tolerated.

PLUM LEAF-CURLING APHID

SYMPTOMS The foliage on plums, damsons and gages becomes tightly curled and crinkled during the spring (*see* **p.27**). Green insects, up to 2mm long, and their whitish cast skins can be found on the underside of the leaves. Winged aphids develop during late spring and early summer and fly away to spend the rest of the summer on various herbaceous plants. After that time, the infestation on plums dies out and trees start to produce normal foliage again.

CAUSE A sap-feeding aphid, *Brachycaudus helichrysi*, which migrates back to plums in late summer to lay over-wintering eggs near the buds.

CONTROL Small trees can be sprayed with thiacloprid in early spring.

PLUM MOTH

SYMPTOMS A pinkish-white caterpillar, up to 12mm long, with a brown head feeds inside a ripening plum fruit. Most feeding occurs around the stone and the damaged area becomes filled with the caterpillar's brown excrement pellets (*see* **p.48**). Damaged fruits tend to ripen prematurely and often have a depressed area on the surface where the fully-fed larva has eaten its way out before seeking an over-wintering site under a loose flake of bark.

CAUSE Caterpillars of a moth, *Cydia funebrana*, which usually has one generation a year, but may have two in hot summers.

CONTROL There are no insecticides available to amateur gardeners which will control plum moth. Pheromone traps (*see p.86*) are available for detecting the presence of this pest by capturing male plum moths. On isolated plum trees, pheromone traps may disrupt the females' mating success sufficiently to reduce the number of maggoty fruits.

PLUM POX (SHARKA)

SYMPTOMS The fruit develops discoloured, usually dark brown streaks or rings which are easiest to see on red-fleshed plums. On darker-coloured or yellow fruits there is clear pitting on the surface. Affected fruits may fall early and have an unpleasant and very acidic taste. Pale green blotches or rings develop on the foliage. Although very damaging, this disease is much less common in Britain than the rest of Europe. It may be confused with physical injury caused by very heavy rainfall or hail.

CAUSE The plum pox virus which is transmitted by aphids and may be introduced on plum rootstocks.

CONTROL None available. In Britain, any suspected outbreaks must be reported to the Department for Environment, Food and Rural Affairs.

PLUM RUST

SYMPTOMS The upper leaf surface of plums or anemones is covered with bright orange-yellow spots from mid-summer onwards. In severe cases the whole leaf may appear yellow. Numerous rust-brown, fungal pustules full of brown spores develop in corresponding areas beneath the leaf (*see p.22*). Towards late summer and early autumn spores darken and appear black. Affected leaves fall prematurely and if severe infections occur for several years in succession the tree may be weakened.

CAUSE The fungus *Tranzschelia pruni-spinosae*, which is encouraged by damp weather. Despite the link between the disease on plums and on anemones, it may appear in the same garden on one without any sign of it on the other.

CONTROL Rarely necessary on plums; rake up and dispose of fallen infected leaves. Feed the tree to maintain vigour. Remove and destroy any infected anemones nearby. Avoid growing the plum 'Victoria', which seems particularly susceptible.

PLUM SAWFLY

SYMPTOMS Unlike plum moth, this pest has caterpillar-like larvae that attack plum fruitlets, rather than the mature fruits. The larvae are creamy-white with brown heads and are up to 10mm long. Damaged fruitlets have a round hole bored in them, from which black excrement pellets may protrude. These fruits fall from the tree in late spring and early summer.

CAUSE Larvae of a sawfly, *Hoplocampa flava*, which lays eggs on the blossoms during mid-spring.

CONTROL Plum sawfly is of local occurrence and so preventive measures are only worthwhile if this pest has been troublesome in previous years. If control is necessary, spray seven to ten days after petal fall with pyrethrum. On small trees, remove damaged fruitlets before the larvae leave to go into the soil where they will overwinter. Some plums, such as 'Victoria' and 'Czar', are more susceptible than others.

POCKET PLUM

SYMPTOMS As infected fruits develop, they become deformed, twisted and often banana-shaped. They contain no stone, are hollow like pockets, and pale green and smooth. A white spore layer then develops over the surface and the plums start to shrivel.

CAUSE The fungus *Taphrina pruni*, which appears to be capable of overwintering in twigs. The spores are carried on air currents or by water splash.

CONTROL None available. Remove affected plums promptly, preferably before they turn white. Although causing a loss of crop, this disease rarely affects all the fruit in one year, and may not re-appear for several years.

POLLEN BEETLES

SYMPTOMS Black or bronze-green beetles, measuring 1–2mm in length, cluster in flowers (*see p.43*) in the spring and again in mid-summer. Yellow flowers, such as daffodils, are affected in the spring but many others, including sweet peas, runner beans, marrows, roses, dahlia and shasta daisy, attract these beetles during the summer months. They can be a nuisance, especially when flowers are cut for display in the house, but in general little damage is caused. When abundant, however, the beetles will sometimes cause damage by eating their way into unopened rose buds, and they may also discolour cauliflowers by feeding on the developing flowerhead.

CAUSE There are many pollen beetles, *Meligethes* species, most of which are entirely harmless, as they feed on wild rather than cultivated flowers. The adult beetles feed on pollen, but do not eat enough to interfere with pollination; they may even assist with pollination if they move from one flower to another and carry pollen on their bodies. The species that are of concern to gardeners are generally those that develop as larvae in the flower buds of brassica seed crops. These include the widely grown agricultural crop, oil seed rape. In mid-summer, adult pollen beetles that have bred in oil seed rape fields disperse into the countryside and gardens, where they seek out flowers. They will not breed in garden flower buds, but they can be a temporary nuisance during the summer, and again in the spring when the over-wintered adults re-emerge.

CONTROL Pollen beetles on flowers in the garden have to be tolerated. Insecticides should not be used on them, as they will also kill bees, butterflies and other desirable flower visitors. Spraying would in any case need to be done at very frequent intervals in order to cope with the continual migration of the beetles into gardens. When cut flowers are taken indoors, pollen beetles have an annoying habit of leaving the flowers and crawling about the room. However, placing cut flowers in a shed or garage for a few hours during the day before they are brought into the house, can have the effect of persuading the beetles to leave the flowers without being a nuisance.

POOR POLLINATION OF RASPBERRIES AND STRAWBERRIES

SYMPTOMS Fruits develop but are distorted and may bear dry areas. On raspberry, individual drupelets or small groups of them may be dry and brown. Adjacent drupelets may be perfectly normal. The overall growth of the plant is normal and it may bear other, perfect fruits.
CAUSE Poor pollination, which results in uneven swelling and development of the fruit. A lack of bees, or weather conditions which prevent their flying and pollinating activities.
CONTROL Grow susceptible plants in an area which is provided with shelter. Grow plenty of other plants and flowers to attract bees and other pollinating insects.

POPLAR CANKER

SYMPTOMS In spring a dense, off-white slime is exuded from cracks in young branches. The damaged areas may be girdled and so die back. On older branches, the slime may also be apparent, and roughened areas of bark develop, up to 3cm in diameter. Larger, more "open" cankered areas may develop and stand out clearly from the branch as they become surrounded by raised callus.
CAUSE The bacterium

Xanthomonas populi, which is believed to enter the tree via wounds, such as as those around bud scales, leaf scars, frost crack and damage caused by insects feeding. The precise means of spread is not known, but rain splash and insects are probably the main ones.
CONTROL Prune out infected areas where feasible, in an attempt to limit the spread of the bacteria.

POPLAR YELLOW BLISTER

SYMPTOMS In late spring or early summer raised, rounded blisters develop on upper, and occasionally lower, leaf surfaces. The concave, lower surface of each blister turns bright yellow. In damp weather young shoots may also be attacked. Although leaves are disfigured, the overall vigour of the tree is not affected.
CAUSE The fungus *Taphrina populina*, which overwinters as spores on the bud scales.
CONTROL No control measure is available, and none is justified.

POTASSIUM DEFICIENCY

SYMPTOMS Most frequently poor flowering and/or poor fruit set. Flowers may be undersized. On tomatoes this deficiency is at least partially responsible for the disorder known as blotchy ripening. Potassium is also important in the ripening of wood and in the structure of the stem, and too little may increase the risk of frost or similar damage. Leaf tips may appear scorched around the edges and show brownish-purple spotting beneath (see *p.24*).
CAUSE A deficiency of

potassium in the soil, which is most common on light, sandy soils and those with a low clay content.
CONTROL Apply sulphate of potash in spring and autumn.

POTATO BLACK LEG

SYMPTOMS Foliage is chlorotic (yellow) and stunted; the leaves are small and slightly incurled. These symptoms are easily confused with other problems, but with black leg the stem base is blackened and rotted at ground level. The vascular strands within the stem are similarly discoloured, showing as distinct black spots when the stem is cut across. The parent tuber is completely rotted. The plant may be killed before a crop is produced.
CAUSE The bacterium *Erwinia carotovora* var. *atroseptica*, which is encouraged by wet soil conditions. It is often introduced on symptomless but mildly affected seed tubers. When these are planted they cause affected plants to appear scattered through the crop of other perfectly healthy plants. The bacteria may enter tubers via wounds either whilst in the soil, or at lifting.
CONTROL Take great care at lifting time to keep damage to a minimum. If possible, do not lift crops during wet weather as infection is more likely then. Store only those tubers which appear quite healthy. Avoid growing 'Majestic', 'Arran Pilot', 'Maris Bard', 'Desirée', and 'Estima', as these cultivars seem to be particularly susceptible.

POTATO BLIGHT

SYMPTOMS Necrotic brown patches develop on the

leaves, largely on the leaf tips and around the edges (see *p.28*). As the spots enlarge, the leaves wither and die. In wet weather or under humid conditions slightly fluffy, white fungal growth may be visible around the edges of the spots. This is most common on lower leaf surfaces. The haulms of infected plants develop brownish-black patches and may collapse. Infected tubers develop slightly sunken, dark patches on the skin with a reddish-brown discoloration beneath. The discoloration may then spread into the flesh (see *p.56*). Secondary organisms often invade, turning the dry, brown areas into unpleasant-smelling, slimy wet rot. The whole tuber is then rotted.
CAUSE The fungus *Phytophthora infestans*, also responsible for blight on tomatoes. Spores are produced from the leaf lesions and carried by rain splash or on air currents. For infection to occur, the fungus needs two consecutive 24-hour periods, each with a minimum temperature of 10°C (50°F) and each with a minimum of at least 11 hours where the relative humidity is 89% or more. Spores from the infected top growth are washed down through the soil and may infect tubers.
CONTROL Spray foliage with a fungicide such as Bordeaux mixture, copper oxychloride or mancozeb. In seasons where weather conditions are favourable to blight, spray before the symptoms appear. If a "Blight Infection Period" is mentioned on the radio for your area, it is advisable to spray. Reduce the chances of the infection developing on the tubers by

earthing up deeply. Remove haulms as soon as they show signs of infection. Grow cultivars which show some resistance, such as 'Cara', 'Estima', 'Kondor', 'Lady Balfour', 'Maris Peer', 'Pentland Crown', 'Record', 'Rembrandt', 'Romano', 'Sante', 'Sarpo Axona', 'Sarpo Mira', and 'Valour'. Do not compost any plant remains.

POTATO COMMON SCAB

SYMPTOMS Raised, roughened scabby patches develop on the skin of the tuber (*see p.56*). The skin ruptures, leaving the scabs with ragged edges. Damage may be superficial or result in cracking. The flesh is usually undamaged, although just beneath the scabby patches it may become discoloured.
CAUSE *Streptomyces scabies*, which is particularly common on sandy, light soils with a low organic matter content. It is present naturally in most soils and potatoes planted on ground that was previously grassland are particularly prone to attack. Soil which has been limed is also likely to have more *Streptomyces* in it, whereas acid soil has less.
CONTROL Improve the organic matter content of the soil. Water regularly, as there appears to be a link between this disease and dry soil conditions. Do not lime soil prior to planting potatoes. Use acidic materials such as sulphate of ammonia and superphosphate to reduce scabbing. No fungicide treatment is available. Grow resistant cultivars such as 'Arran Comet', 'Arran Pilot', 'Golden Wonder', 'King Edward', 'Lady Balfour', 'Maris Peer' and 'Rembrandt'.

POTATO CYST EELWORM

SYMPTOMS Potato plants die prematurely and hence produce a poor crop of small potatoes. Affected plants die during mid- to late summer, with leaves progressively yellowing and drying up from the bottom of the stems upwards. When cyst eelworms first occur in a garden, small patches within the potato rows show symptoms. These patches gradually enlarge each time potatoes are grown in that part of the garden until eventually it is impossible to grow a worthwhile crop of potatoes. Tomato plants are also attacked. If infested plants are carefully dug up, spherical eelworm cysts, up to 1mm in diameter and white, yellow or brown in colour can be seen on the roots (*see p.55*). These are the bodies of mature female eelworms and each one can contain up to 600 eggs.
CAUSE Two species of cyst eelworm attack potato and tomato. The golden cyst eelworm, *Globodera rostochiensis*, has female cysts which are white at first, but turn yellow then brown when mature. The white cyst eelworm, *G. pallida*, changes from white to brown without an intermediate stage. Both species develop inside the roots, where their feeding disrupts the uptake of water and nutrients; when mature, the females become swollen and burst through the root wall.
CONTROL There are no effective pesticides available to amateur gardeners for controlling eelworms. Potato cyst eelworms are particularly difficult because their eggs remain viable inside the cysts for many years. The eggs are

stimulated to hatch by chemicals that are secreted into the soil by the roots of suitable host plants. Adopting a system of crop rotation can delay the build-up of damaging infestations, but once a serious eelworm problem has arisen, the usual three or four-year rotation is not long enough to eliminate the pest. Some potato cultivars have some resistance to the golden cyst eelworm. These include 'Accent', 'Nadine', 'Pentland Javelin', 'Rocket', 'Swift' (earlies); and 'Alhambra', 'Cara', 'Nicola', 'Maris Piper', 'Picasso', and 'Stemster' (main crop). In addition there are the maincrop cultivars 'Maxine' and 'Sante' which are resistant to golden cyst eelworm and have some tolerance of the white cyst eelworm. The latter will, however, breed successfully in the roots, so infestations will increase.

POTATO DRY ROT

SYMPTOMS The skin of the potato tuber becomes wrinkled at one end, followed by rapid shrinking, the development of concentric rings of wrinkles, and discoloration (*see p.57*). The affected tissues become black, grey or brown. Pink, white or blue-green pustules of fungal spores develop on the shrivelled tissues. The symptoms are likely to be found on stored tubers only, though it may also cause poor growth after planting.
CAUSE A species of the fungus *Fusarium*, which infects tubers whilst they are still in the ground or as they are being lifted. The fungus enters through the lenticels (breathing pores), eyes, or wounds caused by abrasions, pests or scab infection.

CONTROL Handle potatoes very carefully at lifting and when storing. Minimize other problems such as wireworm, slugs and scab infections. Ensure that the potatoes are properly mature at harvest time and that they are stored in a frost-free but cool and dry place. Do not grow 'Arran Pilot', which is notably susceptible, but consider 'King Edward' or 'Pentland Crown' instead, which are notably resistant.

POTATO EARLY BLIGHT

SYMPTOMS Despite its name, this disease usually occurs late in the season in Britain and Europe, though it is an early season infection in North America. Dark brownish-black spots, often angular in outline and showing concentric ringing, appear on the foliage of both potato and tomato. Sunken spots may develop on the stems and a dark rot and associated furry fungal growth may occur at the sepal end of the fruits. Inadequately fed plants may also be more susceptible. Although causing some damage, this disease rarely affects cropping.
CAUSE The fungus *Alternaria solani*, which can carry over from year to year on infected plant debris.
CONTROL Rarely necessary, but the treatments applied to control potato or tomato late blight (*Phytophthora infestans*) should give incidental control.

POTATO GANGRENE

SYMPTOMS Small areas of damage, slightly sunken with clearly defined edges, develop on the tubers after lifting (*see p.57*). The skin may become slightly wrinkled. The affected tubers rot, turning wet and pale pink, and then darken. There is no visible fungal growth. If planted, a gangrene-infected potato produces a poor, gappy crop or no crop at all.

CAUSE The fungus *Phoma exigua* var. *foveata*.

CONTROL Avoid damaging tubers on lifting and inspect all tubers carefully before planting and throughout storage. Dispose of any tubers showing symptoms.

POTATO HOLLOW HEART

SYMPTOMS The tubers may look perfectly normal on the outside, as does the plant itself. When cut open, the tuber has an internal cavity – this may be simple and surrounded by corky, somewhat darkened tissue; it may be star-shaped, or it may consist of two or more interconnecting cavities. Large potatoes are particularly susceptible; small ones are rarely affected.

CAUSE This disorder occurs when potatoes have been growing rapidly (perhaps due to overwatering, or excessive fertilizer use) and are then subjected to a sudden check in growth, usually caused by lack of water. Conversely, it may also occur if the plants are subjected to an over-dry soil, followed by a sudden increase in the soil moisture due to heavy rain or watering. In some instances it appears to be induced by extremely high temperatures.

CONTROL Grow plants in uniform stands and ensure regular but not excessive watering, particularly during dry weather.

POTATO INTERNAL RUST SPOT

SYMPTOMS The tubers and the whole plant look perfectly normal, but when cut open, the flesh of the tuber is marked with many irregular, small rust-brown spots.

CAUSE The precise cause is unknown, but unsuitable soil conditions are probably largely responsible. It may occur on sandy and clay soils but is usually most severe on light soils or loam soils with a low organic matter content. Low potassium and lime levels may also be responsible.

CONTROL Improve the condition of the soil prior to planting – incorporate plenty of humus and water regularly as the crop grows. The soil could also be limed but this must be done well in advance of planting. Grow 'King Edward' or 'Arran Consul', which appear to show a degree of resistance.

• *see also* POTATO COMMON SCAB *p.165.*

POTATO POWDERY SCAB

SYMPTOMS Small, nearly circular scabby patches develop on the tuber. Each has a raised margin. The scabs later burst open and produce masses of brown spores which are released into the soil (*see* **p.56**). Occasionally a canker form of this disease occurs when affected tubers are seriously deformed. They may then be incorrectly assumed to have potato wart disease.

CAUSE The fungus *Spongospora subterranea*, which is particularly common in wet seasons and on heavy soils, especially those which have grown many crops of potatoes.

CONTROL Dispose of infected tubers; do not incorporate them into the compost heap. Do not grow potatoes on that site for at least three years. Improve soil aeration before planting potatoes. Avoid growing 'Pentland Crown', as this is particularly susceptible.

• *see also* POTATO WART DISEASE *below.*

POTATO SILVER SCURF

SYMPTOMS Inconspicuous silvery markings, usually lines, develop on the skin of the tuber (*see* **p.56**). The flesh of the tuber is not affected and this disease is of little practical consequence unless the tubers are to be used for showing. The condition usually only becomes apparent when the tubers are stored, particularly if the storage conditions are too humid. The markings may turn black if the fungus produces spores.

CAUSE The fungus *Helminthosporium solani*.

CONTROL None necessary.

POTATO SPRAING

SYMPTOMS The flesh of tubers is marked by tan-brown or red-brown rings or arcs of discoloration (*see* **p.56**). There may be development of some corky tissue associated with the discoloration, too. Tubers may, sometimes, be distorted and the stems and foliage may be mottled yellow.

CAUSE Tobacco rattle virus and occasionally potato mop top virus, which are spread by free-living nematodes (eelworms) in the soil. Infection may be started if infected tubers are planted.

CONTROL No chemical control available. Remove affected plants, as tobacco rattle virus has a wide host range and may affect sweet pepper, hyacinth, gladiolus, tulip, China aster, tobacco plants and many common weeds. Practise good weed control. Grow potatoes on a fresh site. 'Pentland Dell' appears to be particularly susceptible to this problem, so other cultivars should be grown in preference.

POTATO VIRUSES

SYMPTOMS These vary with the precise combination of viruses involved, the potato variety and the growing conditions. Leaf symptoms include yellow flecking, streaking or mosaicing, dark spotting and leaf distortion, upward rolling and stiffening of the leaflets. If home-saved seed potatoes are used for many years in succession, there may be a gradual but noticeable decline in the plant vigour and cropping and this is often due to the build-up of viral diseases.

CAUSE Many different viruses, including potato virus X, potato virus Y and potato leaf roll virus.

CONTROL None is available. Always grow potatoes on a fresh site and do not save tubers for seed. Buy tubers certified as virus-free.

POTATO WART DISEASE

SYMPTOMS The tubers are covered in dense, warty outgrowths, at first off-white, and later black. Affected tubers may rot. Tubers may only start to show symptoms in store. Occasionally small, pale green growths rather like those of crown gall may develop on the stem at soil level. This disease is rare,

but very damaging. The symptoms may be confused with the distorted form of powdery scab, or even with fusarium rot of potato.
CAUSE *Synchytrium endobioticum*, which is usually found in wet years or on heavy soils, and infects tubers through the eyes. Spores are produced by the wart-like growths. These spores are very resilient and can remain viable in the soil for about thirty years, even in the absence of a potato crop. They are easily spread from one area to another in particles of soil adhering to tools and footwear.
CONTROL Infected tubers should be reported to the local Department for Environment, Food and Rural Affairs. Potatoes must not be grown on infested land.
• *see also* POTATO POWDERY SCAB *p.166.*

POWDERY MILDEWS

SYMPTOMS White, powdery fungal growth develops on the leaf surface (*see **p.25***). Upper leaf surfaces are usually affected first and most severely but the symptoms may spread to the lower leaf. The mildew may also develop on all other above ground parts of the plant; the precise location of its growth depends on the host plant and the species of mildew involved. Affected leaves or other plant parts may yellow and become distorted. Distortion is particularly common if infection occurs on young foliage. Affected fruits may crack and split because they are unable to expand normally (*see **p.49***). Sometimes buff or even pale brown fungal growth may be seen; this is most common on rhododendron

(where the mildew is often restricted to the lower leaf surface), gooseberry and laurel. The mildew may kill off small areas of leaf tissue which then drop away, causing a shot-hole effect. Growth may be poor and in extreme cases dieback or even death may occur following premature leaf fall.
CAUSE Various fungi, in particular many species of *Oidium, Microsphaera, Podosphaera, Uncinula, Erysiphe* and *Phyllactinea*. Each usually only infects a single genus or closely related host plants. These fungi are often encouraged by plants growing in dry soils and by humid or damp air around the top growth.
CONTROL Remove infected leaves promptly. Spray with a suitable fungicide such as myclobutanil, or plant oils, where appropriate. Keep plants adequately watered but avoid overhead watering. Where available, grow resistant varieties.
• *see also* APPLE POWDERY MILDEW *p.101;* BEGONIA POWDERY MILDEW *p.107;* GERANIUM DOWNY MILDEW *p.132;* HYDRANGEA POWDERY MILDEW *p.140;* OAK POWDERY MILDEW *p.153;* PANSY POWDERY MILDEW *p.155;* PEA POWDERY MILDEW *p.157;* PHLOX POWDERY MILDEW *p.160;* QUINCE POWDERY MILDEW *p.168;* RHODODENDRON POWDERY MILDEW *p.171;* ROSE POWDERY MILDEW *p.173.*

PRIVET THRIPS

SYMPTOMS During the summer months, privet and lilac leaves develop a dull green appearance, becoming increasingly silvery brown on their upper surface (*see **p.22***). Elongate, narrow-bodied insects up to 2mm long occur on leaves. The adults are black with three

white bands across wings that are folded back over the body. The immature nymphs are similar in shape but are creamy-yellow and wingless.
CAUSE Sap-feeding insects, *Dendrothrips ornatus*, which are more troublesome in hot, dry summers.
CONTROL Light infestations will have little effect on the plant, but if damage starts becoming noticeable, especially before late summer, spray with thiacloprid, acetamiprid, bifenthrin, or pyrethrum.

PROLIFERATION

SYMPTOMS Only the flower or bud is affected. Buds develop within the centre of developing buds (*see **p.44***). These may never mature further or may produce fully grown flowers which grow out from the centre of the original flower. Occasionally a "triple-decker" effect is seen, with three flowers on top of each other. Stems and foliage may also appear between flowers. On fruiting plants, each flower may develop into a fruit.
CAUSE Most commonly, physical injury such as that caused by a late frost which damages developing buds. Occasionally virus infection is responsible. If symptoms occur all over the plant and for several successive years, then virus would seem a likely cause and the affected plant should be removed.
CONTROL Rarely necessary, though affected shoots could be pruned out.

PYRACANTHA LEAF MINER

SYMPTOMS An oval, silvery-brown patch develops on the upper leaf surface of *Pyracantha*, or firethorn,

where the moth's caterpillar has eaten the internal tissues. As the caterpillar matures, it spins silk threads that cause the leaf to fold in half upwards so that the mine is no longer visible. Leaves at the bottom of the plant are often more heavily attacked than those elsewhere.
CAUSE Caterpillars of a tiny moth, *Phyllonorycter leucographella*, which can have three generations a year, with leaf damage being most obvious over the winter period. It was first recorded in Britain in 1989, but is now widespread in England and is continuing to spread. It originates from southern Europe, but occurs in gardens in many northern European countries.
CONTROL Pyracanthas can tolerate heavy infestations, so control measures are often unnecessary. This is fortunate as there are no effective insecticides for leaf miner.

PYRACANTHA SCAB

SYMPTOMS Greyish-black or dark khaki-coloured, scabby patches develop on the leaves and berries (*see **p.49***). Infected leaves become chlorotic and may fall prematurely. Infected berries may remain small, start to crack or split and fall.
CAUSE The fungus *Spilocaea pyracanthae*, which is encouraged by damp weather. It does not appear to overwinter on fallen leaves (unlike apple and pear scab), but overwinters on infected leaves which remain on the plant, and as tiny pustules on the shoots.
CONTROL Prune out severely infected areas, and twigs bearing the pustules. Some cultivars show a degree of resistance to this infection (and also to fireblight) and

these include 'Orange Charmer', 'Shawnee', and 'Golden Charmer'.

PYTHIUM ROOT ROT

SYMPTOMS The whole plant wilts and dies back. Seedlings are particularly prone, but older plants may also succumb if otherwise lacking in vigour. Roots are killed and may disintegrate, giving the plant the appearance of having only a few roots.
CAUSE The fungus *Pythium*, which is soil- or water-borne.
CONTROL Remove infected plants promptly, together with the soil or compost in the immediate vicinity of the roots. Water seeds and seedlings with a copper-based fungicide such as Bordeaux mixture, Cheshunt compound or copper oxychloride. Ensure strict hygiene and do not water seedlings with water from a water butt or use unsterilized compost or soil for raising seedlings.

QUINCE LEAF BLIGHT

SYMPTOMS Numerous irregular, small spots appear on the foliage. These are dark red at first but soon turn black and may join together (*see* **p.48**). The leaves turn yellow and then brown and fall prematurely. Occasionally similar spotting develops on the fruits and these may show some distortion. Shoot tips may also be infected and then develop dieback.
CAUSE The fungus

Diplocarpon mespili (syn. *Fabrea maculata*), which may overwinter on infected shoots.
CONTROL Rake up and dispose of fallen leaves. Prune out infected stems.

QUINCE POWDERY MILDEW

SYMPTOMS Powdery, white deposits on the upper leaf surface and occasionally on the lower surface.
CAUSE The fungus *Podosphaera leucotricha*, which can also affect apple.
CONTROL Keep quinces well watered, as dry soil seems to encourage mildew development. Apply a mulch to the moist soil to preserve soil moisture. Prune out severely infected stems. Prune established trees to improve air circulation within the crown.
• *see also* APPLE POWDERY MILDEW *p.101*.

RABBITS

Rabbits (*Oryctolagus cuniculus*) can feed on a very wide range of plants. Herbaceous plants may be grazed down to ground level, while the foliage and soft shoots of woody plants are eaten up to a height of about 50cm. Rabbits also gnaw the bark from the base of trunks, especially on young apple trees, and the tree dies if bark is lost from most or all of the trunk's circumference. Bark feeding occurs at any time of year, but trees are particularly at risk when snow or frost makes

other food items unavailable. Rabbits are inquisitive animals that often feed on newly planted plants, even if they have previously ignored similar established plants in the same garden. They were greatly reduced by a viral disease, myxomatosis, during the 1950s and 60s, but they now have greater resistance and are often numerous in rural areas.
CONTROL Direct control measures, such as shooting, trapping and gassing of rabbit warrens, are usually not suitable in domestic gardens. A better solution is to keep rabbits away from plants by netting. Fences around the perimeter of a garden need to be at least 120–140cm tall with a further 30cm sunk below soil level, and angled outwards to discourage rabbits from burrowing underneath. The wire netting needs a maximum mesh size of 25mm, otherwise young rabbits may squeeze through. Where individual plants or small flower beds are being protected, it is usually sufficient to enclose them with netting 90cm high without the need to bury part of the netting. Gates to gardens or enclosed flower beds also need to be rabbit-proof and kept closed when not in use. Tree bark can be protected by placing wire netting or spiral tree guards around the base of the trunks. Animal repellent products cannot be relied on to give long-term protection. Although most plants can be eaten, there are some which are usually left alone; these feature in various published lists of "rabbit-proof plants". Looking at other nearby gardens will also provide suggestions for plants that local rabbits ignore.

RASPBERRY BEETLE

SYMPTOMS All cane fruits, including raspberry, blackberry, loganberry and tayberry, are susceptible. The ripe fruits have dried-up patches at the stalk end of the berries (*see* **p.50**). When fully grown, the larvae are up to 8mm long and are creamy-white with pale brown markings on their upper surface. The grubs feed initially at the base of the berries but later feed in the inner core or plug. They are often seen crawling around the bowl after the fruits have been picked.
CAUSE The larval stage of a small greyish-brown beetle, *Byturus tomentosus*. Eggs are laid on the flowers during early and mid-summer.
CONTROL Timing of control measures is important in order to control the newly-hatched grubs before extensive damage has occurred. On raspberries, spray when the first pink fruits are seen; on loganberries and other hybrid berries, at 80 per cent petal fall; and on blackberries, when the first flowers open. Spray with pyrethrum at dusk to minimize the risk to bees and other pollinating insects.

RASPBERRY CANE BLIGHT

SYMPTOMS The shoots or the whole cane start to die back during the summer. This is due to an infection at the base of the canes, which causes them to become brown and the bark to rupture. They also become very brittle and may snap off readily at or just above ground level. Minute, black, pinprick-sized, fungal fruiting bodies develop on dead areas and exude spores.

Other spore forms develop later and these forcibly eject their spores into the air.

CAUSE The fungus *Leptosphaeria coniothyrium*, which is carried both by rain or watering splash and on air currents. Infection most frequently takes place through wounds, such as those caused by cane midge attack, late spring frosts, or following pruning. The fungus can persist in the soil on infected plant debris.

CONTROL Prune and train the plant carefully to avoid infection. Protect from frost and control any pests which may cause wounds through which the fungus can gain entry. If infection occurs, prune out affected canes and cut away any discoloured wood from the crown.

RASPBERRY CANE SPOT

SYMPTOMS Purple spots with silvery-white central areas develop on the stems of raspberries and hybrid cane fruits. They may also appear on the leaves and flower stalks (*see p.34*). On loganberries, the fruits may be infected. The bark may split as the spots enlarge and whole canes may be killed.

CAUSE The fungus *Elsinoe veneta*, which usually appears first on the canes in early summer, and may then spread to the foliage and the fruits.

CONTROL Prune out infected canes. Avoid growing the raspberry cultivars 'Norfolk Giant' and 'Lloyd George' as these are particularly prone to this infection. Spray with copper oxychloride.

RASPBERRY LEAF AND BUD MITE

SYMPTOMS Pale yellow, rounded blotches develop on the upper surface of raspberry leaves (*see p.18*) from late spring onwards. On the underside of the leaves, corresponding with the yellow blotches, are slightly darker areas, compared with the usual silvery-green colour. By mid-summer the foliage may be extensively discoloured and leaves at the shoot tips may be distorted. These symptoms can be confused with those of some virus diseases. Canes affected by the mite normally grow to their usual height and produce an adequate crop, whereas virus-infected plants lack vigour and crop badly.

CAUSE Microscopic, sausage-shaped mites, *Phyllocoptes gracilis*, which suck sap from the underside of leaves. In the autumn they hide in or near buds but cause no damage in the winter.

CONTROL None of the insecticides available to amateur gardeners is effective against this mite. Fortunately, its impact on the plant's health and cropping ability is not as great as the symptoms might suggest. Raspberry cultivars vary in their susceptibility; 'Malling Jewel' is often attacked, while 'Leo', 'Glen Prosen', 'Glen Lyon', and 'Terrie Louise' have some resistance. Plants growing in warm, sheltered places are also likely to have heavier infestations.

RASPBERRY RUST

SYMPTOMS Bright orange, raised pustules full of fungal spores develop on leaves of raspberries and blackberries in spring or early summer. They develop first on the upper leaf surface. Towards the end of summer or in early autumn, dark brown-black spores develop on the lower surface. Premature leaf fall occurs (*see p.40*).

CAUSE The fungus *Phragmidium rubi-idaei*, which can overwinter on infected fallen leaves and on the canes.

CONTROL Rake up and dispose of infected leaves. The raspberry cultivar 'Glen Prosen' shows a degree of resistance. Avoid growing 'Glen Moy', 'Glen Clova, 'Malling Delight', 'Malling Jewel', and 'Malling Joy'.

RASPBERRY SPUR BLIGHT

SYMPTOMS Dark purplish patches develop around the buds of new canes. These increase in size and spread up and down the canes, causing extensive discoloration. In autumn and winter they change colour, becoming greyish-silver and are covered in numerous minute raised, black, pinprick-sized, fungal fruiting bodies. In spring, infected canes bear few viable buds and produce weak shoots which wither and die, or persist but crop very poorly.

CAUSE The fungus *Didymella applanata*, which is usually most troublesome in wet weather, but occurs in most seasons regardless of the weather. Plants that have been fed excessive quantities of nitrogen or which are overcrowded are more prone to infection and encourage rapid disease spread.

CONTROL Prune out any badly infected canes.

RASPBERRY VIRUSES

SYMPTOMS Yellow patterning of the foliage (*see p.18*), most commonly mosaic patterns often combined with slight distortion, stunting and reduction in vigour and cropping. There may also be downward curling of the leaves. The foliage symptoms may be confused with those caused by raspberry leaf and bud mite, but the other symptoms are not present.

CAUSE Various viruses, either alone or in combination with each other. The most common are raspberry mosaic, curly dwarf and raspberry yellow dwarf.

CONTROL Remove and dispose of infected plants. Control aphids and other sap-feeding pests as these may spread the virus. Do not grow replacement raspberries on the same site, as some viruses are transmitted by soil-living nematodes.

RATS

Rats cause damage to stored fruits and vegetables, and also to growing root crops. Rats also spoil foodstuffs by contaminating it with their excrement and urine. Rats often carry in their urine the bacterium that causes Weil's disease; this can cause a serious illness in humans. The brown rat, *Rattus norwegicus*, is widespread in both Britain and the rest of Europe. Local authorities will often undertake rat control in homes and gardens. Otherwise they can be controlled with rat traps or poisoned baits based on bromadiolone, coumatetralyl, or difenacoum. Poisoned baits and traps need to be placed where rats are active, but they must always be covered up so that young children and pets do not have access to them.

RED BERRY MITE

SYMPTOMS Blackberry fruits ripen unevenly, with some parts of the berries remaining red and hard (see *p.49*). This problem is more frequently seen in hot summers. The first fruits to develop usually ripen normally, but later fruits show progressively more severe symptoms.

CAUSE Microscopic gall mites, *Acalitus essigi*, that feed by sucking sap from the flowers, foliage and fruits. Only the fruits suffer damage, however, which is due to a toxic saliva secreted by the mites as they feed.

CONTROL None of the insecticides currently available to amateur gardeners is effective against this mite. Some of the overwintering mites can be eliminated by pruning out to the ground and burning the old fruiting canes.

RED SPIDER MITE

See BOX RED SPIDER MITE *p.110*; CONIFER RED SPIDER MITE *p.120*; FRUIT TREE RED SPIDER MITE *p.130*; GLASSHOUSE RED SPIDER MITE *p.133*.

RED THREAD

See TURF RED THREAD *p.186*.

REPLANT PROBLEMS

See ROSE SICKNESS *p.173*.

REVERSION

See CURRANT REVERSION *p.122*.

RHODODENDRON BUD BLAST

SYMPTOMS Flower buds fail to open, and turn brown and dry (see *p.41*). They may remain on the plant for several years. The surface of the bud is covered in numerous tiny, black, bristle-like fungal outgrowths. These are the spore-bearing parts of the fungus, and it is the presence of these which differentiates between this and dry bud with which it is often confused. Quite often infected buds and healthy ones are seen growing side by side. Leaf, stem and general growth are unaffected.

CAUSE The fungus *Pycnostysanus azaleae*, spread by rhododendron leafhoppers as they lay their eggs in the bud scales.

CONTROL The leafhoppers first appear in mid- to late summer, so pick off and dispose of infected buds before that time of year. Control of the leafhopper is possible but rarely fully effective in controlling bud blast spread, because the pest simply re-invades from nearby gardens.

• *see also* BUD DROP *p.112*; BUD, DRY *p.112*.

RHODODENDRON LACEBUG

SYMPTOMS During the summer the upper leaf surface of rhododendrons, especially those growing in sunny positions, develop a coarse yellowish mottling. The underside of the leaves is covered in rusty-brown excrement spots deposited by the lacebugs (see *p.21*). These are present between late spring and early autumn. The wingless, immature nymphs are yellowish-brown, while the adults are 4mm long, brownish-black and with wings that are folded flat over their backs. The transparent wings have many veins, giving them a lace-like appearance – hence the name lacebug.

CAUSE Sap-feeding insects, *Stephanitis rhododendri*, which have one generation a year. In late summer the adult lacebugs lay overwintering eggs in the midribs of the leaves.

CONTROL Avoid planting rhododendrons in warm, sunny spots. If the pest does occur, spray the foliage with bifenthrin, thiacloprid, acetamiprid, or pyrethrum. Damaged leaves do not regain their green colour and can persist on the shrub for more than a year; new growth develops normally once the pest has been dealt with.

• *see also* PIERIS LACEBUG *p.161*.

RHODODENDRON LEAF SPOT

SYMPTOMS Spots with clearly defined edges develop on the leaves. The older leaves are usually worst affected and, if severely damaged, some may fall prematurely. The spots are brownish-purple, often with a distinct black or very dark purple ring around the edge (see *p.14*). The spots often have concentric ringing with numerous pinprick-sized, raised, black fungal fruiting bodies on the dead areas.

CAUSE The fungus *Gleosporium rhododendri*, which, although causing unsightly damage to leaves, is usually only likely to reach serious levels on a plant lacking vigour or in poor health.

CONTROL Remove severely affected leaves. Take steps to improve the general vigour of the plant by regular feeding, watering and mulching.

RHODODENDRON LEAFHOPPER

SYMPTOMS The creamy-yellow immature nymphs are present on the underside of rhododendron leaves from late spring to mid-summer, but cause no noticeable damage. The adult insects are more readily seen, as they often sun themselves on the upper leaf surface between mid-summer and mid-autumn. They are 8–9mm long and turquoise-green with two orange stripes on each forewing. When disturbed, they jump off the leaf and fly a short distance before returning to the plant.

CAUSE A sap-feeding insect, *Graphocephala fennahi*, which causes no direct damage to rhododendrons. The females, however, insert overwintering eggs into the flower buds and this makes entry wounds for the fungal disease rhododendron bud blast (see *above*).

CONTROL Reduce the incidence of bud blast by spraying regularly against the adult leafhoppers during late summer and early autumn. Bifenthrin, thiacloprid, or pyrethrum will control leafhoppers, but more may fly in from nearby gardens, especially during hot weather, which increases leafhopper activity.

RHODODENDRON PETAL BLIGHT

SYMPTOMS Small spots develop on the petals – these are white on coloured flowers and pale brown on white blooms (see *p.42*). The spots increase in size and appear water-soaked. The entire flower rapidly collapses and becomes slimy and shapeless. Affected flowers do not fall but remain on the plant. Dry weather may then cause the dead flowers to dry out and they may remain on the plant in this state until the following spring.

CAUSE The fungus *Ovulinia azaleae*, which is encouraged by mild, humid growing conditions. The spores develop on the petals as they deteriorate and are spread by insects and on air currents.
CONTROL Remove and dispose of infected flowers promptly to minimize spread of the fungus.

RHODODENDRON POWDERY MILDEW

SYMPTOMS Yellow blotches develop on the upper leaf surface. Beneath each is a corresponding patch of felty, buff-coloured fungal growth (*see p.15*). Premature leaf fall occurs. Occasionally the fungal growth occurs on the upper surface and is then more off-white and powdery than buff and felty.
CAUSE The precise identity of the fungus or fungi is as yet undetermined, but the most important is probably a species of *Erysiphe*. Dry soil and wet conditions around the foliage in late summer and autumn encourage both its development and spread.
CONTROL Pick off severely affected leaves. Spray with a suitable fungicide such as myclobutanil. Keep plants well watered and mulched but avoid overhead watering. Avoid growing *R. cinnabarinum* and its hybrids, and hybrids with species from the subsections *Campylocarpa* or *Fortunea* as their parentage.

RHODODENDRON RUST

SYMPTOMS The upper leaf surface develops bright orange-yellow spots with corresponding yellow and brown fungal pustules appearing beneath. The disease is not very common and may be confused with the damage caused by rhododendron lacebug (*see opposite*).
CAUSE The fungus *Chrysomyxa ledi* var. *rhododendri*, which persists within the affected leaves as long as they remain on the plant. It is encouraged by damp conditions.
CONTROL Remove infected leaves promptly, pruning out entire shoots if necessary. Some success may be achieved by spraying with a suitable fungicide such as myclobutanil or penconazole, though this treatment does not work as reliably as it does on most other rust infections.

RHUBARB GUMMING

SYMPTOMS The sticks develop normally but some are spoiled by drops of resin-like exudation which appear along their length. Sticks may crack around the areas of gumming, allowing the entry of secondary organisms which then cause rotting.
CAUSE The precise cause is not known but poor growing conditions, in particular an erratic supply of moisture, are believed to be involved. Inadequate feeding may also play a part.
CONTROL Improve growing conditions and pay particular attention to watering and feeding the plant.

ROBIN'S PIN CUSHION

SYMPTOMS Wild hedgerow roses, sucker growth on grafted roses and some species roses develop roughly spherical galls, up to 60mm across, on stems (*see p.39*) and occasionally leaves in mid- to late summer. The gall consists of a hard central core in which many white grubs live and feed. The exterior of the gall is covered in a dense mass of moss-like growths which are reddish or yellowish-green in colour. In the autumn the mossy covering dries up, leaving the inner core in which the grubs overwinter before pupating in the spring.
CAUSE A small gall wasp, *Diplolepis rosae*, which lays eggs in buds in mid-summer. The galls that subsequently develop are also sometimes known as bedeguar galls.
CONTROL Apart from creating the galls, this insect has little adverse effect on the plant and so control measures are unnecessary. If the galls are considered unsightly they can be pruned out.

ROOT APHIDS

SYMPTOMS Infested plants tend to have reduced vigour and are more likely to wilt during sunny weather. The aphids, up to 2–3mm long, usually a dirty cream colour but sometimes bluish-green, live amongst the roots or at the stem bases of host plants (*see p.55*). Root aphids often secrete a white, powdery or fluffy wax from their bodies and this gives infested roots and nearby soil particles a white coating.
CAUSE Several different species of root aphids occur in gardens. These include *Smynthurodes betae* on runner and French beans, *Pemphigus bursarius* on lettuce (*see* LETTUCE ROOT APHID *p.146*), *Thecabius auriculae* on auriculas, *Trama troglodytes* on Jerusalem artichoke, *Maculolachnus submacula* on roses (*see* ROSE ROOT APHID *p.173*), *Aphis sambuci* on pinks, *Dysaphis crataegi* on carrots and parsnips.
CONTROL Root aphids are more difficult to control than those that feed on the foliage because they are hidden by the soil. With edible plants, crop rotation helps avoid infestation by aphids that have overwintered in the soil on the remains of the previous year's crops.

ROOT KNOT EELWORMS

SYMPTOMS Many plants can be attacked by root knot eelworms. These microscopic animals live within the roots and cause knobbly swellings (not to be confused with the nitrogen-fixing nodules normally found on the roots of legumes). They disrupt the uptake of water and nutrients, so that plants lack vigour and have poor leaf colour. In northern Europe this pest is mainly encountered in greenhouses, but can attack garden plants on light sandy soils.
CAUSE Several *Meloidogyne* species may infest a wide range of plants.
CONTROL There are no effective chemical controls available to gardeners. Infested plants should be destroyed, together with the soil around the roots.

ROOT MEALYBUGS

SYMPTOMS Root mealybugs are mainly a problem on pot plants, particularly cacti, succulents, pelargonium, ferns, African violet and fuchsia, causing them to lack vigour. The mealybugs are up to 2mm long and they cover themselves and the roots with a white, waxy powder (see *p.55*). They have flatter, more elongate bodies than root aphids (see *p.171*), which have a globular body shape.

CAUSE Several *Rhizoecus* species occur in greenhouses and on the roots of house plants. They tend to be more troublesome on plants which are normally grown in a dryish potting compost.

CONTROL Root mealybugs are difficult to eliminate. Drenching the potting compost with thiacloprid, acetamiprid, or thiamethoxam can reduce infestations. With many plants, however, it is better to discard the old infested plants.

ROSE APHIDS

SYMPTOMS The younger leaves, shoot tips and flower buds are often covered in dense colonies of green, yellowish-green or pink insects up to 2–3mm long (see *p.43*). Heavy infestations cause stunted growth and poor quality blooms, while the foliage becomes sticky with a sugary honeydew excreted by the aphids. Sooty mould may develop on the honeydew, which also traps the white cast skins shed by the aphids as they grow.

CAUSE Several species of aphid occur on roses; the most troublesome is usually *Macrosiphum rosae*.

CONTROL Look for aphids from mid-spring onwards, and spray with bifenthrin, acetamiprid, or thiacloprid before heavy infestations have a chance to develop. Organic sprays, such as pyrethrum, fatty acids, and plant oils, are also effective but may need to be used more frequently.

• see also ROSE ROOT APHID, *p.173*.

ROSE BALLING

SYMPTOMS Flower buds fail to open or start to do so but never open fully (see *p.44*), generally in damp weather. The outer petals become pale brown, dry and papery. The inner petals are perfectly normal but hidden beneath the outer casing of dead ones. If damp weather persists, the bud rots off, usually becoming covered in grey mould growth. The rose's general growth is not affected; healthy buds and flowers may appear among those showing balling.

CAUSE Most commonly rain followed by bright or hot sun on the petals. The petals become slightly scorched and form an outer casing which prevents undamaged petals from appearing.

CONTROL Little can be done to prevent this, but do not water roses on hot, bright days. Water instead during the evening and directly at the base of the plant. Prune out blooms showing the symptoms before they have a chance to develop grey mould, as this secondary infection may then result in dieback.

ROSE BLACK SPOT

SYMPTOMS Diffuse purple-black spots or blotches develop on the leaves (see *p.13*), shortly followed by yellowing, and then premature leaf drop. The spots do not have sharply defined edges and can be seen to be made up of tightly packed strands of fungal growth growing through the surface of the leaf. The spots enlarge and may sometimes join together. If black spot occurs several years in

succession, particularly if infection occurs early in the summer, the rose may be severely weakened. Much smaller purple-black spots (usually only a millimetre or two across) may also develop on the stems, particularly on species roses.

CAUSE The fungus *Diplocarpon rosae*, which overwinters on stem lesions, and also on the bud scales and on fallen leaves.

CONTROL Dispose of infected leaves promptly. During spring, prune out stems showing the lesions associated with black spot. Spray with a suitable fungicide either immediately after spring pruning, or just before the leaves start to break, whichever is appropriate to the pruning regime. Repeat applications of fungicide according to the instructions. It is usually necessary to make a number of applications in order to achieve reasonable control. Fungicides could include myclobutanil, penconazole, or mancozeb. Winter treatments of soil tend not to be very effective, because of the way in which the leaves are curled and packed on the ground. Affected leaves should, however, be raked up and disposed of as soon as possible. Check current catalogues and choose roses that are described as showing good disease resistance.

ROSE CANKER AND DIEBACK

SYMPTOMS Stems discolour and die back or fail to produce growth in the spring. They may have a purple or blackish patch of discoloration which becomes uniform if the stem is allowed to die off totally (see *p.40*). In extreme cases,

the rose may be killed. During damp weather fuzzy grey fungal growth may develop from areas damaged by *Botrytis* (grey mould) and minute raised, fungal bodies may develop on the areas killed by the other fungi.

CAUSE Various fungi, in particular *Leptosphaeria Coniothyrium* and *Botrytis cinerea*. These gain entry to the stems, usually at the base, via wounds or other points of injury. These diseases are all encouraged by poor growing conditions such as over-wet soil, too deep planting, and excessively high mulching (when the mulch is placed too close to the stems and at some height up them).

CONTROL Prune out affected shoots completely, cutting right back into healthy tissue. Clear mulch or excess soil away from the base of the plant. Provide good growing conditions – water, feeding and so forth – to ensure good, sturdy replacement growth.

ROSE GREY MOULD

See GREY MOULD *p.135*; ROSE CANKER (*above*).

ROSE LEAF-ROLLING SAWFLY

SYMPTOMS During late spring and early summer, rose leaflets become tightly rolled in a downwards direction (see *p.27*). The rolled leaflets often hang down compared with unaffected parts of the leaf. In the following weeks the leaflets are eaten by pale green caterpillars, up to 10mm long, which are hidden within them.

CAUSE The leaf curling is

brought about by chemicals injected into the leaves by female sawflies, *Blennocampa phyllocolpa*, as they insert their eggs. There is one generation a year, with egg laying occurring in early summer. **CONTROL** It can be difficult to prevent the damage occurring, especially when hot weather coincides with the sawfly's flight period. This increases their activity and the number of eggs laid. Light infestations can be dealt with by picking off rolled leaflets. If many leaves are affected, it is better to leave them on the plant and spray with thiacloprid.

ROSE LEAFHOPPER

SYMPTOMS Rose leaves develop a coarse, whitish mottling on their upper surface (*see* **p.21**) during the summer. Roses growing against a wall or in other sheltered positions are particularly susceptible and may be heavily infested; by late summer much of the green colour will have gone from their leaves. Pale yellow insects, up to 3mm long, live on the underside of the leaves and readily jump off when disturbed. They have narrow bodies which are broadest at the head end and taper to the rear. The immature nymphs are creamy-white and crawl about on the lower leaf surface. **CAUSE** Sap-feeding insects, *Edwardsiana rosae*, which have two or three generations during the summer. **CONTROL** The insecticides used against rose aphids (*see p.172*) also control leafhoppers.

ROSE POWDERY MILDEW

SYMPTOMS Powdery, white fungal growth develops on the upper leaf surface (*see* **p.20**). Leaves may be attacked when very young and extensive distortion then results. On young foliage the fungal growth frequently develops on both leaf surfaces. The fungal growth is superficial and can be rubbed off but the leaf tissue beneath is discoloured. Powdery mildew growth may also develop on the stems, flower buds and thorns. When flower buds are severely infected, the buds may fail to open fully. Affected leaves fall early. **CAUSE** The fungus *Sphaerotheca pannosa*, which overwinters as stem infections and also in dormant buds. The spores are air-borne. Powdery mildew is encouraged by dry conditions around the roots and by moist air around the foliage. Roses growing up against walls are often particularly badly damaged because of the drying effect of the bricks and also the rain-shadow effect. **CONTROL** Prune out badly infected stems and spray with a suitable fungicide such as myclobutanil, mancozeb, penconazole, or sulphur. Keep roses adequately watered, particularly during hot, dry weather, but avoid wetting the foliage. Apply a good bulky mulch to encourage soil moisture retention. Check current catalogues and choose roses described as showing good disease resistance.

ROSE ROOT APHID

SYMPTOMS This dark brown aphid, which measures 3–4mm in length, feeds by sucking sap from rose roots. It would probably go unnoticed if it did not climb up the stems to deposit eggs there in the autumn. Recently-laid eggs are brown, but they soon become shiny black (*see* **p.35**). The eggs are about 1mm long, shaped like rugby balls and may be thickly clustered on the lower part of the stems. They will hatch in early spring and soon afterwards the young aphids go down into the soil. **CAUSE** A sap-feeding aphid, *Maculolachnus submacula*. The root-feeding activities of this aphid seem to have little impact on roses and infestations can vary considerably in numbers from year to year. **CONTROL** Many of the eggs can be removed when roses are being pruned. Where this is not possible, the aphids can be sprayed like other rose aphids (*see p.172*) when they have hatched.

ROSE RUST

SYMPTOMS Bright orange spots develop on the upper leaf surface, with corresponding bright orange spore masses beneath (*see* **p.16**). Severely infected leaves may fall prematurely. The leaf infections develop in early summer and are the most common and most obvious. Later in the summer, or in very early autumn, dark brown spore masses develop beneath the leaf as the winter spores are produced to replace the summer ones. Defoliation at this stage is almost inevitable. Occasionally, spring infections may develop, but these are usually only seen on species roses as a rupturing of the stem and the production of large quantities of bright orange spores which burst out from within the stem. The stem above may then wither as the crack allows secondary organisms to invade. **CAUSE** The fungus *Phragmidium*, in particular *P. tuberculatum* and, less frequently, *P. mucronatum*. Moist air is essential for the development and spread of the infection, as the spores are carried on air currents. The fungus may persist within the stems of plants which have shown spring infections. The spores can also overwinter on the soil surface and on fallen debris, stems, fences and stakes. **CONTROL** Prune out stems showing spring infections. This should be done both promptly and thoroughly, so that no fungus is left within the adjacent tissue. Improve air circulation by pruning and by avoiding planting too closely. Spray with a suitable fungicide such as myclobutanil, penconazole, or mancozeb.

ROSE SICKNESS/ REPLANT PROBLEMS

SYMPTOMS Newly planted roses fail to thrive and may show signs of dieback. There is no obvious cultural problem – such as poor planting, very dry or very wet soil – or any signs of pest attack. The roots may appear poorly developed, compact and dark in colour and there may also be rotting of the finer roots. Roses are most often affected by this, but other plants, in particular apple, cherry, peach, pear, quince and certain plums may be affected too. When planted on a site previously occupied by the same species, plants can fail to thrive. However, if they are removed promptly and replanted on a fresh site they may recover.

CAUSE The precise cause is uncertain and may well vary from plant to plant and also with location. The main causes are soil-living nematodes (eelworms) which may transmit viruses. Soil-borne fungi such as *Thielaviopsis basicola* and *Pythium* may also play a part, as may nutritional factors, such as nutrient depletion.

CONTROL If possible, do not allow the problem to arise, as there is no reliable control other than clearing soil from the roots, then replanting on a fresh site. If necessary, change the soil in the bed or area before planting. Soil must be changed to a depth of at least 45cm, preferably deeper. If replacing individual roses, change a cube of soil measuring at least 45 x 45 x 45cm. Feeding plants with a high nitrogen fertilizer and incorporating materials high in nitrogen into the planting hole may also help minimize damage. Treating the soil with a mycorrhizal fungus may help to reduce the incidence of replant problems.

ROSE SLUGWORM

SYMPTOMS Yellowish-green, caterpillar-like larvae, up to 14mm long with pale brown heads, feed on the foliage during summer. They graze away the lower leaf surface, then the remaining damaged tissues dry up and turn whitish-brown (*see p.18*).

CAUSE Larvae of a sawfly, *Endelomyia aethiops,* which has two generations during the summer. Larvae of the second generation in mid- to late summer are often more numerous and

therefore capable of more extensive damage.

CONTROL Slugworm larvae are relatively easy to control with most insecticides, including thiacloprid, pyrethrum, or bifenthrin.

ROSE VIRUSES

SYMPTOMS Symptoms show up most clearly on leaves, usually as vein clearing, yellow flecking or mottling (*see p.21*). The markings tend to be quite mild, and not as obvious as on most other virus-infected plants, but the plant may show some distortion and stunting. Symptoms may be confused with those caused by weedkiller damage, the latter being far more common than virus symptoms.

CAUSE Various viruses, including strawberry latent ringspot and rose mosaic.

CONTROL None is currently available. Some rose viruses are spread by soil-living nematodes, but other methods may be involved as well. Infected plants showing severe symptoms should be removed and destroyed.

ROSEMARY BEETLE

SYMPTOMS Foliage at the shoot tips and flower spikes of rosemary (*Rosmarinus officinalis*), lavenders (*Lavandula* spp.) and thyme (*Thymus* spp.) are eaten during late summer to early summer. The larvae are greyish white, soft bodied grubs which measure up to 8mm long. The adult beetles are up to 9mm long and are metallic dark green with five purple stripes on each wing case. Both adults and larvae cause damage to their host plants.

CAUSE A leaf beetle, *Chrysolina americana.*

Despite its latin name, this insect originates from southern Europe but is now more widespread in northern Europe. It became established in Britain during the 1990s. It is present on its host plants as non-feeding adult beetles during the mid-summer period. It begins feeding and laying eggs in late summer to autumn, with both larvae and adults being present during the autumn–spring period.

CONTROL Damage only occurs where a heavy infestation has developed. Hand picking can reduce infestations. If necessary spray with bifenthrin, acetamiprid, thiacloprid, or pyrethrum. Only the last two can be used on herbs being used for culinary purposes. Young larvae are more susceptible to insecticides than the adult beetles.

ROSY APPLE APHID

SYMPTOMS Pinkish-grey insects, up to 2mm long, cluster on the young foliage of apples in the spring. Their feeding causes leaves at the shoot tips to become curled and yellowish. The aphids also feed on the young fruitlets and this can result in them staying small, with a pinched appearance around the eye end of the fruits (*see p.48*). The distribution of damaged fruits can be very variable, with some branches producing normal fruits, while others have stunted ones at harvest time.

CAUSE A sap-feeding insect, *Dysaphis plantaginea,* which overwinters on the tree as eggs. These hatch at bud burst and the aphids are active on the tree from then until early or mid-summer, when they migrate to their summer host plants, which

are wild flowers known as plantains.

CONTROL Spray against the newly-hatched aphids as the foliage emerges from the buds, using bifenthrin or thiacloprid. Poor control is likely once the foliage has become curled.

RUSTS

SYMPTOMS The symptoms may develop on the foliage and stems, depending both on the host plant and the rust involved. Spores, either as spore masses or pustules, are usually bright orange or dark brown, and their colour may vary with the time of year. Rusts may produce several distinct spore stages, commonly known as spring, summer, and winter spores. Those produced early in the year are generally orange or bright yellow and those later in the year usually brown. Occasionally the rust produces gelatinous masses containing the spores, or "cluster cups". Infected areas usually discolour and may wither and die off. Certain rusts have alternate hosts: although unrelated, both plants are needed for the fungus to complete its life cycle. On others, the life cycle needs only one host.

CAUSE Various fungi, the most common being species of *Puccinia, Uromyces, Phragmidium, Melampsora* and *Gymnosporangium.* The spores need a moist environment to germinate and infect, and so rust infections are generally at their most severe in damp conditions.

CONTROL Remove infected leaves. Spray plants with a suitable fungicide, such as penconazole, mancozeb, or myclobutanil (check the labels for the suitability of each chemical for the host plant that you have in mind). Improve air circulation. Where available, grow resistant cultivars.

• see also Bean Rust *p.105*; Bluebell Rust *p.109*; Fuchsia Rust *p.130*; Hollyhock Rust *p.137*; Iris Rust *p.141*; Leek Rust *p.145*; Periwinkle Rust *p.160*; Plum Rust *p.163*; Rhododendron Rust *p.171*; Rose Rust *(p.173)*.

Salt Damage

Symptoms Leaf browning and curling, often associated with the leaf veins. Leaves may be undersized and fall prematurely. Dieback may also occur. Most tree species could be affected, but symptoms are particularly common on sycamore, maple, spruce, alder, lime and plane.

Cause Sodium chloride, usually as a result of a coastal situation, excessive use of salt on roads in winter, or to the presence of a salt dump for road salting being placed too close to trees. Occasionally, stormy weather may produce salt-laden winds, which cause salt damage a considerable distance inland, as well as to trees growing by the sea.

Control Avoid contamination and where feasible, alert the local council to potential problems.

Sawflies

Sawflies have caterpillar-like larvae, but they belong to the insect order called the Hymenoptera. They are therefore more closely related to ants, bees and wasps than to butterflies or moths, which belong to the Lepidoptera order. Sawfly larvae can be distinguished from moth and butterfly larvae by counting the clasping prolegs on the abdomen. Both types of larvae have three pairs of jointed legs on the thorax, but sawfly larvae have at least seven pairs of prolegs, while Lepidoptera caterpillars have five or less. Sawfly larvae are mainly leaf-feeders and many of the pest species are capable of causing extensive defoliation. Other feeding habits include:

boring into developing fruits – see Apple Sawfly *p.101*; Plum Sawfly *p.163*;

causing galls or leaf distortion – see Rose Leaf-rolling Sawfly *p.172*; Willow Bean Gall Sawflies *p.192*;

leaf-mining – see Geum Sawflies *p.132*.

• see also Aquilegia Sawfly *p.102*; Aruncus Sawfly *p.102*; Berberis Sawfly *p.107*; Geranium Sawfly *p.132*; Gooseberry Sawfly *p.135*; Pear and Cherry Slugworm *p.158*; Rose Slugworm *p.174*; Solomon's Seal Sawfly *p.179*.

Scale Insects

Scale insects are sap-feeding pests that attack the foliage and stems of many plants in greenhouses and gardens. The insects are hidden underneath waxy shells or scales that they secrete over their bodies. Newly-hatched scales crawl around looking for suitable places to feed, but they soon settle down and remain immobile for most of their lives. The colour of the protective scale is usually brown or greyish white and it may be flat or domed. The size of the scale varies according to the species and is in the range of 1–6mm. Some scale insects excrete a sugary liquid called honeydew, which makes the foliage sticky and permits the growth of sooty moulds. Most scales lay their eggs under the protection of their own bodies, but the cushion scales deposit eggs among white, waxy fibres secreted by the female scales. The stage in the life cycle most vulnerable to insecticides, such as thiacloprid, fatty acids, or plant oils, is that of the newly-hatched nymphs. With most outdoor species of scales, nymphs are present in mid-summer, but on indoor plants breeding may be continuous throughout the year.

• see also Beech Bark Scale *p.106*; Brown Scale *p.112*; Cushion Scale *p.122*; Euonymus Scale *p.127*; Fluted Scale *p.129*; Hemispherical Scale *p.137*; Horse Chestnut Scale *p.139*; Hydrangea Scale *p.140*; Juniper Scale *p.142*; Mussel Scale *p.151*; Oleander Scale *p.154*; Soft Scale *p.179*; Viburnum Scale *p.187*; Wisteria Scale *p.193*; Woolly Vine Scale *p.194*.

Sciarid Flies

See Fungus Gnats *p.131*.

Sclerotinia

Symptoms Affected tissues become discoloured, brown and wet as they rot. Large quantities of fluffy, cotton wool-like fungal growth develop and are scattered with large, black sclerotia (fungal resting bodies) which may be several millimetres long. The fungal growth and sclerotia may also develop within the stems. This disease may affect most above ground plant parts, but fruit and stem base infections are most common (*see p.36*). In addition, it may develop on corms and tubers kept in store.

Cause The fungus *Sclerotinia sclerotiorum*, which is encouraged by cool, damp conditions. The sclerotia fall into the soil where they remain dormant until spring. They then produce spore-producing, cup-shaped fungal growths (called apothecia) which are responsible for the following year's infections.

Control Remove and burn affected plant parts promptly. Dispose of all debris at the end of the season. Do not grow susceptible plants on that site for at least four years following an outbreak of the disease.

Scorch

Symptoms Pale, scorched-looking patches develop which are usually pale brown or bleached. Damaged areas usually dry out, becoming crispy and brown. Leaves or petals are most commonly affected but occasionally other above-ground plant parts may be damaged, too. Stems or trunks of trees and shrubs may show scorching only on the exposed side. Other growth remains normal.

Cause Hot or bright sun is the most common cause. The problem is exacerbated by droplets of moisture on the leaf surface – the droplets help to further magnify the sun's rays and increase the damage. Soft- or hairy-leaved plants are often most severely affected. Similar damage may occur with chemical contamination;

either an unsuitable pesticide is used, or as a result of injury by a contact weed-killer. Less frequently, damage is due to the presence of moisture of any sort on the foliage, particularly if the leaves remain wet overnight as temperatures drop. On bark, the injury is usually the result of the scorching effect of bright sun on thin or young bark.

CONTROL Take care when watering, and water in early evening when leaves are not exposed to direct sunlight, yet have time to dry off before nightfall. Ensure that chemicals are selected properly and used with great care. In greenhouses and conservatories, provide adequate shading during the summer months.

• see also HORSE CHESTNUT MARGINAL LEAF SCORCH *p.139;* IRIS SCORCH *p.140;* NARCISSUS LEAF SCORCH *p.152.*

SERIDIUM CANKER

SYMPTOMS Foliage of cypresses (*Cupressus*), junipers (*Juniperus*), Leyland cypresses (× *Cupressocyparis leylandii*) and thuyas (*Thuja*) loses colour, becoming slightly yellowed and losing its lustre. It then turns brown and dies (*see p.35*). This deterioration is due to the development of a canker on the affected stem. The bark becomes slightly roughened and resin is produced from around the cankers. Close inspection reveals black, raised, pinprick-sized fungal bodies around the cankered area. These may be rounded or may have burst, leaving minute craters. The infection is often first noticed as many small areas of dying foliage, but it may attack larger branches and cause

extensive dieback. The tree may be rendered very unsightly and can even be killed.

CAUSE The fungus *Seiridium cardinale* (syn. *Coryneum cardinale*), which enters branches through twig crotches or fine cracks in the bark, and attacks most readily when the tree is growing slowly. The spores are produced from within the fungal bodies on the cankers and are carried on the wind. The fungus kills by girdling the affected stems as the canker enlarges.

CONTROL Prune out infected areas promptly, as this may prevent further damage occurring. There is no chemical control available. Check other susceptible trees in the area, such as *Cupressus macrocarpa, C. sempervirens* var. *horizontalis, C. glabra* and *Thuja plicata,* and remove any infected areas.

SEMPERVIVUM LEAF MINER

SYMPTOMS The outer leaves on the rosettes of house leeks (*Sempervivum* species) become discoloured and collapse due to the inner tissues being eaten by the larvae of a hoverfly. This pest has two generations in early and late summer, when it makes the plants look as though they are rotting off.

CAUSE The larvae of a leaf-mining hoverfly, *Cheilosia caerulescens,* which has spread across Europe in recent years and is now established in Britain. The adult flies are 8–10mm long and have black bodies with short white hairs. They lack the yellow and black stripes on the abdomen found on many of the hoverflies that have aphid predator larvae.

When fully grown, the legless creamy white maggots are 8–10mm long. Each larva will feed on several leaves before it is ready to pupate in the soil.

CONTROL On small plantings, it is feasible to search for mined leaves and remove the grubs before they cause too much damage. Compost drench pesticides used for vine weevil control, such as thiacloprid and acetamiprid, could be applied to container-grown house leeks in early and late summer.

SHANKING OF GRAPES

SYMPTOMS Individual berries within the bunch fail to colour up normally – those of black cultivars turn red, those of white cultivars remain translucent. They then become wrinkled and, if left on the bunch, may start to resemble raisins. They have a watery or unpleasantly sour taste.

CAUSE A disorder which may result from many different cultural problems. Under- or over-watering are the most common causes; stagnant soil conditions and under-feeding may also be factors.

CONTROL Cut out the affected berries and spray the foliage regularly with a foliar feed to boost the vine's vigour. Ensure adequate but not excessive watering and feeding. Check that drainage is not impeded.

SHARKA

See PLUM POX *p.163.*

SHIELDBUGS

When viewed from above, these insects have a shield-like shape and a broad, flattened beetle-like

appearance, with sucking mouthparts. Some prey on other insects, but the majority feed on plant sap, although their impact on cultivated plants is negligible. Several species occur in gardens; the more common types include green shieldbug (*Palomena prasina*), hawthorn shieldbug (*Acanthosoma haemorrhoidale*), birch shieldbug (*Elasmostethus interstinctus*) and dock shieldbug (*Coreus marginatus*). They measure up to 12mm long and are green, brown or a combination of yellowish-green and reddish-brown. They are most often seen between late summer and early autumn, when they tend to sun themselves on plant leaves before seeking sheltered places in which to overwinter. Eggs are laid between mid-spring and early summer after the adult insects have become active again. The immature nymphs have a more rounded body shape, but increasingly come to resemble the adults as they mature.

CONTROL Not required for the above.

• see also SOUTHERN GREEN SHIELDBUG *p.179*

SHOTHOLE

SYMPTOMS Holes develop on the leaves (*see p.30*), usually with an inconspicuous brown ring around the edge of each hole. Brown spots of dead tissue (*see p.18*) are present on the leaf.

CAUSE Various fungal and bacterial infections, including bacterial canker and occasionally powdery mildews, may be responsible. The pathogens kill off areas of leaf,

producing leaf spots. The damaged areas then fall away, leaving holes.
CONTROL Determine the primary cause of the leaf damage and take the appropriate action.

SHOT-HOLE BORERS

SYMPTOMS Branches and sometimes the trunk of fruiting and ornamental forms of plums, cherry, almond, and less frequently, apple and pear, have many round holes, 1–3mm in diameter, in the bark. Underneath the bark is a maze of tunnels (*see p.33*) where white, legless grubs have been feeding. Some species also bore into the heart wood. Sawdust may be seen coming from the shot-holes in the bark, where the dark brown, cylindrical adult beetles, 2–4mm long, have emerged in early summer.
CAUSE Several species are collectively known as shot-hole borers on fruit trees. These include *Scolytus rugulosus*, *S. mali* and *Xyleborus dispar*. All are likely to attack branches or trees that are already in poor condition, and they are not a problem on vigorous trees.
CONTROL Chemical control gives poor results since much of the life cycle is protected under the bark. Prune out and burn infested branches and improve the tree's health by feeding in the spring and keeping it watered in dry summers. Unsuitable growing conditions, such as poorly drained soil or a weak root system arising from bad planting, will make trees prone to dieback and shot-hole borers.

SILVER FIR ADELGID

SYMPTOMS The smaller stems on silver firs such as *Abies procera*, *A. grandis* and *A. balsamea* develop lumpy growths where small black insects covered with white, waxy fibres are feeding.
CAUSE A sap-feeding insect, *Adelges piceae*, which is active between late spring and autumn.
CONTROL Little can be done on large trees, but those small enough to be sprayed can be treated with bifenthrin or pyrethrum in the spring.

SILVER LEAF

SYMPTOMS Single or small numbers of branches develop foliage which has a silver sheen. On some hosts, such as rhododendron, the silvering is almost imperceptible, or non-existent. If affected stems more than 2.5cm in diameter are cut across, they have a central brownish stain. Other limbs appear perfectly normal, but may show symptoms at a later date. Affected limbs die off or fail to leaf up the following spring. The symptoms may be confused with those caused by other problems, such as adverse weather conditions, drought, malnutrition or insect attack. Fruiting bodies, often grouped together and adhering fairly closely to the bark, may develop on dead wood. Their exposed surface is a dark purple-grey colour.
CAUSE The fungus *Chondrostereum purpureum*, a fresh-wound parasite that invades freshly created wounds on deciduous trees. Pruning cuts or pest attack are common points of entry. The spores are produced from the fruiting bodies found on infected limbs of living trees, or on fallen timber. The infection is spread on air currents or on pruning tools. The fungus produces a toxin which causes the upper leaf surface to separate from the main leaf blade. The air which accumulates between the two layers alters the light-reflecting qualities of the leaf, causing it to appear silvered. The leaves do not carry the infection.
CONTROL There is no control available, but occasionally mildly affected trees may recover. Generally, the infection spreads throughout the tree unless infected branches are removed to a point at least 15cm past the point at which the staining in the wood ends. Infection is least likely to occur during the summer months, so all pruning of susceptible trees should be carried out at this time. Although no longer recommended in most cases, trees which are particularly prone to silver leaf infection should have all pruning wounds painted with a wound treatment. This should be carried out at the same time as the pruning. Locate any possible sources of infection and dispose of them. Plums and cherries are the trees showing greatest susceptibility to this disease, 'Victoria' plum being the most susceptible cultivar. The rootstock may also influence the likelihood of infection occurring: some, such as 'Brompton', are very prone, whereas others, such as 'Pixie', show a good degree of resistance.

SLIME FLUX AND WETWOOD

SYMPTOMS Watery substances, sometimes slightly viscous and orange, pink or white in colour, ooze from the stems of affected plants. The slime flux usually first appears a few inches above soil level and may have an unpleasant odour. The bark beneath may be killed and the upper parts of the stem wilt and die. Not all stems on the plant are affected at the same time.
CAUSE Injury to the stem early in the season when the sap pressure is high causes sap to leak out. The high sugar content of the sap means that it is rapidly colonised by various yeasts, fungi and bacteria which are responsible for the thickening and discoloration.
CONTROL Cut out affected stems, pruning back to perfectly sound wood, even if this means going below soil level. Attempt to prevent stem injury.

SLIME MOULD

SYMPTOMS Grey, off-white, yellow or orange growth develops, most commonly on stems, or on other plant parts. It is usually in the form of numerous tiny spheres which may appear to dissolve when wet, as the spores are released. The plant's growth is not affected, however, and if the slime mould is washed or scraped off, the plant beneath is generally unharmed. Slime moulds are particularly common on grass (*see p.58*).
CAUSE Various species of slime mould – non-parasitic organisms, believed to be most closely related to fungi.
CONTROL No chemical control is available, but none is justified, as the slime moulds are unlikely to cause much damage. Wash off with a strong jet of water.

SLUGS

Several species of slugs occur in gardens. These include the grey field slug (*Deroceras reticulatus*), the large black slug (*Arion ater*), garden slug (*A. hortensis*) and keeled slugs (*Milax* species). They can damage a wide range of plants and are present throughout the year, continuing to feed in the winter if temperatures are above 5°C (40°F). Irregular holes are eaten in foliage (*see **p.29***), flowers (*see **p.43***) and stems (*see **p.36***), but examination of plants by torchlight on a mild evening may be necessary to find slugs feeding. Slugs secrete a slimy mucilage from their bodies, which may leave a distinctive silvery deposit where they have been feeding. Some slugs live mainly underground and damage potato tubers (*see **p.57***) and tulip bulbs.

CONTROL Slugs can never be eliminated from gardens, so control measures should be concentrated on protecting vulnerable plants, such as seedlings and soft new growth on herbaceous plants. Slug pellets containing metaldehyde or ferric phosphate can be scattered thinly amongst the plants. Poor results will be achieved during cold or dry weather, when slugs are relatively inactive. Metaldehyde causes slugs to produce an excessive amount of slime and die of dehydration, but they may recover under wet conditions. Slug pellets can kill cats and dogs, so take care in their use. Less toxic to pets are slug killers based on aluminium sulphate. This is mainly active against young slugs and is best applied in mild, damp weather in the spring and autumn. Non-chemical controls include hand-removal after dark on mild evenings, and use of beer traps. Jam jars or similar containers half filled with beer are sunk in the ground almost to ground level. Slugs, attracted by the odour, fall in and drown. Biological control in the form of a nematode, *Phasmarhabditis hermaphrodita*, can be used. It needs to be applied to soils which are moist, well-drained and above 5°C (40°F); the best times for use are spring and autumn. The microscopic nematodes penetrate the bodies of slugs that have gone into the soil to seek shelter during the day, and release bacteria which cause a fatal disease in infected slugs. Under suitable conditions it gives a significant reduction in slug numbers for at least six weeks. This is a useful method of tackling underground slugs which damage bulbs and potato tubers, since these slugs come to the soil surface infrequently and are poorly controlled by slug pellets. Damage to potato tubers can be reduced by lifting the crop as soon as the tubers have matured. There is some variation in the susceptibility of potato cultivars to slugs. The cultivars 'Maris Piper', 'Cara', 'Rocket', 'Kondor' and 'Maris Bard' are all prone to be heavily attacked, while 'Pentland Ivory', 'Pentland Dell', 'Wilja', 'Stemster', 'Charlotte', and 'Estima' are less susceptible. Barriers that slugs and snails are reluctant to cross are available. These generally consist of moisture absorbent minerals for surrounding plants, and copper tapes for putting around plant pots.

SMALL BULB FLY

See NARCISSUS BULB FLY *p.150*.

SMALL ERMINE MOTHS

SYMPTOMS The stems and foliage of various plants, including hawthorn, blackthorn, bird cherry, willow, apple, *Euonymus* and *Sedum*, become covered with dense, whitish, silk webbing in summer (*see **p.38***). Whitish-yellow caterpillars with a row of black dots down each side of their bodies, up to 20mm long, feed on leaves. Infestations can be heavy enough to cause severe defoliation. The adult moths are silvery-white with many black spots on their wings; they emerge from pupae in the webbing in late summer.

CAUSE Caterpillars of several species of small ermine moths, *Yponomeuta* species. Most have one generation a year but *Y. vigintipunctata*, which attacks *Sedum telephium* and its hybrids, has two.

CONTROL Light infestations can be dealt with by pruning out the webbed shoots. Otherwise spray forcefully to penetrate the webbing with bifenthrin or pyrethrum.

SMUTS

SYMPTOMS Affected areas of the plant are discoloured and may be distorted (*see **p.35***). In a few cases, such as dahlia smut (*see p.120*), slight leaf spotting is all that is seen. In other plants, swellings develop and erupt to reveal large masses of soot-black spores (*see **p.44***). Stems, leaves, flowers and seeds may be affected. Some smut infections (for example, anther smuts) are systemic within the plant, but most are restricted only to discrete areas of the plant.

CAUSE Various species of the fungi *Entyloma*, *Ustilago* and *Urocystis*, which are spread by rain or water splash from nearby infected plant matter, or from the soil surface.

CONTROL Remove affected areas from plants as soon as seen, preferably before the spores are released. Avoid splashing water around plants and clear up all debris at the end of the season. Grow new plants on a fresh site.

• *see also* DAHLIA SMUT *p.123*; DIANTHUS ANTHER SMUT *p.124*; WINTER ACONITE SMUT *p.192*.

SNAILS

The most common snail pest is the garden snail, *Helix aspersa*. Banded snails, *Cepaea* species, are less troublesome. Many plants are eaten by snails between spring and autumn. Irregular holes are rasped in the foliage (*see **p.29***), and the stems and flowers (*see **p.43***), particularly of annuals and herbaceous plants. Snails are mostly active after dark or in wet weather. They secrete a slimy mucilage from their bodies and this often dries to leave a silvery deposit. Snails are less numerous in areas with acid soils, which lack the calcium salts necessary to form a snail's shell.

CONTROL Snails can be controlled by the methods used against slugs (*see above*). Biological control is less effective as snails live mainly above soil level and are less likely to be infected by the bacterium-carrying nematodes.

SNOWDROP GREY MOULD

SYMPTOMS Fuzzy, grey fungal growth, typical of a grey mould type of infection, develops around the foliage and/or flower stems of infected snowdrops. The plant starts to rot and as the infection spreads back, the bulb may be killed. Infection often first appears as the foliage comes above ground. Small black sclerotia (fungal resting bodies) develop around the bulb and foliage. The sclerotia drop off and overwinter in the soil.
CAUSE The fungus *Botrytis galanthina*, which is encouraged by mild winters and wet spring weather. The spores are spread by rain splash and on air currents.
CONTROL Remove infected plants and the soil in their immediate vicinity. Do not grow snowdrops on that site for at least five years.

SOFT ROT

See BACTERIAL SOFT ROT *p.104*.

SOFT SCALE

SYMPTOMS Yellowish-brown, flat, oval scales, measuring up to 3 or 4mm long, occur on the underside of leaves and are often clustered along the larger leaf veins. Bay, citrus, ivy, ferns, *Ficus* species and *Schefflera* are frequently attacked, but this pest can infest a wide range of indoor plants. They excrete a sugary substance called honeydew that makes the upper leaf surface sticky and allows the growth of sooty mould.
CAUSE A sap-feeding insect, *Coccus hesperidum*, which breeds continuously throughout the year on house plants and in heated greenhouses.

CONTROL Spray the underside of the leaves of non-edible plants with acetamiprid or thiamethoxam. Culinary herbs, like bay, and ornamental plants can be sprayed with thiacloprid, fatty acids, or plant oils. Heavily infested plants may need three applications of the last two at two-week intervals in order to control nymphs as they hatch from eggs that are concealed beneath the adult scales' bodies. Alternatively, in greenhouses in the summer, a parasitic wasp, *Metaphycus helvolus*, can be introduced for biological control.

SOLOMON'S SEAL SAWFLY

SYMPTOMS Greyish-white caterpillar-like larvae with black heads and up to 20mm long eat the foliage of Solomon's seal in early to mid-summer, often reducing plants to bare stalks (*see* **p.31**). The adult insects have black bodies and two pairs of smoky-grey wings. They emerge when the plant is coming into flower, and females lay rows of eggs in the stems, causing purplish scars up to 25mm in length.
CAUSE Larvae of a sawfly, *Phymatocera aterrima*.
CONTROL In spite of severe leaf loss, affected plants normally make adequate growth and flower the following year, so control may not be necessary. The larvae can be removed by hand or sprayed with thiacloprid, pyrethrum, or bifenthrin when damage begins in early summer.

SOOTY MOULD

Sooty mould is a black or sometimes greyish-green non-parasitic fungus that

grows on foliage and other surfaces that have an accumulation of honeydew on them. Various sap-feeding insects, including aphids, whiteflies, mealybugs and some scale insects and suckers, excrete honeydew. This falls down onto the upper surface of leaves below where the pests are feeding. Sooty mould can therefore be found on plants that are themselves free of pests, but which are growing beneath trees that are infested with aphids. Heavy coatings of sooty mould spoil a plant's appearance and reduce the amount of light and air reaching the foliage. There is no direct control for sooty mould, but dealing with the pest that is causing the problem will allow new growth to remain clean.

SOUTHERN GREEN SHIELDBUG

SYMPTOMS Pale green shield-shaped insects about 10–12mm or their immature nymphs are found clustered on the seed heads of ornamental plants, runner beans, tomatoes and raspberries, mainly in late summer to early autumn. The nymphs have a more rounded body shape than the adult and they are black or green with many rounded spots, which may be white, pink or yellow, on the upper surface. The spotty nature of the nymphs can result in them being misidentified as ladybirds. Their feeding can affect the development of bean pods and fruits on tomato and raspberry.
CAUSE A sap-feeding plant bug, *Nezara viridula*, that is native to southern Europe but has been established in south east England

since 2003. There is a native green shieldbug, *Palomena prasina*, that is more widespread in Britain and which does not cause damage to garden plants. The nymphs of the latter are green with black markings but no rounded spots. Adult native green shieldbugs are pale green with a distinctive blackish brown diamond-shaped area at the rear end of the body. Southern green shieldbugs are uniformly pale green when adult.
CONTROL Britain is probably at the northern edge of this bug's range and the climate may have to get warmer still before this becomes a significant pest. At present, the bugs do not become numerous until early autumn, by which time its host plants are coming to the end of their growing and cropping season. If control is required on ornamental plants, spray with thiacloprid, acetamiprid, or bifenthrin. The last mentioned can also be used on beans and tomato.

SPARROWS

The house sparrow, *Passer domesticus*, will peck the flowers of crocus (*see* **p.46**), primrose and polyanthus to shreds. Sparrows also eat the flower buds of many other plants, including runner bean and wisteria, although loss of flower buds on these plants is more frequently due to dry soil conditions than bird damage. Freshly sown grass seed is also liable to be eaten, and sparrows can cause an uneven soil surface by dust-bathing in seed beds. Seedling lettuce and beetroot plants can be defoliated.
CONTROL Netting is the only certain means of protecting flowers, seeds, and seedlings

from damage caused by sparrows. An alternative is to place sticks in the soil and criss-cross the area with black cotton. Sparrows have difficulty in seeing the cotton and are threfore deterred from feeding. Encourage rapid germination and growth through the vulnerable seedling stage by sowing in warm soil and keeping the young plants watered in dry weather.

SPLITTING

SYMPTOMS Fruits and occasionally stems may show cracking. This usually runs longitudinally and initially the rest of the plant appears normal. The splitting allows other potential pathogens to enter, and so dieback or rotting may occur later. The cracks may not necessarily become infected and may dry or heal over.
CAUSE An erratic supply of water, or great fluctuations in temperature are the most common causes. Erratic availability of nutrients may be involved.
CONTROL Concentrate on improving growing conditions for plants by regular watering and feeding, and use a mulch to help prevent soil drying out completely. Keep a watch on damaged areas, and in the case of affected fruits, remove them so they are not infected by secondary organisms.

SPRINGTAILS

SYMPTOMS Active white insects, up to 2mm long, live in the compost of pot plants (*see p.54*), especially those growing in peat-based types. They are most frequently seen when plants are being watered, as this washes them out of the bottom of the pot or flushes them up onto the soil surface.
CAUSE There are many species of springtails, but those associated with pot plants are generally *Onychiurus* species. These are wingless insects with a prominent pair of antennae on their heads.
CONTROL Springtails feed on decaying plant material and fungi. No control measures are required, as they will not harm the roots of established plants.

SQUIRRELS

The grey squirrel, *Sciurus carolinensis*, can be delightful to watch as it scampers around a garden, but squirrels can be very destructive. They dig up and eat crocus corms and tulip bulbs, eat flower buds and shoot tips, take nuts (*see p.52*), strawberry and tree fruits, and also strip the bark from trees such as sycamore, beech, and ash. If bark is lost from all or most of the circumference of the trunk, the growth above that point will dry up and die. Squirrels can also cause annoyance by taking food put out for birds.
CONTROL Shooting, traps and poisoned baits are used in forests against squirrels; however, these methods are impractical and undesirable in gardens. They are also unlikely to be effective unless used over a wider area than an individual garden. Squirrels are very mobile animals and others will soon move in to occupy the territory of those which have been removed. Netting can give protection to fruits and flowering shrubs such as magnolia and camellia during periods when squirrels are showing an interest in them. Wire netting is better used for permanent structures such as fruit cages, as squirrels can easily bite through plastic. Netting can also be placed over areas where bulbs and corms have been planted, to deter squirrels from digging them up. Squirrel-proof bird feeders are available from most garden centres.

STACHYS CASE-BEARING CATERPILLAR

SYMPTOMS In early summer, the foliage of *Stachys* species develops a number of brown blotches where the internal tissues have been eaten out. The small caterpillars responsible for this damage spin silk cases around their bodies. Hairs and other fragments of the host plant's leaves are incorporated into these cases. The caterpillars are almost entirely hidden but, when feeding, the larvae will partly emerge in order to bore into the leaf. The feeding area is limited to how far the caterpillar can reach inside the leaf without leaving its protective case.
CAUSE Caterpillars of a tiny moth, *Coleophora lineola*, which feed on various types of *Stachys* and related plants. It is most frequently seen on lambs' ears, *Stachys byzantina* (formerly *S. lanata*), where the caterpillars incorporate leaf hairs into the case to give them a silvery-white, furry covering measuring up to 4mm long and 2–3mm wide.
CONTROL The caterpillars can be removed by hand or controlled by spraying with bifenthrin or pyrethrum when damage is detected in late spring.

STEM AND BULB EELWORM

This microscopic nematode, *Ditylenchus dipsaci*, occurs as several biological races which have different ranges of host plants. The types that are most likely to be encountered in gardens are narcissus eelworm (*see p.151*), onion eelworm (*see p.154*) and phlox eelworm (*see p.160*). These pests cause distinctive forms of distortion on the host plants indicated, and often cause their death.

STORAGE ROTS

SYMPTOMS Fruits, roots and tubers kept in store may begin to rot. In some cases the infection may already have taken place while the plant was growing, but the storage conditions are such that rotting is encouraged. In other instances, previously perfectly healthy plants or plant parts may actually become infected whilst they are in store, most often from neighbouring diseased material. The rotting may also be accompanied by the development of fungal growth on the surface of the fruit, roots or tubers or spreading within the plant
CAUSE A very wide range of fungi and bacteria, some of which are described as secondary, enter only through injured areas; others, which are primary pathogens, cause damage to previously sound and uninjured tissue.
CONTROL Store only perfect fruits, roots and tubers. Inspect stores frequently and remove anything showing signs of deterioration, however slight. Provide storage conditions as close as possible to those advised.

STRAWBERRY BLACK EYE

SYMPTOMS Strawberry flowers develop a distinct dark brown or black centre. The petals and growth of the rest of the plant usually appear quite healthy and normal.

CAUSE Frost damage to the young flower which injures the centre. Flowers damaged in this way do not later form fruit.

CONTROL Use cloches, horticultural fleece or similar material to protect crops, particularly early cultivars, from frost. Pinch out affected flowers.

STRAWBERRY GREEN PETAL

SYMPTOMS The flowers are reduced in size and bear green petals. Plants produce few, if any, fruits and those which are formed are very small and often distorted. The plant itself is generally somewhat stunted and distorted and may show yellow markings on the leaves. Once flowering is over, the older leaves may turn red.

CAUSE A mycoplasma, known as strawberry green petal, which is transmitted by leafhoppers.

CONTROL Remove and burn infected plants promptly and do not propagate from them. Spray against leafhoppers.

• see also LEAFHOPPERS *p.144.*

STRAWBERRY GREY MOULD

SYMPTOMS Fruits develop normally, but as they start to ripen and colour, pale brown necrotic patches develop on their skin. The fruit then deteriorates rapidly, becoming soft and brown. Shortly afterwards a fuzzy, grey fungal growth develops over the surface (*see p.50*). Numerous spores are given off if the fruit is disturbed.

CAUSE The fungus *Botrytis cinerea*, which is very common on both living and dead plant material. The spores enter the open flowers but remain dormant until the fruits start to ripen. The spores are then readily spread by either rain or water splash and on air currents. Strawberries which are not infected at the flowering stage, but have formed healthy fruits which have been injured in some way (by slugs, for example), may subsequently become infected by spores from these rotting fruits. The fungus may carry over from year to year either on plant debris or as sclerotia in the soil.

CONTROL Control is very difficult, because the fungus is so widespread. Remove all dead and injured plant parts and fruits before they become infected. Remove infected fruits promptly. Do not leave plant debris lying around the garden. Avoid wetting the strawberry plant foliage and flowers when watering.

STRAWBERRY LEAF SPOT

SYMPTOMS Reddish-purple spots develop on the leaves (*see p.12*), each with a grey centre. Off-white fungal growth and, occasionally, numerous tiny, raised, black fungal fruiting bodies may develop on the spots. The spots may join together and cover much of the leaf surface (*see p.17*). They may also then spread to the flower and leaf stems. Although it spoils the appearance of the foliage, this infection usually has little effect on the vigour of the plant – it may occur in spring but more often appears fairly late in the season.

CAUSE The fungus *Mycosphaerella fragariae*, which forms sclerotia (black fungal resting bodies) which may overwinter in the soil and then cause the earlier infections in the spring.

CONTROL Remove badly infected leaves. Clear up all infected debris at the end of the season.

STRAWBERRY MITE

SYMPTOMS Strawberry plants make poor growth in the latter half of the summer, with new leaves appearing puckered and failing to expand to their full size. Similar symptoms can be caused by some virus diseases. The strawberry mites, which are whitish-brown and less than 1mm long, live in the crown of the plant and feed on the developing leaves. A x10 hand lens is necessary to see them.

CAUSE A tarsonemid mite, *Phytonemus pallidus*, which feeds by sucking sap. It tends to be more troublesome in hot, dry summers.

CONTROL None of the pesticides available to amateur gardeners will control strawberry mite. Destroy affected plants and start a new bed with certified stock strawberry plants which will be free of problems such as strawberry mite and virus diseases.

STRAWBERRY RED CORE

SYMPTOMS Groups of stunted plants appear in late spring, bearing reddish-brown, strangely stiff leaves. The roots are discoloured and have a red inner core.

CAUSE The fungus *Phytophthora fragariae*, which is encouraged by heavy, wet soils and produces resistant spore stages from the deteriorating root material, releasing these into the soil where they may persist for more than ten years. The fungus is easily introduced on contaminated plants or on soil adhering to plants, footwear, tools, etc.

CONTROL None available. Remove infected plants immediately, together with the soil in the vicinity of their roots, and burn them. Do not grow strawberries on the same site again and avoid moving soil from that area to other parts of the garden.

STRAWBERRY SEED BEETLES

SYMPTOMS Seeds are removed from the surface of ripening strawberry fruits, causing small, dried-up brown patches. The culprits are fast-moving black beetles, about 15mm long, with slender reddish-brown or black legs. They are mainly active after dark. In addition to eating the seeds, they sometimes eat into the fruits, but it is difficult to distinguish this form of damage from that caused by slugs.

CAUSE Carabid or ground beetles, *Pterostichus* species and *Harpalus rufipes*. At other times of the year these beetles will feed on weed seeds.

CONTROL Keep the strawberry bed free of weeds to discourage a build-up of seed beetles. Jam jars sunk into the soil will act as pitfall traps, but they will also capture beneficial ground beetles that are predatory in their feeding habits. Because ripe strawberries have such thin skins, chemical controls are undesirable while ripe fruit is present but, if necessary, plants with unripe fruits can be sprayed with bifenthrin at dusk.

STRAWBERRY VIRUSES

SYMPTOMS Stunting and distortion of the whole plant, usually combined with stunting, distortion and yellowing of the leaves. The discoloration is seen as distinct patterns – mosaic, ring spot, streaking or yellowing around the leaf edge. Infected plants fail to thrive and may not flower or fruit properly.
CAUSE Various viruses, most commonly strawberry yellow edge or little leaf virus, crinkle (spread by aphids), arabis mosaic, strawberry ring spot and tomato black ring (spread by eelworms).
CONTROL None currently available. Remove and burn infected plants as soon as the symptoms are noticed. Control aphids and leaf-hoppers with appropriate insecticides. Do not grow strawberries on that site again. Do not propagate from infected plants, as their runners will also be infected, even if without symptoms. Always buy certified stocks of strawberries, guaranteed to be virus-free when purchased. Certain cultivars are known to be particularly susceptible to particular viral infections and these include 'Royal Sovereign' and 'Redgauntlet' (yellow edge).
• see also APHIDS p.99; LEAFHOPPERS p.144.

STRAWBERRY YELLOWS

SYMPTOMS Leaves turn yellow. The entire plant may be affected or it may appear blotched yellow with part of the leaves remaining green. Symptoms usually appear most severe in mid-summer.
CAUSE The precise cause is not known, but it is thought to be largely the result of unsuitable soil conditions. A soil which is excessively wet or which is lacking adequate nutrients will generally encourage these symptoms.
CONTROL Determine which adverse soil conditions are involved and attempt to improve problems such as poor drainage. Apply a fertilizer containing trace elements.

SUCKERS

These are small sap-feeding insects that cause damage on various garden plants during the spring and/or summer months. They are also known as psyllids and have very distinctive immature nymphs that appear to be squashed flat when viewed from the side. The adult insects have a more rounded body shape and resemble winged aphids. They are 2–3mm long when adult. An immature nymph has wing pads on its upper surface, which become larger each time the nymph sheds its skin, until the adult stage is reached with fully formed wings. Some suckers cause gall formation on their host plants while others live exposed on the leaf surface.
• see also ALDER SUCKER p.98; APPLE SUCKER p.102; ASH SUCKER p.103; BAY SUCKER p.104; BOX SUCKER p.110; ELAEAGNUS SUCKER p.126; EUCALYPTUS SUCKER p.127; PEAR SUCKER p.159.

SWEETCORN SMUT

SYMPTOMS Individual kernels on the cob become hugely swollen and deformed (see **p.51**). Each is pale grey and ruptures to release large quantities of powdery black spores. These spores may become mixed with rain to produce a black liquid which runs down the plant. Occasionally, other parts of the plant may be affected. The infection is not systemic and healthy and infected cobs may develop on the same plant.
CAUSE The fungus *Ustilago maydis*, which is quite common when summer temperatures are high. The spores are carried on air currents and by rain and water splash, and either infect other sweetcorn plants directly or persist in the soil.
CONTROL None available. Remove infected cobs before the swellings have a chance to rupture. Remove and burn all infected plant debris at the end of the season. Do not grow sweetcorn on that site for at least five years.

SWIFT MOTHS

SYMPTOMS Long, slender, white caterpillars, up to 35mm long, with brown heads are found in the soil (see **p.53**). They feed on plant roots, especially of herbaceous perennials, and can also damage bulbs, corms and rhizomes. Heavy infestations cause plants to lack vigour; they are sometimes killed. Some species take two years to complete their larval development, so the caterpillars may be unearthed at any time of year. The adult moths emerge and lay eggs from late spring to late summer.
CAUSE There are several species of swift moth, but the most troublesome in gardens are the ghost swift moth, *Hepialus humuli*, and the garden swift moth, *H. lupulinus*.
CONTROL Good weed control can help avoid infestations, as the moths often deposit eggs where there is a dense cover of vegetation. Any caterpillars found during digging or division of perennials should be crushed and thrown to the robins. Chemical control is difficult, as the older and more damaging larvae are more tolerant of pesticides.

SYCAMORE SOOTY BARK DISEASE

SYMPTOMS Leaves wilt and die back on infected branches, but remain attached to the tree. If the bark is pared off, green or dark yellow staining may be visible (see **p.36**). Although these are the first symptoms, the problem may only become apparent at a later stage, when black spore masses or blister-like outgrowths develop on the bark of affected stems and when the tree does not come into leaf in the spring. These outgrowths both enlarge and elongate and may be associated with the extensive death and breaking up of the bark. Death of the tree usually occurs within a couple of years.
CAUSE The fungus *Cryptostroma corticale*, which appears to be encouraged by warm

temperatures and, in particular, by hot summers.
CONTROL Remove infected trees promptly in order to limit the spread of the disease.

TAR SPOT OF ACER

SYMPTOMS Large, slightly raised, shiny black blotches develop on the leaves (*see **p.13***), each of which can measure up to 15mm in diameter.
CAUSE The fungus *Rhytisma*, in particular *R. acerinum*, which can overwinter on fallen infected leaves. In spring it produces spores in a gelatinous coating which adhere to young leaves as they are produced. Leaves may fall prematurely, but this is not always the case and the vigour of the tree is rarely significantly affected.
CONTROL None available, but it is not necessary. Rake up and dispose of affected leaves as they fall.

TARSONEMID MITES

SYMPTOMS Mainly greenhouse and house plants are attacked by minute, whitish-brown mites that infest the developing leaves at shoot tips and the flower buds. Their feeding activities cause distorted flowers with white or brown flecks on the petals. The new foliage fails to expand to its full size and becomes progressively more stunted and distorted as the infestation develops (*see **p.27***). In severe attacks, growth comes to a complete

halt and the stems have a brownish scarring. Tarsonemid mites can be a particular problem on cyclamen, begonias, pelargoniums, busy Lizzies and many other pot plants.
CAUSE Microscopic mites that feed by sucking sap. These include broad mite, *Polyphagotarsonemus latus*, and cyclamen mite, *Phytonemus pallidus*. The latter also occurs as races that attack some garden plants – *see* MICHAELMAS DAISY MITE *p.149* and STRAWBERRY MITE *p.181*.
CONTROL None of the pesticides available will control tarsonemid mites. Infested plants should be destroyed. This pest usually comes into a greenhouse when an infested plant is acquired. Care should be taken to avoid this by purchasing plants from reputable suppliers.

THRIPS

Thrips are narrow-bodied, elongate insects up to 2mm long. They feed by sucking sap and generally cause a fine silvery-white mottled discoloration of the upper leaf surface. Thrips will readily feed from the upper leaf surface, unlike most sap-feeding pests that remain on the underside of leaves. Adult thrips are usually black with two pairs of wings that are heavily fringed with hairs. The wings are folded back along the body when not in use and so are not readily visible. Some thrips have black and white areas on their wings, which give the insects a banded appearance. The immature nymphs are wingless and creamy-yellow in colour. Some species, like western

flower thrips, can spread certain plant virus diseases on their mouthparts. To control specific thrips commonly found in gardens and greenhouses, *see* BANDED PALM THRIPS *p.104*; GLADIOLUS THRIPS *p.133*; GLASSHOUSE THRIPS *p.133*; ONION THRIPS *p.155*; PEA THRIPS *p.157*; PRIVET THRIPS *p.167*; WESTERN FLOWER THRIPS *p.191*.
Others may be controlled by spraying with pyrethrum, thiacloprid, acetamiprid, thiamethoxam, or bifenthrin when signs of activity are seen.

THUNDERWORMS

Thunderworms derive their common name from the fact that they are often seen after heavy rain, particularly between late spring and mid-summer. Under damp conditions the nematodes, *Mermis* species, emerge from the soil and climb onto plants where they can be seen writhing about. They are less than 1mm in diameter, but can be up to 60mm long. They are creamy-white, although blackish body contents can often be seen through the semi-transparent skin. Eggs are deposited on foliage and if the eggs and leaf are subsequently eaten by a suitable animal, such as a caterpillar, grasshopper or earwig, the eggs will hatch and develop as parasites in the insect's gut. Eventually the fully-formed nematode emerges and goes into the soil. Thunderworms can only develop inside insects and so pose no threat to humans or pets.

TOADSTOOLS

See TURF TOADSTOOLS *p.186*.

TOMATO BLIGHT

SYMPTOMS Brown patches develop on the leaves, causing them to dry and curl. The stems may also show blackening. Infected fruits develop a brown discoloration and may appear to shrink inwards. They may rot, either on the plant or a few days after being picked. Outdoor tomatoes are affected more frequently than those grown under glass. Potatoes may also be attacked (*see* POTATO BLIGHT *p.164*). Tomatoes usually succumb slightly later than potatoes.
CAUSE The fungus *Phytophthora infestans*, which is encouraged by moist or wet weather, particularly if it occurs in late summer. The spores are carried on air currents or by water splash, and need a minute film of moisture on the foliage if they are to cause damage.
CONTROL Preventive spraying is worthwhile, particularly if the tomatoes are growing in an area where blight is common and if there are other potato or tomato crops in the vicinity. In a wet season this should be carried out as soon as the first tomatoes have set and then repeated according to the manufacturer's instructions. A copper-based fungicide or one based on mancozeb should be used.

TOMATO BLOTCHY RIPENING

SYMPTOMS Patches of hard, yellow or green flesh remain unripened (*see **p.50***). These areas may be anywhere on the fruit and, unlike tomato greenback (*see p.184*), this disorder does not follow a regular pattern. The patches are not clearly demarcated.

The damage is only visible on mature fruits, and those on the lower trusses are generally the most frequently affected.

CAUSE This disorder is believed to be caused by some form of malnutrition, and a deficiency of potassium is probably the single most important factor. Dry soil or compost or a poorly functioning root system may also play a part. High temperatures in the greenhouse also increase the damage.

CONTROL Keep the plants well fed, in particular ensuring that a high potash feed is used. Keep the greenhouse well ventilated and the plants well watered.

TOMATO CYST EELWORM

See POTATO CYST EELWORM *p.165*.

TOMATO FRUIT SPLITTING

SYMPTOMS Tomatoes develop normally but start to split, usually just before picking (*see* **p.50**). The splits may become infected by secondary organisms such as grey mould (*see p.135*), and may then rot. Occasionally the split surface temporarily dries up and the fruit does not rot.

CAUSE Erratic growth and fruit swelling. This is usually caused by an irregular supply of water or wildly fluctuating temperatures.

CONTROL Ensure that plants are adequately watered at all times and are not subjected to temperature extremes. In greenhouses, provide shading and ventilation. Remove affected fruits before secondary organisms invade.

TOMATO GHOST SPOT

SYMPTOMS On unripe fruits, pale green or yellow rings of discoloration appear. As the fruits develop and colour, the rings gradually turn yellow or pale orange. The fruit's development is not affected and it is perfectly edible.

CAUSE The fungus *Botrytis cinerea*. Spores are readily spread by rain or water splash and on air currents. The fungus is widespread on both living and dead plant matter. The rings of discoloration result from a hypersensitive reaction to the fungal spores.

CONTROL As affected fruits rarely deteriorate, there is no need to remove them, but ensure you eliminate sources of infection in the area. Spraying plants with carbendazim may help.

TOMATO GREENBACK

SYMPTOMS As greenhouse tomatoes begin to ripen, a partial or complete ring of clearly demarcated leathery tissue remains around the stalk end. It does not ripen properly, remaining either green or yellow (*see* **p.50**). The rest of the plant develops normally.

CAUSE A deficiency which is under genetic control, and is brought about by several different factors either alone or, more commonly, in combination. Low levels of potassium and phosphorus can be involved but hot, bright sunlight, which causes heat injury, is probably the most important single factor.

CONTROL Provide adequate shading for the plants, particularly once the fruits start to develop. Do not remove too much of the plant's foliage, as this acts as a good shading material. Ventilate the greenhouse to prevent excessively high temperature levels. Keep the plants well fed, ensuring adequate supplies of potash and phosphate in particular. Choose tomatoes which show resistance to this disorder. The following are suitable, but in some cases it appears that the resistance may be breaking down, and even these may show some symptoms in very hot weather: 'Cherry Wonder', 'Craigella', 'Estrella', 'Eurocross BB', 'Golden Sunrise', 'Grenadier', 'Matador', 'Moneymaker', 'Shirley', 'Tigercross', 'Tigrella', 'Totem', and most modern F1 hybrids.

TOMATO LEAF MOULD

SYMPTOMS Yellow blotches develop on the upper leaf surface. Beneath each spot is a grey or faintly purple velvety fungal growth. The lower leaves are usually attacked first, but all leaves may eventually be affected. The leaves turn brown, wither and dry but remain hanging on the plant. A secondary infection of grey mould may mask the symptoms of leaf mould. While greenhouse tomatoes may be frequently affected, those that are grown out of doors rarely develop this disease.

CAUSE The fungus *Fulvia fulva*, which spreads rapidly in warm, humid conditions. The spores are carried on air currents and may also be spread by insects, and on clothing. The fungus overwinters on infected plant debris and on greenhouse structures.

CONTROL Remove and burn all infected plants at the end of the season and thoroughly clean out the greenhouse. Ensure that the ventilation is adequate and attempt to minimize excessively humid conditions by spacing plants out well and by carefully removing some of the foliage. Grow tomatoes which show resistance to this disease, such as 'Cyclon', 'Dombello', 'Estrella', 'Eurocross BB', 'Grenadier', 'Seville Cross', 'Shirley', and 'Sonatine'.

TOMATO MOTH

SYMPTOMS The foliage and fruits of tomatoes are eaten by caterpillars. Newly-hatched larvae feed together by grazing away the lower leaf surface, causing the remaining upper surface to become papery-white. In greenhouses, such damage can often be seen first on leaves which are growing close to the glass. As they grow larger, the caterpillars disperse and eat holes right through the foliage, and also bore into green and ripening fruits. When fully grown, the caterpillar can be up to 35mm long and either brown or pale green in colour, with a yellow line – marked by many tiny white dots – along each side of the body.

CAUSE Caterpillars of a moth, *Lacanobia oleracea*, which is also sometimes known as the bright-line brown-eye moth.

CONTROL Look for leaves bearing newly-hatched caterpillars and remove them immediately. Otherwise spray with an insecticide such as pyrethrum or bifenthrin when caterpillars are seen.

TOMATO PITH NECROSIS

SYMPTOMS The upper leaves wilt and turn yellow as the fruits start to ripen. Dry, dark brown to black patches develop on the stems. If cut longitudinally, the internal tissues of the stem can be seen to be discoloured, usually black in the pith cavity or in the pith itself.

CAUSE The bacterium *Pseudomonas corrugata*, which is probably spread by rain and water splash and by handling, and possibly on any tools used when removing side shoots.

CONTROL Remove and burn infected plants promptly to reduce the risk of the infection spreading.

TOMATO VIRUSES

SYMPTOMS Several different viruses may affect tomatoes and the symptoms they produce vary, not only with the virus involved, but also depending on the tomato cultivar and the prevalent growing conditions. Common leaf symptoms include yellow mottling or mosaicing, distortion and the production of unusually small leaves. With tomato mosaic, in addition to these symptoms, the youngest leaves may curl downwards. With fern leaf, the leaf blades are extensively narrowed and malformed and may be mistakenly assumed to be damaged by weedkiller. Other symptoms of virus infection may include yellow streaking, stunting and the bronzing of the young fruits, yellow streaking of the fruits, or failure of the plant to set fruit.

CAUSE The viruses most commonly involved, either alone or in combination, are tomato spotted wilt, streak, tomato mosaic, potato virus X and cucumber mosaic. All these viruses are readily spread by handling and in some cases by aphids or other sap-feeding pests. Many of these viruses have very wide host ranges, so they may be introduced from unrelated plants in the garden, or may be spread to these from infected tomatoes.

CONTROL Remove and dispose of infected plants promptly. Avoid handling plants unnecessarily and always wash hands thoroughly after doing so. Control aphids and other sap-feeding pests. Remove and burn all crop debris at the end of the season. Grow tomato cultivars which are described as showing resistance. At present, the following cultivars are listed as showing resistance to tomato mosaic virus: 'Cumulus', 'Cyclon', 'Dombello', 'Estrella', and 'Shirley'.

TORTRIX MOTHS

See CARNATION TORTRIX MOTH, *p.115.*

TULIP FIRE

SYMPTOMS As the foliage emerges above ground it may be withered, distorted and covered in numerous buff-coloured flecks (*see **p.22***), or scorched. It soon becomes covered in fuzzy, grey fungal growth. Elliptical grey-brown marks also develop on the stalks. The plants may fail to mature, though some may flower, or flower buds may develop but fail to open. Those which open normally may develop numerous small, bleached spots on the petals, each one being caused by the plant's hypersensitive reaction to the spores. The flower stems may topple over and in wet weather the whole plant may become covered in the fungal growth and then rot rapidly. Bulbs may rot or may appear firm but bear black sclerotia (fungal resting bodies) clustered around the neck (*see **p.55***).

CAUSE The fungus *Botrytis tulipae*, which overwinters as pinhead-sized sclerotia in the soil or on infected bulbs. The spores produced on all the above-ground parts of the plant are readily spread on air currents and by water splash.

CONTROL Remove infected foliage promptly. Do not plant tulips on the same ground for more than two years in succession, as this may encourage build-up of the sclerotia in the soil. Remove and dispose of severely infected plants, together with the soil in the immediate vicinity of the roots. Check bulbs carefully before planting and only use those which appear to be perfectly healthy and free from signs of sclerotia. Late planting may also decrease the chances of the disease developing. On sites where the disease has been identified, do not replant with tulips for at least three years. At the end of the season lift tulip bulbs and dust them with sulphur.

TULIP GREY BULB ROT

SYMPTOMS Bulbs fail to emerge above ground, leaving gaps in the bed. If shoots are produced, they are extensively distorted and soon wither and die off. There are no signs of fungal growth on the affected foliage. Infected bulbs turn grey and dry as they start to rot. The roots and the basal plate may be all that remain. Soil tends to adhere to infected bulbs, especially at the noses. A dense, white or grey felty fungal growth develops between the scales and on the outside of the bulb. Black sclerotia (fungal resting bodies) develop in amongst the fungal growth and may reach up to 7mm in size, remaining viable in the soil for anything up to five years.

CAUSE The fungus *Rhizoctonia tuliparum*, which may attack many bulbs, including alliums, amaryllis, brodiaeas, chionodoxas, colchicums, crocus, eranthis, fritillaries, gladioli, hyacinths, iris, ixias, lilies, daffodils and narcissi, snowdrops, scillas and tulips.

CONTROL Remove and burn affected plants, together with the soil around them. Do not grow susceptible bulbs in that soil for at least five years.

TURF DOLLAR SPOT

SYMPTOMS Grass dies off in patches, becoming straw-yellow (*see **p.58***). Patches usually measure between 7.5–10cm in diameter.

CAUSE The fungus *Sclerotinia homeocarpa*, which is most prevalent during warm, damp spells in early autumn, and on lawns which have become compacted.

CONTROL Improve drainage by spiking the lawn in autumn. Avoid using grass seed mixtures containing a high proportion of *Festuca* species (the fescues) as these are particularly prone to this infection.

TURF FAIRY RINGS

SYMPTOMS Rings develop on the turf, increasing in size as the fungus spreads. Around the periphery of the ring one or more zones of very lush, green grass develop. There is also an inner ring of similarly luxuriant grass. The area between these two rings is dead and brown, or the soil may be completely bare. Occasionally, incomplete rings or other formations develop. The fungus grows just beneath the soil surface and around the bases of the grasses, forming a dense, off-white fungal growth or mycelium. Pale brown, slender toadstools, each 4–10cm tall, may develop at the outer part of the middle zone in summer or autumn (*see* **p.59**).

CAUSE The fungus *Marasmius oreades*. The death of the middle zone of grass is caused by drought – the dense fungal mycelium is so water-repellent that water cannot penetrate to the grass roots in this area.

CONTROL There is no chemical control available on the gardening market. Rake up and dispose of all toadstools, preferably before their caps open to release spores. Mow the infected area of grass separately and collect up and burn mowings from this area in order to limit the risk of the fungus spreading. Dig out the affected grass and topsoil to a depth of at least 30cm. In addition, dig out the grass in the area, extending about 30cm beyond the edges of the ring to the same depth. Dispose of all affected grass and soil, taking care not to contaminate surrounding areas of lawn or grass. Fill the trench with topsoil and re-seed.

TURF OPHIOBOLUS PATCH

SYMPTOMS Bleached patches develop and spread. Fungal growth may be visible in among dead, brown roots and stolons of the grasses.

CAUSE The fungus *Gaeumanomyces graminis*, which is very damaging on lawns composed largely of fine grasses It is also very prevalent on chalky or limey soils or on lawns which have been limed.

CONTROL Remove affected areas and several inches of soil beneath the remains of the grasses. Use fresh topsoil to fill in, and then re-seed. Avoid liming the soil.

TURF RED THREAD

SYMPTOMS Patches of either reddish or bleached grass develop, up to 7.5–8cm in diameter. In amongst the grasses and attached to them, dark pink, horn-like, gelatinous, branching strands of fungus develop (*see* **p.59**). These strands of fungus subsequently turn pale pink and slightly fluffy. The grass is rarely killed, but it is weakened and its appearance spoiled.

CAUSE The fungus *Corticium fuciforme*, which is most common on lawns composed mainly of fine lawn grasses such as the fescues. Perennial rye grass and annual meadow grass are less susceptible and are rarely affected. Inadequate nitrogen levels and poor aeration encourage the disease. It is usually most prevalent in the year following a drought, because the dry weather will have made the usual fertilizing routine very difficult.

CONTROL Improve drainage by aeration and ensure the grass is adequately fertilized.

Apply nitrogen as sulphate of ammonia and this may cure the problem. Treat with trifloxystrobin.

TURF SNOW MOULD

SYMPTOMS Small patches of dying grass develop, gradually turn brown, then increase in size and may even join together. In damp weather the dying patches may become covered in a pale pink, fluffy fungal growth (*see* **p.59**). Snow mould is most commonly seen in the autumn months of the year and during mild spells in winter.

CAUSE The fungus *Monographella nivalis* and occasionally other fungi. Snow mould is often encouraged by poor aeration and by damp conditions. If the grass is walked on in winter after a fall of snow, the disease frequently develops on these areas.

CONTROL Carry out regular routine spiking, and use any other methods for improving aeration. Treat with trifloxystrobin. Avoid walking on the grass during wet or cold weather.

TURF THATCH FUNGAL MYCELIUM

SYMPTOMS Bleached, yellowed or occasionally reddish patches of grass develop on the lawn: the grasses are severely stressed or may be killed (*see* **p.58**). A dense mat of fungal growth grows around the bases of the grasses.

CAUSE Various fungi, often described as thatch fungi, which are not directly pathogenic but live on organic debris, including thatch (remains of old grasses, mowings and other debris) which is found at the base of the grasses. These fungi form a strongly water-repellent layer, which causes the grass to die from drought stress. Lawns which have not been properly maintained, and newly-laid turf, are the most susceptible.

CONTROL Carry out all necessary lawn maintenance: scarification, spiking and weed and moss control. Do not allow mowings to be redistributed over the lawn. Buy turf only from a reliable, reputable supplier and check it carefully on delivery. Where necessary, the only solution may be the removal of badly infested areas followed by re-seeding or re-turfing.

TURF TOADSTOOLS

SYMPTOMS The grass grows normally, but toadstools develop in random arrangements on the lawn. Toadstools may also develop in straight lines, following the path of tree roots or other buried woody debris.

CAUSE Various fungi – including ink caps, field mushrooms and puff balls – which do not attack the grasses but feed on organic debris in the soil.

CONTROL Brush off toadstools as soon as they appear, preferably before they release their spores. If the problem is recurrent, the only long-term solution may be to excavate beneath the grass and remove the tree root or other material on which the fungus is feeding.

TURNIP GALL WEEVIL

SYMPTOMS The adult beetles lay eggs on the roots of turnip, swede and leafy brassicas such as cabbage, cauliflower, Brussels sprouts and broccoli. The feeding activities of the white, legless grubs cause the plant to produce rounded swellings that enclose the grubs. This can be confused with roots swollen by infection with clubroot disease (*see p.119*). They can be distinguished by cutting the swellings in half. Turnip weevil galls are hollow and contain grubs, or have circular exit holes if the grubs have left to pupate in the soil, whereas clubroot galls are solid and less regularly rounded.

CAUSE Larvae of a beetle, *Ceutorhynchus pleurostigma*. Infestations on the roots of leafy brassica plants have little impact on their growth, but this pest is more troublesome when it is developing on edible roots such as turnip and swede.

CONTROL There is no effective chemical control and so the presence of galls has to be tolerated.

TWO-SPOTTED MITE

See GLASSHOUSE RED SPIDER MITE *p.133*.

VAPOURER MOTH

SYMPTOMS Hairy caterpillars, up to 30mm long, occur during late spring to late summer on a wide range of trees and shrubs (*see p.31*). They are greyish-black with numerous orange-red spots on the body; there are four clumps of pale yellow hairs, resembling shaving brushes, on the upper body surface and long tufts of black hairs pointing forward from the head end, and backwards from the rear end. The

caterpillars generally occur in small numbers, but are sometimes abundant enough to cause significant damage.

CAUSE Caterpillars of a moth, *Orgyia antiqua*, which emerge in mid- to late summer. The female is wingless and after mating she lays her eggs on the silk cocoon from which she has emerged. The eggs are the overwintering stage.

CONTROL Hand picking will deal with light infestations. This pest is rarely so abundant that it is necessary to use pyrethrum or bifenthrin.

VERTICILLIUM WILTS

SYMPTOMS Plants show wilting of the foliage (*see p.29*). Most or all of the leaves on an affected branch or stem show the symptoms, but the whole plant is unlikely to be affected simultaneously. The leaves may show interveinal discoloration (usually as yellowing or browning between the veins) and they then die. Stem death follows shortly, but it may be several years before a large shrub is killed completely. Smaller shrubs or herbaceous perennials and annuals may be killed within a single season. If the bark is removed from an affected stem, the vascular tissue (conducting tissue) is seen to be discoloured, usually by brown or purple-brown streaks which run the length of the stem (*see p.33*), but are usually most apparent towards the base. Roots have a central core of discoloured conducting tissue.

CAUSE *Verticillium dahliae* and *V. albo-atrum*. These are both found in infested soil, on plant debris and within the plant sap. *V. dahliae*

forms fungal resting bodies which may persist in the soil. The host ranges are wide, and as well as many cultivated plants, several common garden weeds may harbour the infection too.

CONTROL None available. Remove affected plants promptly, together with the soil surrounding their roots. As the infection can be spread on pruning tools, always thoroughly clean any tools after they have been used on an infected plant. Avoid replanting any of the following on the same site: maples, Indian bean tree (*Catalpa bignonioides*), Judas tree, smoke bush (*Cotinus*), golden rain tree (*Koelreuteria paniculata*), stag's horn sumach, lime, berberis, daphnes, quince, apples, pears or *Prunus*.

VIBURNUM BEETLE

SYMPTOMS The foliage of viburnums, particularly *Viburnum tinus*, *V. opulus* and *V. lantana*, is reduced to a lacework of damaged leaves during the summer. Most of the damage is caused by creamy-yellow, black-spotted beetle larvae, up to 7mm long, during late spring to early summer. The greyish-brown adult beetles, 5–6mm long, emerge from pupae in the soil in late summer and cause additional, but less extensive, damage.

CAUSE Adults and larvae of a beetle, *Pyrrhalta viburni*. This pest overwinters on its host plant as eggs, which are placed in small pits in the bark of the stems.

CONTROL Look for signs of damage on the new foliage in late spring, and spray with bifenthrin, thiacloprid, acetamiprid, or pyrethrum while the larvae are still

small and not yet at their most damaging.

VIBURNUM SCALE

SYMPTOMS The upper leaf surface on *Viburnum tinus* and ivies (*Hedera* species) becomes sticky with honeydew excreted by this pest. Black sooty moulds often develop on the honeydew. The scales live on the underside of leaves, usually next to a leaf vein. They are pale yellow, flat, oval creatures up to 4mm long. When fully mature in late spring, the females are pale brown and they cover themselves with a fluffy, white waxy material in which they deposit their eggs.

CAUSE A sap-feeding insect, *Lichtensia viburni*. It is more troublesome on plants growing against a wall or in other sheltered places.

CONTROL Spray thoroughly with thiacloprid, acetamiprid, fatty acids, or plant oils, especially on the underside of the leaves, in mid-summer to control the more vulnerable newly-hatched scale nymphs.

VIBURNUM WHITEFLY

SYMPTOMS The only viburnum affected is laurustinus (*Viburnum tinus*), but this pest also attacks strawberry trees (*Arbutus*). In mid-summer, white-winged insects, 1–2mm long, occur on the underside of the younger leaves. The immature nymphs are flat, oval, yellowish-green, scale-like objects found on the lower leaf surface in late summer. More easily seen is the overwintering pupal stage; these are black, oval objects 1mm long, which are encrusted with a white, waxy substance (*see p.15*). Heavy infestations may cause soiling of the foliage with honeydew and sooty moulds.

CAUSE Adults and nymphs of a sap-feeding insect, *Aleurotrachelus jelinekii*.

CONTROL Light infestations can be ignored, since the pest has little direct impact on its host plant's growth. If necessary, spray the underside of the leaves with thiacloprid, acetamiprid, bifenthrin, pyrethrum, fatty acids, or plant oils, when adult whitefly are present.

VINE ERINOSE MITE

SYMPTOMS During mid- to late spring, parts of grape vine leaves bulge upwards (*see p.15*); the underside of these deformed areas is covered in a dense mat of fine, whitish hairs. As the summer progresses, the hairs darken and become yellowish or brown. The upper surface of the blistered areas may also become discoloured.

CAUSE Microscopic gall mites, *Colomerus vitis*, which suck sap from the foliage and secrete chemicals into the leaf to induce the excessive growth of hairs. Apart from the leaf distortion, the mites seem to have little impact on the plant's vigour and ability to produce grapes.

CONTROL No effective chemicals are available. Removing affected leaves may check infestations, but as the mites are of little consequence, heavy defoliation would be more harmful than tolerating their presence.

VINE WEEVIL

SYMPTOMS Both the adult and larval stages cause damage. The larvae are plump,

creamy-white, legless grubs with brown heads. They are up to 10mm long and have slightly curved bodies. They live in the soil (*see p.53*) and feed on roots, which are either severed or, on woody plants, have the outer bark removed. This results in plants making slow growth, followed by wilting and death when insufficient roots remain to sustain the plant. The grubs also bore into the tubers of plants such as begonia and cyclamen. A wide range of garden and indoor plants is attacked and plants grown in pots or other containers are particularly at risk. Larval damage is mostly seen between early autumn and mid-spring. The adult weevils are 9mm long, with dull black, pear-shaped bodies and antennae that are bent at right angles about half-way along their length. They are slow-moving beetles that emerge at night and eat irregular notches from the leaf margins of many plants (*see p.29*), but especially rhododendrons, *Euonymus japonicus*, and hydrangea. The adults can be found at any time from spring to autumn.

CAUSE Larvae and adults of a beetle, *Otiorhynchus sulcatus*. All the adults are female and capable of laying many hundreds of eggs over a period of several months.

CONTROL Thiacloprid or acetamiprid applied as a compost drench to container-grown ornamental plants will protect them from vine weevil grubs. Biological control with a pathogenic nematode, *Steinernema kraussei*, is an alternative means of protecting plants. Water nematodes into the potting compost or garden soil in late summer, before

the grubs have grown large enough to cause serious damage. Nematodes are ineffective in heavy soils, or those which are dry or colder than 5°C (41°F). Adult weevils can be searched for by torchlight and removed from plants. The leaf damage caused by the adults is rarely sufficient to affect plant growth, but if it is causing concern, plants can be sprayed at dusk on mild evenings with thiacloprid or acetamiprid.

VIOLET GALL MIDGE

SYMPTOMS Leaves become greatly thickened and fail to unfurl (*see p.27*). If broken open, orange-white maggots or white silk cocoons, 2mm long, will be found inside the galls. Galled leaves are present throughout the year, but are often found in the winter after normal foliage has died down.

CAUSE A small fly, *Dasineura affinis*, which has three or four generations during the summer when females deposit their eggs on the developing leaves.

CONTROL Heavy infestations may kill plants, but usually they survive. Affected leaves can be removed by hand, but it is difficult to prevent reinfestation. There are no insecticides recommended for this pest.

VIOLET ROOT ROT

SYMPTOMS Plants are stunted and chlorotic, but the most characteristic symptoms are seen on the affected roots, tubers and rhizomes. Their surface is covered in a dense network of dark purple fungal strands and often have considerable quantities

of soil adhering to them (*see p.57*). Closely packed areas of fungal threads form into sclerotia which may fall into the soil as the tissue beneath starts to rot and turns brown, but this may be largely the effect of secondary organisms.

CAUSE The fungus *Helicobasidion purpureum*, which is most prevalent in wet, acid soils and persists in the soil as sclerotia. It may also remain viable on plant tissue for a considerable time.

CONTROL Lift and burn infected plants to minimize disease carry-over. Where possible, remove infected plants before they start to disintegrate and release sclerotia into the soil. Improve soil drainage. Do not grow susceptible plants on that soil.

VIRUS VECTORS

Some virus diseases affecting plants can be spread by certain pests and these are described as virus vectors. It is mainly sap-feeding insects such as some aphids, leafhoppers and thrips that are involved. When they insert their needle-like mouthparts into virus-infected plants, they will pick up some virus particles which can be transmitted to other plants when they move on to new host plants. The champion virus vector seems to be the peach potato aphid, *Myzus persicae*, which feeds on a wide range of plants and is capable of transmitting more than a hundred viruses. Some soil-dwelling eelworms, such as some *Xiphinema*, *Trichodorus* and *Longidorus* species, can transmit virus diseases when they feed on roots. The spread of viruses can be limited by controlling insect pests where feasible, and by removing plants affected by damaging virus diseases. Raspberries, strawberries and daffodils are susceptible to eelworm-vectored viruses. New plants should be put into fresh ground, since if they are planted where virus-infected plants were previously growing, eelworms may infect them with viruses when they start feeding on the roots.

VIRUSES

SYMPTOMS The most common and widely spread symptoms of viral infection are stunting of the whole plant, and distortion (*see **pp.25, 27***). The leaves and occasionally other above-ground parts of the plant develop markings (*see **p.23***). These are usually yellow, though on some plants such as orchids they may be dark brown or black. The markings are commonly seen as streaks, mosaic patterns, flecks, mottles or ring spots. Flowers may show distortion or fail completely, or they may form but show "flower breaking" where pale streaks of colour develop on dark petals (*see **p.42***), or coloured streaks develop on pale petals. Infected plants generally crop poorly or fail to crop at all and in many cases death is premature. Some viruses may infect certain plants without causing visible symptoms, and this is called latent infection. For example, cucumber mosaic virus (*see p.121*) can exist in a number of common weeds which, although looking perfectly normal, can then act as a source of viral infection for other plants.

CAUSE A wide range of viruses, most of which may be found alone or in combination with other viruses. There are various methods of transmission, the most common being by sap-feeding insects, in particular aphids. Leafhoppers, whitefly, plant hoppers, thrips, scale insects, mites and even beetles may also be involved. In most cases the infection is non-persistent, that is to say the insect can acquire the virus after feeding on an infected plant for only a few seconds, but can then transfer it to other plants only within a short period of time. Cucumber mosaic virus (*see p.121*) is an example of this. Some insect-borne viruses are persistent, that is, the insect cannot transmit the virus for several hours after feeding on an infected plant, but is then capable of transmitting it for a long time afterwards, and sometimes for the duration of its life. Tomato spotted wilt, which is transmitted by western flower thrips (*see p.191*), is an example of this. Some viruses are transmitted by soil-inhabiting nematodes and fungi. Viruses may also be transmitted by handling, by grafting, by tubers, bulbs and corms, and during routine operations such as disbudding or removal of side shoots.

CONTROL None available. Remove and dispose of affected plants promptly. Wash hands and tools thoroughly after handling infected plants. Do not put replacement plants of a similar type on the same site, and control potential vectors. Good weed control is essential too. Do not propagate from infected plants. Where possible, choose virus-resistant cultivars and buy plants certified as virus-free.

VOLES

Voles are rodents with shorter tails and more rounded heads than the mice and rats that they resemble. They feed on a wide range of plant material, including seeds, flower buds, bulbs, corms and plant roots. They also gnaw bark from the roots, trunks and branches of young trees and shrubs (*see **p.36***). If bark is eaten away around the entire stem (girdling), this prevents the uptake of water and nutrients, leading to the death of branches or sometimes the whole plant. Several species occur in gardens, including the mouse-sized bank vole, *Clethriomys glareolus*, and the short-tailed field vole, *Microtus agrestis*. The water vole, *Arvicola amphibius*, is larger, and can be confused with a rat. Vole numbers can fluctuate widely, especially of the short-tailed field vole, and it is usually only in years when they are abundant that they cause serious damage to garden plants.

CONTROL Where field or bank voles are causing problems in a garden, some protection can be given by setting mouse traps in the places where voles are active. Pieces of dessert apple seem to be a particularly attractive bait for the traps. These need to be placed under covers made of logs or bricks to keep birds away and prevent pets from harming themselves. Water voles have become scarce in Britain and are now a legally protected species.

WALNUT BLISTER MITE

SYMPTOMS During early summer, leaves develop oval, upwardly bulging areas up to 20mm long and about 10mm wide. The underside of these blistered areas is covered with a dense mass of creamy-white hairs which darken and become brown later in the summer. The upper surface of galled areas may become purplish-yellow.

CAUSE Microscopic gall mites, *Aceria erineus*, which suck sap from the underside of the foliage and cause the distorted growth. Most walnuts are affected and sometimes a large proportion of leaves will have galls by late summer.

CONTROL No effective insecticides are available to gardeners. Fortunately, the mite has no impact on the tree's vigour or cropping.

WALNUT BLOTCH

SYMPTOMS Leaves develop brown, necrotic blotches, and premature leaf fall occurs. Similar lesions may develop on the fruits, turning them from green to black.

CAUSE The fungus *Marssonina juglandis*, which overwinters on fallen leaves.

CONTROL Rake up and dispose of affected leaves.

WALNUT SOFT SHELL

SYMPTOMS The walnuts appear to develop normally, but the shell fails to form properly at one end. It may be thin or totally absent, so the nut within may become damaged or rotted by organisms which can gain entry through the faulty area. **CAUSE** The precise cause is unclear, but soft shell is believed to be largely the result of poor feeding and an erratic supply of water at that stage when the nuts are developing. **CONTROL** Keep trees fed, particularly if they have been cropping well. Mulch to help soil moisture retention.

WASPS

Social wasps, such as *Vespula vulgaris* and *V. germanica*, and tree nesting species, such as *Dolichovespula media* and *D. sylvestris*, are important as plant pests because of their interest in ripe fruits. On relatively soft-skinned fruits, such as grapes and plums, they are capable of initiating damage (*see* **p.49**), but on apple and pears they generally only enlarge on damage started by birds. Wasps eat the soft inner tissues and gradually hollow out large cavities (*see* **p.47**). Some species of wasp make their papery nests above ground level in the branches of trees and shrubs. This can make hedge-cutting hazardous. **CONTROL** If wasp nests can be located, they can be controlled by placing some insecticide dust, such as bendiocarb, in the nest entrance at dusk when wasps have stopped flying. Aerial nests are best dealt with by using an aerosol containing phenothrin + tetramethrin. There is no point in spraying fruit trees against wasps, since any killed will soon be replaced by others. Ripening fruits can be protected by enclosing some of the better fruit

trusses in bags made from old nylon tights or muslin. See p.86 for information on making a wasp trap.

WATER CORE

See APPLE WATER CORE *p.102.*

WATER LILY APHID

SYMPTOMS Olive-green or brownish insects, about 2mm long, cluster on the upper surface of water lily leaves and on the flowers during the summer. Other floating and emergent pond plants are also liable to be attacked. Heavy attacks result in poor growth and flowering, with water lily leaves appearing dirty due to the accumulation of aphids and their whitish cast skins. **CAUSE** A sap-feeding aphid, *Rhopalosiphum nymphaeae*. It overwinters as eggs on wild and cultivated plums and cherries, feeding on the foliage of these plants in the spring before migrating to water lilies in early summer. **CONTROL** Although many insecticides will control aphids, none of them can be used on pond plants, as they will kill fish, frogs and other pond wildlife. Forceful spraying with plain water will reduce infestations by knocking aphids off the plants.

WATER LILY BEETLE

SYMPTOMS Damage starts in late spring and continues throughout the summer. Yellowish-brown beetles, 7mm long, eat irregularly shaped slots in the lily pads (*see* **p.30**). In mid-summer the larvae, which have elongate bodies, measuring

up to 8mm in length, that are black on the upper surface but pale yellow underneath, continue this damage. Infested leaves are are not only unsightly but also deteriorate rapidly as secondary rots develop on the damaged tissues. The adult beetles also feed on the flowers. **CAUSE** Adults and larvae of a beetle, *Galerucella nymphaeae*. The entire life cycle of eggs, larvae, pupae and adults takes place on the upper surface of water lily leaves, and there are usually two generations during the summer. **CONTROL** Insecticides cannot be used, as they will be harmful to fish and other animals in the pond. With small ponds, adult beetles, larvae and pupae can be removed by hand. Forceful spraying with water will dislodge the larvae, some of which will fail to find their way back onto the leaves.

WATER LILY LEAF SPOT

SYMPTOMS Red-brown spots, sometimes with a distinct yellow band around the edge, develop on the leaves. The upper leaf surface is generally most affected, but spotting may also occur beneath the leaves. Concentric zoning and numerous tiny, black pinprick-sized fungal bodies may also be visible. As the spots mature, they darken, turning a brownish-black colour. The centre of the spot may disintegrate, leaving a hole. **CAUSE** The fungus *Ramularia nymphaerum*, which is carried by water splash. **CONTROL** Remove affected leaves promptly. No chemical control is available

and none should be attempted, as it may pose a threat to pond life.

WATERLOGGING

SYMPTOMS The plant appears to be suffering from drought, with wilting foliage (*see* **p.34**) and, in extreme cases, flowers (*see* **p.71**). The leaves may be yellowed and may fall early. Inspection of the roots may reveal that they have disintegrated, often peeling readily, to leave just the dead vascular strands. **CAUSE** Overwatering, excessive rain or poor drainage may all be involved. **CONTROL** Attempt to improve soil conditions and avoid over-application of water. On heavy soils ensure that a large area of soil is lightened by the incorporation of plenty of grit, bulky organic matter and other materials before planting. The application of a foliar feed to any remaining green leaves may help to counteract the effects of slight waterlogging; it stimulates root production to replace some of those roots which have been lost. Where feasible, plant on a slight mound to improve drainage.

WEEDKILLER DAMAGE – BROADLEAVED

SYMPTOMS The leaf blades are distinctly narrowed (*see* **p.27**), often to such an extent that the leaf veins are so closely drawn together that they appear parallel. Petioles and stems may show spiral twisting, and the foliage may become cupped (*see* **p.27**). Fruits may not form or if they do, they may be distorted. Tomatoes produce plum-shaped fruits which are hollow with few,

if any, seeds and little flesh. The stems of brassicas develop roughened gall-like growths at or just above soil level and these may appear to be adventitious roots (*see p.39*). Affected plants may live, but they can fail to crop normally.
CAUSE The plant has taken up broadleaved weedkiller, probably applied to control weeds nearby, into its tissues. Contamination may occur in many ways, such as inadvertent spraying, spray drift, use of watering cans or spraying equipment which have been insufficiently washed out, and mulching with lawn mowings which contain traces of weedkillers.
CONTROL Take great care when spraying and never attempt to do it on hot, windy or gusty days. Protect nearby plants with polythene sheeting or bags. Where possible, use a watering can and dribble bar rather than a sprayer, as these produce larger droplets that are less likely to cause drift problems. Keep equipment used for applying weedkillers clearly labelled and use it for this purpose only. Always wash it out thoroughly after use. Prune out severely damaged areas on trees or shrubs and feed and water to promote new replacement growth.

WEEDKILLER DAMAGE – CONTACT

SYMPTOMS Brown scorched spots develop on the foliage (*see pp.23, 34*) and any other part of the plant on which the chemical lands. If the weedkiller is applied on bulbs before the leaves have died down properly, bleached yellow or white foliage is produced the following season, which soon withers and dies. It

may take several years for the bulb to recover fully.
CAUSE Contamination by the the contact weedkiller diquat.
CONTROL As for broadleaved weedkiller damage.

WEEDKILLER DAMAGE – GLYPHOSATE

SYMPTOMS Stunted rosettes of short shoots and malformed, strap-like leaves develop in place of normal side shoots. The symptoms resemble those caused by a witches' broom infection (*see p.193*).
CAUSE The widely-used weedkiller glyphosate, which may cause problems on a range of plants, in particular those which form suckers.
CONTROL As for broadleaved weedkiller damage (*above*).

WEEVILS

Weevils are beetles that are characterized by having clubbed antennae that can be bent at right angles about halfway along their length. The front of the head projects forward to form a snout or rostrum which may be long, as in the nut weevil (*see p.151*), or short, as in vine weevil (*see p.186*).
• see also APPLE BLOSSOM WEEVIL *p.100;* BEECH LEAF-MINING WEEVIL *p.106;* FIGWORT WEEVILS *p.128;* LEAF WEEVILS *p.145;* PEA AND BEAN WEEVIL *p.156;* TURNIP GALL WEEVIL *p.186.*

WESTERN FLOWER THRIPS

SYMPTOMS Mainly house plants and greenhouse plants are affected, but this insect can also attack outdoor plants during the summer. On some host plants, such as tomato and cucumber, the thrips feed mainly on the foliage, causing a silvery

mottling of the upper leaf surface. On ornamentals such as African violet, streptocarpus, gloxinias, chrysanthemums, verbenas, pelargoniums, fuchsia, cyclamen and busy Lizzies, it is mainly flowers that are attacked (*see p.42*). They cause a pale flecking on the petals, and heavy infestations cause rapid deterioration in the blooms or prevent the buds from opening. Adult western flower thrips are 2mm long and have narrow, elongate, brown bodies. The immature nymphs have a similar shape and are pale yellow. This pest can transmit tomato spotted wilt virus, which causes dieback and distorted growth on many plants.
CAUSE A sap-feeding insect, *Frankliniella occidentalis*, which originates from North America, but became widely established in Europe after 1986. It breeds continuously in heated greenhouses and has many generations a year.
CONTROL Many insecticides used against other thrips have little impact on this species. Biological control is available with a predatory mite, *Amblyseius degenerans*, but this has to be introduced before heavy infestations arise. Alternatively, spray with thiacloprid, acetamiprid, or thiamethoxam. When buying house plants, check the flowers carefully and reject any with thrips or signs of feeding damage on petals.

WHIPTAIL OF BRASSICAS

SYMPTOMS Leaves become mottled yellow and growth is reduced.
CAUSE An inadequate supply of molybdenum (a trace element) in the soil or

compost. This is common on acid soils and is most likely to affect seedlings, especially brassicas.
CONTROL Lime the soil. Apply a proprietary mixture of trace elements, including molybdenum.

WHITE BLISTER

SYMPTOMS Raised, shiny white fungal pustules develop, primarily on the lower leaf surface of brassicas (*see p.14*). Occasionally these may develop on other above-ground plant parts. The pustules are arranged either singly or in concentric rings. Affected leaves may become puckered and distorted, and the upper leaf surface develops yellow, slightly sunken pits.
CAUSE The fungus *Albugo candida*, which is encouraged by humid or moist air. The spores are spread on air currents or by water splash, or occasionally by insects. A resilient resting spore stage may also be formed and this may persist in the soil.
CONTROL Remove infected leaves. Remove and dispose of severely infected plants. Where feasible, grow resistant varieties of vegetable such as the Brussels sprout 'Saxon'.

WHITEFLIES

Whiteflies are sap-feeding insects, about 2mm long, which, when adult, have white wings. They live on the underside of leaves and readily fly up when the plant is disturbed. They lay eggs and these hatch into flat, oval, scale-like nymphs. Like the adults, these nymphs excrete a sugary excrement called honeydew. This drops down onto the upper

surfaces of leaves below, which become sticky and covered with black sooty moulds. The final nymphal stage, sometimes called a pupa, is thicker, and often has hairs or white, waxy secretions on the upper surface. Some garden plants, such as rhododendron, honeysuckle and *Phillyrea*, are occasionally attacked by whiteflies specific to these plants. If necessary, control them as for azalea whitefly.
• see also AZALEA WHITEFLY *p.103*; CABBAGE WHITEFLY *p.114*; GLASSHOUSE WHITEFLY *p.134*; VIBURNUM WHITEFLY *p.187*.

WILLOW ANTHRACNOSE

SYMPTOMS Leaves curl, turn yellow and fall prematurely. Extensive defoliation may occur by mid-summer. Pinhead-sized, dark brown spots may develop on infected leaves and these produce spores. Raised, rough lens-shaped cankers develop on affected stems and these may cause dieback if they enlarge sufficiently to girdle the stem. Although symptoms may appear alarming, and recurrent attacks will weaken the tree, this disease is unlikely to cause the death of an otherwise healthy tree.
CAUSE The fungus *Marssonina salicicola*, which is encouraged by moist conditions. The spores from the leaves and the cankers are carried by water splash. The fungus overwinters in the stem cankers, on bud scales and on fallen leaves.
CONTROL Rake up and dispose of fallen infected leaves. Prune out severely cankered stems, if possible, without spoiling the weeping habit of the tree. Consider growing the resistant Peking willow *Salix babylonica* var. *pekinensis* 'Pendula' or the hybrid *S. × sepulcralis*.

WILLOW BARK APHID

SYMPTOMS Dense colonies of brownish-black aphids, which are up to 5mm long, form on the bark of willow trunks and branches, mainly in late summer (*see p.36*). The upper surface of the aphids is marked with small black dots, with two larger tubercles towards the rear of the abdomen. Large quantities of sugary honeydew are excreted by the aphids and this makes the stems and ground below sticky. Sooty moulds often grow on the honeydew, which also attracts wasps and flies.
CAUSE A sap-feeding insect, *Tuberolachnus salignus*.
CONTROL On willow trees small enough to be sprayed, this aphid can be controlled using bifenthrin, thiacloprid, acetamiprid, or pyrethrum.

WILLOW BEAN GALL SAWFLIES

SYMPTOMS Hard swellings form in the leaf blades during the summer months (*see p.17*). They are yellowish-green or red, and their rounded oval shape resembles that of baked beans. The galls are initiated when the female sawfly lays her eggs in the leaf. She secretes chemicals that induce the leaf to produce a gall around each egg. The egg later hatches into a caterpillar-like larva that gradually hollows out the gall.
CAUSE Several sawflies, *Pontania* species, cause this type of gall on willows, and there are often two generations occurring between late spring and early autumn.

CONTROL Apart from causing galls on the foliage, these sawflies have no harmful effect on willows. Affected leaves can be removed from small trees if they are unsightly, but spraying would not be justified.

WILLOW BLACK CANKER

SYMPTOMS In late spring or early summer angular, red-brown spots develop on the leaves; they spread, causing the leaves to shrivel and turn black, but remain hanging on the stem. The fungus then spreads into the stem and may produce lens-shaped, flattened cankers. It may be confused with willow scab (*see below*) or willow anthracnose (*see above*).
CAUSE The fungus *Glomerella miyabeana*, encouraged by warm temperatures and high humidity.
CONTROL Prune out affected shoots.

WILLOW LEAF BEETLES

SYMPTOMS Both the adult and larval stages damage the leaves of willow, poplar and aspen. The adult beetles are about 4mm long and metallic blue-black or brassy green in colour. The larvae have elongate bodies up to 6mm long, and are mainly black on the upper surface, but creamy-yellow below. Both the larvae and adult beetles graze away part of the leaf surface, causing the remaining damaged tissues to dry up and turn brown or white (*see p.18*). The larvae are gregarious and often feed side by side in groups of up to 30 or more.
CAUSE Several beetles and their larvae, including *Phyllodecta vitellinae*,

P. vulgatissima and *Plagiodera versicolora*, cause damage to willows and poplars from late spring to early autumn. There are usually two generations.
CONTROL Only young trees suffer significant damage, although the foliage on older trees is often disfigured. Small trees can be sprayed with bifenthrin, acetamiprid, or thiacloprid, particularly if infested in early summer.

WILLOW SCAB

SYMPTOMS Similar to willow anthracnose and willow black canker (*see above*). In early spring irregular spots develop on the young foliage, spreading and causing it to turn black and die. Olive-brown pustules develop on the lower leaf surface and on affected stems. Black lesions develop on the stems and may cause girdling (*see p.41*).
CAUSE The fungus *Fusicladium saliciperdum*, which overwinters as pustules on the stems.
CONTROL Prune out the affected stems.

WILLOW WATERMARK DISEASE

SYMPTOMS In mid- to late spring, young leaves turn red and wither. Ooze, which contains the bacteria, is produced from the stems. Leaves deteriorate further and affected branches die back. The tree loses much of its growth, but is rarely killed. The wood develops a watermark-like stain internally and infected cut surfaces show complete or partial red or black rings. The disease is not common, occurring only in certain parts of the country. *Salix alba* var. *caerulea* and other white willows may be

parts of the country. *Salix alba* var. *coerulea* and other white willows may be affected, but it is uncertain whether other species are also susceptible.

CAUSE The bacterium *Erwinia salicis*, which persists in the wood and produces toxins which cause the leaf discoloration and dieback. The bacteria enter the tree via wounds, and are probably spread by birds, insects and on air currents.

CONTROL Remove and burn affected trees.

WILTS

SYMPTOMS Wilting of the foliage followed by dieback. The whole plant is rarely affected simultaneously and may take several seasons to die. Smaller or herbaceous or annual plants may be killed in one season, however. Staining may develop, associated with the vascular or conducting tissues in stems and roots.

CAUSE Various fungi, including *Verticillium*, *Fusarium* and *Ceratocystis* species. Much damage is caused by the blocking of vascular elements and the subsequent deterioration of stems and leaves. The fungi may be spread by pruning, root contact or grafting, or by contaminated sap coming into contact with that of a healthy plant.

CONTROL None available. Where feasible, grow resistant plants. Remove and dispose of affected plants with the soil in the vicinity of the roots.

• see also ASTER WILT *p.103*; CLEMATIS WILT *p.119*; DROUGHT *p.125*; DUTCH ELM DISEASE *p.125*; FUSARIUM WILT *p.131*; PEONY WILT *p.160*; VERTICILLIUM WILTS *p.187*; WATERLOGGING *p.190*.

WIND DAMAGE

SYMPTOMS Leaves appear scorched. One side or face of the plant is generally more severely affected than the rest. The most damage occurs on the side facing the prevailing wind. Flowers may be similarly affected (*see **p.45***). If winds are salt-laden, the effect is usually much more severe and the whole plant may suffer extreme dieback or even be killed. Trees on particularly exposed sites may develop a lop-sided growth habit, known as "krumholtz", where they grow away from the prevailing wind.

CAUSE Wind. Often made worse if wind tunnels are created or if a windbreak is removed.

CONTROL Where possible, plant or erect windbreaks to prevent wind damage. If stems are killed, prune these out to prevent secondary organisms invading.

WINTER ACONITE SMUT

SYMPTOMS Raised, blister-like grey outgrowths develop on stems and leaves, erupting to release large quantities of black, powdery spores. Severely affected parts die back and general growth and vigour is reduced.

CAUSE The fungus *Urocystis eranthidis*, which appears to be systemic (that is, found throughout all parts of the plant and not just on the area showing the symptoms) within the plant.

CONTROL Remove and dispose of affected plants promptly.

WINTER MOTH

SYMPTOMS Yellowish-green caterpillars with paler lines along their bodies feed on the foliage (*see **p.31***) of many fruit trees and other deciduous trees and shrubs between bud burst and early spring. The caterpillars, which measure up to 25mm in length, walk with a looping action, as they have only two pairs of clasping legs on their abdomens. They bind leaves together loosely with silk threads and are often hidden from view. The caterpillars also eat fruit blossom and make holes in apple fruitlets, causing them to develop as badly misshapen fruits.

CAUSE Caterpillars of a moth, *Operophtera brumata*. The adult moths emerge between late autumn and mid-winter. Only males have normal wings; the female have swollen bodies with tiny wings and are incapable of flight. When they emerge from pupae in the soil, the females have to crawl up the tree trunk in order to lay their eggs on the branches.

CONTROL Place sticky grease bands around tree trunks, especially those of fruit trees, in mid-autumn to prevent female moths reaching the branches. Control newly-hatched caterpillars by spraying with either bifenthrin or pyrethrum as the leaves emerge from the buds.

WIREWORMS

SYMPTOMS Wireworms are slender, orange-brown beetle larvae that have three pairs of short legs near the head end (*see **p.54***). They can be up to 25mm long and have a small protuberance on the underside of the rear end of the abdomen. Wireworms live in the soil and can kill seedlings by biting into the stems just below soil level. In late summer, vegetables such as potato tubers (*see **p.56***) and carrots can be extensively tunnelled by the larvae.

CAUSE The larval stage of click beetles (*see p.119*), including *Agriotes lineatus*, *A. obscurus*, *A. sputator* and *Athous haemorrhoidalis*. They are found in the greatest numbers in newly-dug grassland; their numbers will gradually decline over the next three or four years if the ground is kept in cultivation.

CONTROL Where wireworms are troublesome, dig up potato tubers as soon as they have matured to limit damage. There are no soil insecticides for wireworm control.

WITCHES' BROOMS

SYMPTOMS Extensively branched twigs develop, clustered together to form "witches' brooms" in the crown of the tree (*see **p.41***). In winter they may at first be mistaken for birds' nests and can spoil the outline of the tree or hedge, but in summer they appear as dense, leafy areas and so are less obtrusive.

CAUSE Various fungi, in particular species of *Taphrina*, that are spread by water splash and on air currents. Occasionally other factors, in particular mite infestations, may be responsible.

CONTROL Prune out the growths, if desired. Spread seems to be very slow, but removal will help to limit the problem. No chemical control is available.

• see also BIRCH WITCHES' BROOMS *p.107*; HORNBEAM WITCHES' BROOMS *p.138*.

mistaken for birds' nests and can spoil the outline of the tree or hedge, but in summer they appear as dense, leafy areas and are less obtrusive.

CAUSE Various fungi, in particular species of *Taphrina*, that are spread by water splash and on air currents. Occasionally other factors, in particular mite infestations, may be responsible.

CONTROL Prune out the growths, if desired. Spread seems to be very slow, but removal will help to limit the problem. No chemical control is available.
• *see also* BIRCH WITCHES' BROOMS *p.107*; HORNBEAM WITCHES' BROOMS *p.138*.

WOODLICE

SYMPTOMS Grey or pinkish-brown animals (*see* **p.53**), with a hard segmented covering over their bodies. During the day, woodlice hide in dark places, such as under pots, logs and stones, or in the surface layers of compost heaps. They can damage seedlings, but not established plants.

CAUSE Several species of woodlice are common in gardens, including *Oniscus asellus*, *Porcellio scaber* and *Armadillidium vulgare*. The last-mentioned rolls itself up into a ball when disturbed and is known as a pillbug.

CONTROL Woodlice are so abundant in gardens that there is little prospect of achieving any lasting reduction in their numbers. Fortunately, they feed mainly on dead plant material rather than growing plants. Good hygiene in greenhouses will help to keep their numbers down.

WOOLLY APHID

SYMPTOMS A fluffy, white, waxy substance develops on the bark of apple trees and ornamental *Malus*, pyracantha and cotoneaster in summer (*see* **p.35**). The wax is secreted from the bodies of pinkish-brown aphids. During mid- to late spring they are mainly seen around old pruning cuts and in splits in older bark. Later in the summer the aphids spread to young branches where their feeding causes soft, knobbly swellings to develop. These galls often split during frosty weather creating wounds that can become infected with apple canker (*see p.100*).

CAUSE A sap-feeding insect, *Eriosoma lanigerum*. Most aphids overwinter as eggs. This species, however, overwinters as immature nymphs which are found under loose bark.

CONTROL Watch for signs of woolly aphid during spring. If present on ornamental plants, spray with thiacloprid or acetamiprid. On edible apples, some reduction can be achieved with thiacloprid or bifenthrin. Heavy infestations in late summer are difficult to control.
• *see also* BEECH WOOLLY APHID *p.106*.

WOOLLY VINE SCALE

SYMPTOMS This pest attacks many woody plants, including grape vine, alder, currants, gooseberry and pyracantha. When mature in late spring, females deposit eggs in a thick pad of white, waxy fibres (*see* **p.39**). The scales are dark brown and up to 6mm long. If they are removed by hand, the egg masses can be drawn out in long waxy threads. Heavily infested plants show a lack of vigour and the foliage is soiled by sooty moulds.

CAUSE A sap-feeding scale insect, *Pulvinaria vitis*. Nymphs hatch in early summer and feed first on the underside of leaves. Before leaf fall, however, they move back to the bark where they complete their development and overwinter as adults.

CONTROL Ornamental plants can be sprayed with thiacloprid or acetamiprid in mid-summer against the newly-hatched nymphs. On edible plants, pick off the egg masses and spray with either fatty acids, plant oils or thiacloprid (not grapes).

WORM CASTS

See EARTHWORMS *p.125*.

WORMS

See CUTWORMS *p.122*; EARTHWORMS *p.126*; EELWORMS *p.126*; ENCHYTRAEID WORMS *p.127*; FLATWORMS *p.128*; THUNDERWORMS *p.183*; WIREWORMS *p.193*.

YELLOW-TAIL MOTH

SYMPTOMS The adult moth has white wings and body, except for the end of the abdomen, which is covered with yellow hairs. It lays eggs on many trees and shrubs in late summer. It overwinters as young caterpillars which feed during spring and early summer. The hairy caterpillars, up to 30mm long, are black with an orange-red double stripe running down their upper body and have white markings along their sides. They eat both the foliage and flowers of their hosts.

CAUSE Caterpillars of a moth, *Euproctis similis*.

CONTROL Infestations are generally light and caterpillars can either be picked off or may be tolerated on established plants. If heavier attacks occur it may be necessary to spray with bifenthrin or pyrethrum.

YUCCA LEAF SPOT

SYMPTOMS Circular brown spots develop on affected leaves. On variegated yuccas, the spots are usually most prevalent on the variegated parts of the leaf (*see* **p.18**). The spots show concentric ringing and numerous, pinprick-sized, raised, black fruiting bodies develop on them. The spots enlarge and may join up.

CAUSE The fungi *Cercospora concentrica* and *Coniothyrium concentricum*, which produce similar symptoms. The spores are spread by water splash.

CONTROL Remove severely affected leaves.

ZUCCHINI YELLOW MOSAIC VIRUS

SYMPTOMS Bright yellow mosaic patterns appear on the leaves of marrows, courgettes and squashes. Foliage is also frequently distorted and the whole plant stunted. If plants reach their flowering and fruiting stage, fruits are very knobbly and distorted and the skin is blotched dark green.

CAUSE A mosaic virus, spread by aphids.

CONTROL Remove infected plants immediately.

INDIVIDUAL PLANT PROBLEMS

COMMON PROBLEMS TO which different garden plants are subject are listed here grouped by plant type, then under the common name of the plant within its type. Where common names are not known, the index (*pp.216–222*) lists these under plants' botanical names, for ease of access. For each individual plant problem, cross-references are given both to its detailed entry in the A–Z of Pests, Diseases and Disorders, and to any illustrations of the problem's effects in the Gallery of Symptoms.

GARDEN TREES

ONCE ESTABLISHED, trees are usually sufficiently vigorous for many problems to seem insignificant. However, if seriously attacked by a pest or a disease, an established tree's size can make it difficult – sometimes impossible – to treat. Many problems can be avoided by careful maintenance. With deciduous trees, the autumn leaf-fall also helps, removing many pathogens on foliage.

A number of common problems may affect a wide range of unrelated trees (*see right*). Some, however, are more host-specific; trees that are particularly prone to certain problems are listed below. Resistant species or cultivars may be suggested in the entry for the problem in question in the *A–Z of Pests, Diseases and Disorders*.

COMMON PROBLEMS

The most common non-host-specific tree ailments are:

ALGAE 98
APHIDS 99
BRACKET FUNGI 110
CATERPILLARS 115
CORAL SPOT 120
DROUGHT 125
FROST DAMAGE 130
GANODERMA 132
HONEY FUNGUS 138
IRON DEFICIENCY AND LIME-INDUCED CHLOROSIS 141
LEAF SPOTS (FUNGAL) 144
LIGHTNING DAMAGE 70
MAGNESIUM DEFICIENCY 147

PHYTOPHTHORA 161
POWDERY MILDEWS 167
SCORCH 175
VERTICILLIUM WILTS 187
VIRUSES 189
WATERLOGGING 190
WEEDKILLER DAMAGE 191
WIND DAMAGE 193

A

Acacia
FLUTED SCALE 129

Alder (*Alnus*)
ALDER SUCKER 32, 98
GALL MITES 131
PHYTOPHTHORA 161

Almond (*Prunus dulcis*)
APHIDS 99
BACTERIAL CANKER 35, 104
BLOSSOM WILT 41, 109
BROWN ROT 47, 49, 112
BULLFINCHES 46, 113
FASCIATION 43, 127
HONEY FUNGUS 34, 138
IRON DEFICIENCY AND LIME-INDUCED CHLOROSIS 24, 141
PEACH LEAF CURL 26, 157
PHYTOPHTHORA 161
POWDERY MILDEWS 25, 167
SHOT-HOLE 18, 30, 176
SHOT-HOLE BORERS 33, 177
SILVER LEAF 177

Amelanchier *see* SHRUBS AND CLIMBERS 198–9

Ash (*Fraxinus*)
ASH CATKIN GALL MITE 102
ASH SUCKER 102
BACTERIAL CANKER 35, 104
DIEBACK 124
GANODERMA 132
INONOTUS HISPIDUS 140
NECTRIA CANKER 152

B

Bay (*Laurus nobilis*)
BAY SUCKER 26, 104
CARNATION TORTRIX MOTH 20, 115

FROST DAMAGE (FOLIAGE AND STEMS) 28, 130
HORSE CHESTNUT SCALE 40, 139
POWDERY MILDEWS 25, 167
SOFT SCALE 179

Beech (*Fagus*)
BEECH BARK DISEASE 35, 106
BEECH BARK SCALE 35, 106
CANKER 114
CORAL SPOT 41, 120
FELT GALL MITE 19, 127
FOMES ROOT AND BUTT ROT 37, 129
GALL MIDGES 131
GANODERMA 132
HONEY FUNGUS 34, 138
LEAF MINING WEEVIL 24, 106
MERIPILUS GIGANTEUS 39, 148
PHYTOPHTHORA 161
POWDERY MILDEWS 25, 167
SQUIRRELS 52, 180
WOOLLY BEECH APHID 35, 106

Birch (*Betula*)
APHIDS 99
BRACKET FUNGI 37, 38, 110
FELT GALL MITE 19, 127
FOMES ROOT AND BUTT ROT 37, 129
HONEY FUNGUS 34, 138
LEAF SPOT (FUNGAL) 12, 13, 144
LEAF WEEVILS 145
PHYTOPHTHORA 161
POWDERY MILDEWS 25, 167
RUSTS 174
SAWFLIES 175
WITCHES' BROOMS 41, 193

C

Catalpa
MAGNESIUM DEFICIENCY 22, 147
PHYTOPHTHORA 161
POWDERY MILDEWS 25, 167

VERTICILLIUM WILTS 33, 187

Cedar (*Cedrus*)
BRACKET FUNGI 37, 38, 110
CEDAR APHID 32, 116
HONEY FUNGUS 34, 138

Cercidiphyllum japonicum
FROST DAMAGE (FOLIAGE AND STEMS) 28, 130
VERTICILLIUM WILTS 33, 187

Cherry, flowering *see* **Prunus, ornamental**

Cherry laurel *see* **Laurel, cherry**

Crab apple (*Malus*)
APPLE CANKER 36, 100
APPLE SCAB 16, 101
BLOSSOM WILT 41, 109
BROWN ROT 49, 112
CORAL SPOT 41, 120
FIREBLIGHT 35, 128
GANODERMA 132
HONEY FUNGUS 34, 138
LEAF MINERS 19, 101
LEAF WEEVILS 145
PAPERY BARK 40, 156
POWDERY MILDEWS 25, 167
SILVER LEAF 177
WINTER MOTH 31, 193
WOOLLY APHID 35, 194

Cypress, Leyland cypress (*Cupressus, x Cupressocyparis leylandii*)
CROWN GALL 38, 120
CYPRESS APHID 122
FOMES ROOT AND BUTT ROT 37, 129
HONEY FUNGUS 34, 138
JUNIPER SCALE 37, 142
KABATINA SHOOT BLIGHT 142
PHOMOPSIS CANKER 161
PHYTOPHTHORA 161
SEIRIDIUM CANKER 176

D

Dogwood (*Cornus*)
DOGWOOD ANTHRACNOSE 124
HORSE CHESTNUT SCALE 40, 139
PHYTOPHTHORA 161
POWDERY MILDEWS 25, 167

Douglas fir *see* **Fir, Douglas**

E

Elder *see* SHRUBS AND CLIMBERS 198–9

Elm (*Ulmus*)
BARK BEETLES 33, 104
CORAL SPOT 41, 120
DUTCH ELM DISEASE 124
GALL MITE 131
GANODERMA 132
HONEY FUNGUS 34, 138
HORSE CHESTNUT SCALE 40, 139
LEAF SPOT (FUNGAL) 12, 13, 144
PHYTOPHTHORA 161
WITCHES' BROOMS 41, 193

Eucalyptus
EUCALYPTUS GALL WASP 127
EUCALYPTUS SUCKER 127
OEDEMA 19, 154
PHYTOPHTHORA 161
SILVER LEAF 177

F

False acacia (*Robinia pseudoacacia*)
LAETIPORUS SULPHUREUS 38, 143
PHYTOPHTHORA 161

SAWFLIES 175
VERTICILLIUM WILTS 33, 187

Fir, Douglas (*Pseudotsuga menziesii*)
DOUGLAS FIR ADELGID 124
FOMES ROOT AND BUTT ROT 37, 129
PHYTOPHTHORA 161

Fir, silver (*Abies alba*)
HONEY FUNGUS 34, 138
SILVER FIR ADELGID 177

H

Hawthorn (*Crataegus*)
APHIDS 99
APPLE CANKER 36, 100
COTONEASTER WEBBER MOTH 37, 120
CROWN GALL 38, 120
FIREBLIGHT 35, 128
FOMES ROOT AND BUTT ROT 37, 129
HAWTHORN GALL MIDGE 136
HONEY FUNGUS 34, 138
LEAF SPOT (FUNGAL) 12, 13, 144
PHYTOPHTHORA 161
POWDERY MILDEWS 25, 167
RUSTS 174
SILVER LEAF 177

Hazel (*Corylus*) *see* FRUITS AND NUTS 206–7

Holly (*Ilex*)
CUSHION SCALE 15, 122
HONEY FUNGUS 34, 138
LEAF MINERS 17, 136
LEAF SPOT (FUNGAL) 12, 13, 144
PHYTOPHTHORA 161

Honey locust (*Gleditsia*)
CORAL SPOT **41**, 120
GLEDITSIA GALL MIDGE **27**, 134
MERIPILUS GIGANTEUS **39**, 148

Hornbeam (*Carpinus*)
APHIDS 99
CORAL SPOT **41**, 120
LEAF SPOT (FUNGAL) **12**, **13**, 144
WINTER MOTH **31**, 193
WITCHES' BROOMS **41**, 193

Horse chestnut (*Aesculus hippocastanum*)
BLEEDING CANKER **23**, **38**, 108, 138
GANODERMA 132
HONEY FUNGUS **34**, 138
HORSE CHESTNUT LEAF BLOTCH **20**, 139
HORSE CHESTNUT SCALE **40**, 139
LEAF MINING MOTH 139
LEAF SCORCH 144
PHYTOPHTHORA 161
SLIME FLUX AND WETWOOD 177

J

Judas tree (*Cercis siliquastrum*)
CORAL SPOT **41**, 120
VERTICILLIUM WILTS **33**, 187

Juniper (*Juniperus*)
CONIFER RED SPIDER MITE 120
JUNIPER APHIDS 142
JUNIPER SCALE **37**, 142
JUNIPER WEBBER MOTH **32**, 142
KABATINA SHOOT BLIGHT 142
PESTALOTIOPSIS 160
PHOMOPSIS CANKER 161
PHYTOPHTHORA 161
RUSTS 174

L

Laburnum
BLACKFLY 108
DOWNY MILDEW 125
HONEY FUNGUS **34**, 138
LABURNUM LEAF MINER **18**, 142
LEAF SPOT (FUNGAL) **12**, **13**, 144
PHYTOPHTHORA 161
POWDERY MILDEWS **25**, 167
SILVER LEAF 177
VIRUSES 189

Larch (*Larix*)
BRACKET FUNGI **37**, 110
FOMES ROOT AND BUTT ROT **37**, 129
HONEY FUNGUS **34**, 138
LARCH ADELGID **32**, 143
NEEDLE BLIGHTS AND CASTS OF PINES **32**, 152

Laurel, cherry (*Prunus laurocerasus*)
BACTERIAL CANKER **35**, 104
HONEY FUNGUS **34**, 138
LEAF SPOT (FUNGAL) **12**, **13**, 144
MAGNESIUM DEFICIENCY **22**, 147
PHYTOPHTHORA 161
POWDERY MILDEWS **25**, 167
SHOT-HOLE **18**, **30**, 176
SILVER LEAF 177

Lime (*Tilia*)
APHIDS 99
FELT GALL MITE **19**, 127
GANODERMA 132
HORSE CHESTNUT SCALE **40**, 139
LEAF SPOT (FUNGAL) **12**, **13**, 144
LIME NAIL GALL MITE **14**, 147
PHYTOPHTHORA 161
VERTICILLIUM WILTS **33**, 187

Liquidambar
HONEY FUNGUS **34**, 138
IRON DEFICIENCY AND LIME-INDUCED CHLOROSIS **24**, 141

M

Magnolia
CAPSID BUGS **29**, 115
CORAL SPOT **41**, 120
FROST DAMAGE **28**, **44**, 130
GREY MOULD **45**, 135
HONEY FUNGUS **34**, 138
HORSE CHESTNUT SCALE **40**, 139
LEAF SPOT (FUNGAL) **12**, **13**, 144
IRON DEFICIENCY AND LIME-INDUCED CHLOROSIS **24**, 141
SQUIRRELS **52**, 180
VIRUSES 189

Maple (*Acer*)
ACER GALL MITE **17**, 98
APHIDS 99
CORAL SPOT **41**, 120
FELT GALL MITE **19**, 127
GANODERMA 132
HONEY FUNGUS **34**, 138
HORSE CHESTNUT SCALE **40**, 139
POWDERY MILDEWS **25**, 167
SQUIRRELS **52**, 180
SYCAMORE SOOTY BARK DISEASE **36**, 182
VERTICILLIUM WILTS **33**, 187
See also **Sycamore**

Monkey puzzle (*Araucaria araucana*)
HONEY FUNGUS **34**, 138
PHYTOPHTHORA 161

Mountain ash/rowan (*Sorbus aucuparia*)
CORAL SPOT **41**, 120

FIREBLIGHT **35**, 128
HONEY FUNGUS **34**, 138
LEAF WEEVILS 145
LEOPARD MOTH 146
MOUNTAIN ASH BLISTER MITE 150
NECTRIA CANKER 152
SILVER LEAF 177

Mulberry *see* FRUITS AND NUTS 206–7

O

Oak (*Quercus*)
ACORN GALL WASP **52**, 98
APHIDS 99
FISTULINA HEPATICA 128
GALL WASPS 132
GANODERMA 132
GRIFOLA FRONDOSA 135
HONEY FUNGUS **34**, 138
LAETIPORUS SULPHUREUS 143
LEAF MINERS 144
LEAF SPOT (FUNGAL) **12**, **13**, 144
OAK PHYLLOXERA 153
OAK PROCESSIONARY MOTH 153
POWDERY MILDEWS **25**, 167

P

Pear, ornamental (*Pyrus*)
APPLE AND PEAR CANKER **36**, 100
HONEY FUNGUS **34**, 138
POWDERY MILDEWS **25**, 167

Pine (*Pinus*)
BRACKET FUNGI **37**, **38**, 110
CONIFER APHIDS 120
FOMES ROOT AND BUTT ROT **37**, 129
HONEY FUNGUS **34**, 138
PINE ADELGID **39**, 161
RUSTS 174

Plane (*Platanus*)
GANODERMA 132
INONOTUS HISPIDUS 140
POWDERY MILDEWS **25**, 167
SALT DAMAGE 175

Poplar (*Populus*)
APPLE SCAB **16**, 101
BACTERIAL CANKER **35**, 104
HONEY FUNGUS **34**, 138
LEAF SPOT (FUNGAL) **12**, **13**, 144
PEACH LEAF CURL **26**, 157
POPLAR CANKER 164
RUSTS 174
SILVER LEAF 177
WILLOW LEAF BEETLES **18**, 192

Prunus, ornamental
APPLE LEAF MINER **19**, 101
BACTERIAL CANKER **35**, 104
BLOSSOM WILT **41**, 109
BULLFINCHES **46**, 113
CHERRY BLACKFLY 117
CHERRY LEAF SCORCH 117
CHERRY LEAF SPOT **24**, 117
CROWN GALL **38**, 120
FASCIATION **43**, 127
HONEY FUNGUS **34**, 138
MAGNESIUM DEFICIENCY **22**, 147
PEACH LEAF CURL **26**, 157
PEAR AND CHERRY SLUGWORM **19**, 158
PHYTOPHTHORA 161
POWDERY MILDEWS **25**, 167
ROSE SICKNESS/REPLANT PROBLEMS 173
SHOT-HOLE **18**, **30**, 176
SHOT-HOLE BORERS **33**, 177
SILVER LEAF 177
WINTER MOTH **31**, 193
WISTERIA SCALE 193
WITCHES' BROOMS **41**, 193

R

Rowan *see* **Mountain ash**

S

Silver fir *see* **Fir, silver**

Spruce (*Picea*)
CONIFER APHIDS 120
CONIFER RED SPIDER MITE 120
FOMES ROOT AND BUTT ROT **37**, 129
GREEN SPRUCE APHID 135
HONEY FUNGUS **34**, 138
PINEAPPLE GALL ADELGID **38**, 162
RUSTS 174

Strawberry tree (*Arbutus unedo*)
ARBUTUS LEAF SPOT **14**, 102
PHYTOPHTHORA 161
VIBURNUM WHITEFLY **15**, 187

Sumach *see* SHRUBS AND CLIMBERS 198–9

Sweet chestnut (*Castanea sativa*)
HONEY FUNGUS **34**, 138
LAETIPORUS SULPHUREUS **38**, 143
PHYTOPHTHORA 161

Sycamore (*Acer pseudoplatanus*)
ACER GALL MITE **17**, 98
APHIDS 99
CORAL SPOT **41**, 120
FELT GALL MITE **19**, 127
GANODERMA 132
HONEY FUNGUS **34**, 138

HORSE CHESTNUT SCALE **40**, 139
LEAF SCORCH 144
POWDERY MILDEWS **25**, 167
SQUIRRELS **52**, 180
SYCAMORE SOOTY BARK DISEASE **36**, 182
TAR SPOT OF ACER **13**, 183
VERTICILLIUM WILTS **33**, 187
WINTER MOTH **31**, 193
WISTERIA SCALE 193

T

Thuja
CONIFER APHIDS 120
HONEY FUNGUS **34**, 138
JUNIPER SCALE **37**, 142
KABATINA SHOOT BLIGHT 142
SEIRIDIUM CANKER 176

Tulip tree (*Liriodendron*)
HONEY FUNGUS **34**, 138
LEAF SPOT (FUNGAL) **12**, **13**, 144
MAGNESIUM DEFICIENCY **22**, 147
SLIME FLUX AND WETWOOD 177

W

Walnut (*Juglans*) *see* FRUITS AND NUTS 206–7

Willow (*Salix*)
ANTHRACNOSE 99
APHIDS 99
DAEDALEOPSIS CONFRAGROSA **37**, 122
GALL MITES 131
HONEY FUNGUS **34**, 138
PHELLINUS IGNARIUS 160
RUSTS 174
WILLOW BARK APHID **36**, 192
WILLOW BEAN GALL SAWFLIES **17**, 192
WILLOW BLACK CANKER 192
WILLOW LEAF BEETLES **18**, 192
WILLOW SCAB **41**, 192
WILLOW WATERMARK DISEASE 192

YZ

Yew (*Taxus*)
BIG BUD MITE **41**, 107
CARNATION TORTRIX MOTH **20**, 115
LAETIPORUS SULPHUREUS **38**, 143
PHYTOPHTHORA 161
VINE WEEVIL **29**, **53**, 188

Zelkova
DUTCH ELM DISEASE 125

INDIVIDUAL PLANT PROBLEMS

SHRUBS AND CLIMBERS

CORRECT PRUNING is one of the keys to maintaining healthy shrubs and climbers, thinning overcrowded and old growth to prevent a stagnant environment. Prompt recognition of problems is also vital; the removal of much badly affected growth can spoil an attractively shaped shrub or carefully trained climber.

Some common problems may affect many unrelated plants (*see right*). Others are more host-specific; plants particularly prone to certain problems are listed below. Resistant species or cultivars may be suggested in the entry for the problem in question in the *A–Z of Pests, Diseases and Disorders*.

COMMON PROBLEMS

The most common non-host-specific ailments affecting shrubs and climbing plants are:

ALGAE 98
APHIDS 99
CORAL SPOT 120
DEER 123
FROST DAMAGE 130
HONEY FUNGUS 138
LEAF SCORCH 144
LEAF SPOT (FUNGAL) 144

IRON DEFICIENCY AND LIME-INDUCED CHLOROSIS 141
MAGNESIUM DEFICIENCY 147
PHYTOPHTHORA 161
POWDERY MILDEWS 167
RABBITS 168
VERTICILLIUM WILTS 187
VIRUSES 189
WEEDKILLER DAMAGE 191

A

Actinidia
GLASSHOUSE RED SPIDER MITE **22**, **23**, 134
HONEY FUNGUS **34**, 138

Amelanchier
FIREBLIGHT **35**, 128

Aucuba
FROST DAMAGE (FOLIAGE AND STEMS) **28**, 130
PHYTOPHTHORA 161

B

Bamboo
APHIDS 99

Barberry (Berberis)
APHIDS 99
HONEY FUNGUS **34**, 138
LEAF SPOT (FUNGAL) **12**, **13**, 144
POWDERY MILDEWS **49**, 167
RUSTS 174
SAWFLY 107

Box (Buxus)
BOX BLIGHT 110
BOX RED SPIDER MITE 110
BOX SUCKER **26**, 110
LEAF SPOT (FUNGAL) **12**, **13**, 144
MUSSEL SCALE **36**, 151
PHYTOPHTHORA 161
RUSTS 174

Broom (Cytisus)
BROOM GALL MITE **39**, 111
FROST DAMAGE (FOLIAGE AND STEMS) **28**, 130
HONEY FUNGUS **34**, 138
SILVER LEAF 177
WATERLOGGING **34**, 190

Buddleja
APHIDS 99
CAPSID BUGS **29**, **46**, 115
EARWIGS **46**, 126

FIGWORT WEEVILS 128
GLASSHOUSE RED SPIDER MITE **22**, **23**, 133
HONEY FUNGUS **34**, 138
LEAF AND BUD EELWORM **25**, 143
LEAF SPOT (FUNGAL) **12**, **13**, 144
MULLEIN MOTH 151
VIRUSES **23**, **27**, 189

C

Camellia
APHIDS 99
CAMELLIA GALL **26**, 114
CAMELLIA PETAL BLIGHT 114
CAMELLIA YELLOW MOTTLE **22**, 114
CARNATION TORTRIX MOTH **20**, 115
CUSHION SCALE **15**, 122
HONEY FUNGUS **34**, 138
LEAF SPOT (FUNGAL) **12**, **13**, 144
PESTALOTIOPSIS 160
PHYTOPHTHORA 161
SQUIRRELS **52**, 180
VINE WEEVIL **29**, 188

Carpenteria
LEAF SPOT (FUNGAL) **12**, **13**, 144

Caryopteris
CAPSID BUGS **29**, **46**, 115
HONEY FUNGUS **34**, 138

Ceanothus
DROUGHT **44**, 125
FROST DAMAGE **28**, **44**, 130
HONEY FUNGUS **34**, 138
IRON DEFICIENCY AND LIME-INDUCED CHLOROSIS **24**, 141
MUSSEL SCALE **36**, 151

Clematis
APHIDS 99
CAPSID BUGS **29**, **46**, 115
CATERPILLARS 115

CLEMATIS GREEN PETAL 118
CLEMATIS WILT 119
EARWIGS **46**, 126
GLASSHOUSE RED SPIDER MITE **22**, **23**, 133
LEAF SPOT (FUNGAL) **12**, **13**, 144
POWDERY MILDEWS **49**, 167
SLIME FLUX AND WETWOOD 177
VIRUSES **23**, **27**, 189

Cotinus
POWDERY MILDEWS **49**, 167
VERTICILLIUM WILTS **29**, 187

Cotoneaster
APHIDS 99
BROWN SCALE **36**, 112
COTONEASTER WEBBER MOTH **37**, 120
FIREBLIGHT **35**, 128
HONEY FUNGUS **34**, 138
LEAF SPOT (FUNGAL) **12**, **13**, 144
MUSSEL SCALE **36**, 151
SILVER LEAF 177
VIRUSES **23**, **27**, 189
WOOLLY APHID **35**, 194

D

Daphne
CUCUMBER MOSAIC VIRUS **21**, **25**, 121
FASCIATION **43**, 127
GREY MOULD **34**, 135
HONEY FUNGUS **34**, 138
LEAF SPOT (FUNGAL) **12**, **13**, 144
VERTICILLIUM WILTS **29**, 187
VIRUSES **22**, **23**, 189

Dogwood (Cornus)
DOGWOOD ANTHRACNOSE 124
HORSE CHESTNUT SCALE **40**, 139
MUSSEL SCALE **36**, 151
PHYTOPHTHORA 161
POWDERY MILDEWS **49**, 167

E

Elaeagnus
CORAL SPOT **41**, 120
ELEAGNUS SUCKER 126
LEAF SPOT (FUNGAL) **12**, **13**, 144

Elder (Sambucus)
BLACKFLY 108
GLASSHOUSE RED SPIDER MITE **22**, **23**, 133
HONEY FUNGUS **34**, 138
LEAF SPOT (FUNGAL) **12**, **13**, 144
VIRUSES **23**, **27**, 189

Escallonia
BROWN SCALE **36**, 112
SILVER LEAF 177
VIRUSES **23**, **27**, 189

Euonymus
BLACKFLY 108
CROWN GALL **38**, 120
CUSHION SCALE **15**, 122
EUONYMUS SCALE* **21**, 127
POWDERY MILDEWS **49**, 167
SMALL ERMINE MOTHS **38**, 178
VINE WEEVIL* **29**, 188
VIRUSES **22**, **23**, 189
(*E. japonica only)

F

Fatsia
CAPSID BUGS **29**, **46**, 115

Firethorn (Pyracantha)
BROWN SCALE **36**, 112
CORAL SPOT **41**, 120
FIREBLIGHT **35**, 128
PYRACANTHA LEAF MINER 167
PYRACANTHA SCAB 167
WOOLLY APHID **35**, 194
WOOLLY VINE SCALE **39**, 194

Flowering currant (Ribes)
AMERICAN GOOSEBERRY MILDEW **49**, 98
BROWN SCALE **36**, 112

CORAL SPOT **41**, 120
CROWN GALL **38**, 120
CURRANT AND GOOSEBERRY LEAF SPOT 121
CURRANT BLISTER APHID **27**, 121
DIEBACK 124
GREY MOULD **34**, 135
HONEY FUNGUS **34**, 138
IRON DEFICIENCY AND LIME-INDUCED CHLOROSIS **24**, 141
RUSTS 174
WOOLLY VINE SCALE **39**, 194

Forsythia
BULLFINCHES **46**, 113
CAPSID BUGS **29**, **46**, 115
FORSYTHIA GALL **40**, 130
HONEY FUNGUS **34**, 138
LEAF SPOT (FUNGAL) **12**, **13**, 144
POWDERY MILDEWS **49**, 167

Fuchsia, hardy
CAPSID BUGS **29**, **46**, 115
ELEPHANT HAWK MOTH **30**, 126
FLEA BEETLES **29**, 128
FUCHSIA GALL MITE 130
FUCHSIA RUST **18**, 131
GLASSHOUSE RED SPIDER MITE **22**, **23**, 133
LEAF SPOT (FUNGAL) **12**, **13**, 144

H

Hawthorn (Crataegus)
see GARDEN TREES 196–7

Heathers (Erica, Calluna)
HONEY FUNGUS **34**, 138
LEAF SPOT (FUNGAL) **12**, **13**, 144
PHYTOPHTHORA 161
POWDERY MILDEWS **49**, 167

Hebe
APHIDS 99
CARNATION TORTRIX MOTH **20**, 115
DOWNY MILDEW 125
LEAF SPOT (FUNGAL) **12**, **13**, 144

Honeysuckle (*Lonicera*)
HONEYSUCKLE APHID 138
LEAF SPOT (FUNGAL) **12**, **13**, 144
POWDERY MILDEWS **49**, 167
SILVER LEAF 177
THRIPS 183
WHITEFLIES 191

Hydrangea
APHIDS 99
CAPSID BUGS **29**, **46**, 115
FOOT AND ROOT ROT **55**, 129
FROST DAMAGE (FOLIAGE AND STEMS) **28**, 130
GLASSHOUSE RED SPIDER MITE **22**, **23**, 133
GREY MOULD **34**, 135
HONEY FUNGUS **34**, 138
HYDRANGEA SCALE **15**, 140
IRON DEFICIENCY AND LIME-INDUCED CHLOROSIS **24**, 141
LEAF SPOT (FUNGAL) **12**, **13**, 144
MAGNESIUM DEFICIENCY **22**, 147
POWDERY MILDEWS **49**, 167
VINE WEEVIL **29**, 188
VIRUSES **23**, **27**, 189

Hypericum
RUSTS 174

I

Ivy (*Hedera*)
BLACKFLY 108
BRYOBIA MITES 112
LEAF SPOT (FUNGAL) **12**, **13**, 144
SOFT SCALE 179
VIBURNUM SCALE **15**, 187
VINE WEEVIL **29**, 188
VIRUSES **22**, **23**, 189

J

Japanese quince (*Chaenomeles*)
APHIDS 99
BROWN ROT **47**, 112
BROWN SCALE **36**, 112
GREY MOULD **34**, 135
IRON DEFICIENCY AND LIME-INDUCED CHLOROSIS **24**, 141

Jasmine (*Jasminum*)
GLASSHOUSE RED SPIDER MITE **22**, **23**, 133
GREY MOULD **34**, 135
VIRUSES **23**, **27**, 189

Jerusalem sage (*Phlomis*)
LEAFHOPPERS 144

L

Laurel (*Laurus*)
BACTERIAL CANKER **35**, 104
BAY SUCKER **26**, 104

CARNATION TORTRIX MOTH **20**, 115
HONEY FUNGUS **34**, 138
HORSE CHESTNUT SCALE **40**, 139
LEAF SPOT (FUNGAL) **12**, **13**, 144
POWDERY MILDEWS **49**, 167
SOFT SCALE 179

Lavatera
LAVATERA LEAF AND STEM ROT 143
LEAF SPOT (FUNGAL) **12**, **13**, 144
RUSTS 174

Lavender (*Lavandula*)
CUCKOO SPIT **39**, **43**, 121
GREY MOULD **34**, 135
HONEY FUNGUS **34**, 138
LAVENDER SHAB 143
LEAF SPOT (FUNGAL) **12**, **13**, 144
ROSEMARY BEETLE 174

Lilac (*Syringa*)
HONEY FUNGUS **34**, 138
LEAF SPOT (FUNGAL) **12**, **13**, 144
LILAC BLIGHT **33**, 146
LILAC LEAF MINER **19**, 147
POWDERY MILDEWS **49**, 167
PRIVET THRIPS **22**, 167
SILVER LEAF 177

M

Magnolia *see* GARDEN TREES 196–7

Mahonia
BERBERIS SAWFLY 107
MAHONIA RUST **12**, **24**, 148
POWDERY MILDEWS **49**, 167

Mexican orange (*Choisya*)
CARNATION TORTRIX MOTH **20**, 115
FLUTED SCALE 129
FROST DAMAGE **28**, **44**, 130
GLASSHOUSE RED SPIDER MITE **22**, **23**, 133
SCORCH 175
SNAILS **29**, 178

Mock orange (*Philadelphus*)
BLACKFLY 108
LEAF SPOT (FUNGAL) **12**, **13**, 144
VIRUSES **23**, **27**, 189

Morning glory (*Ipomoea*)
FROST DAMAGE (FOLIAGE AND STEMS) **28**, 130
GLASSHOUSE RED SPIDER MITE **22**, **23**, 133
MAGNESIUM DEFICIENCY **22**, 147
VIRUSES **23**, **27**, 189

N

New Zealand flax (*Phormium*)
LEAF SPOT (FUNGAL) **12**, **13**, 144
PHORMIUM MEALYBUG 161

P

Photinia
APPLE SCAB **16**, 101
FIREBLIGHT **35**, 128
LEAF SPOT (FUNGAL) **12**, **13**, 144
PEAR SCAB **48**, 159

Pieris
PIERIS LACEBUG 161

Privet (*Ligustrum*)
APHIDS 99
HONEY FUNGUS **34**, 138
LEAF SPOT (FUNGAL) **12**, **13**, 144
LILAC LEAF MINER **19**, 147
PRIVET THRIPS **22**, 167
VIRUSES **23**, **27**, 189

R

Rhododendron
APHIDS 99
AZALEA WHITEFLY 103
BUD, DRY 112
CATERPILLARS 115
CROWN GALL **38**, 120
FROST DAMAGE **28**, **44**, 130
GREY MOULD **34**, 135
HONEY FUNGUS **34**, 138
IRON DEFICIENCY AND LIME-INDUCED CHLOROSIS **24**, 141
LEAF SCORCH 144
PHYTOPHTHORA 161
PIERIS LACEBUG 161
RHODODENDRON BUD BLAST **41**, 170
RHODODENDRON LACEBUG **21**, 170
RHODODENDRON LEAFHOPPER 170
RHODODENDRON PETAL BLIGHT **42**, 170
RHODODENDRON POWDERY MILDEW **15**, 171
RUSTS 174
SILVER LEAF 177
VINE WEEVIL **29**, 188

Rock rose (*Helianthemum*)
APHIDS 99
LEAF SPOT (FUNGAL) **12**, **13**, 144
POWDERY MILDEWS **49**, 167

Rose (*Rosa*)
CAPSID BUGS **29**, **46**, 115
CROWN GALL **38**, 120

DEER **41**, 123
DIEBACK 124
DOWNY MILDEW 125
GLASSHOUSE RED SPIDER MITE **22**, **23**, 133
GREY MOULD **34**, 135
HONEY FUNGUS **34**, 138
LEAF-CUTTING BEES **30**, 144
MAGNESIUM DEFICIENCY **22**, 147
NITROGEN DEFICIENCY 152
POTASSIUM DEFICIENCY **24**, 164
RABBITS 168
ROBIN'S PIN CUSHION **39**, 171
ROSE APHIDS **43**, 172
ROSE BALLING **44**, 172
ROSE BLACK SPOT **13**, 172
ROSE LEAF-ROLLING SAWFLY **27**, 172
ROSE LEAFHOPPER **21**, 173
ROSE POWDERY MILDEW **20**, 173
ROSE ROOT APHID **35**, 173
ROSE RUST **16**, 174
ROSE SICKNESS/REPLANT PROBLEMS 173
ROSE SLUGWORM **18**, 174
VIRUSES **23**, **27**, 189
WEEDKILLER DAMAGE – BROADLEAVED **27**, 190
WEEDKILLER DAMAGE – GLYPHOSATE 191
WINTER MOTH **31**, 193

Rosemary (*Rosmarinus*)
BROWN SCALE **36**, 112
CUCKOO SPIT **39**, **43**, 121
FROST DAMAGE (FOLIAGE AND STEMS) **28**, 130
HONEY FUNGUS **34**, 138
LEAFHOPPERS 144
ROSEMARY BEETLE 174

Russian vine (*Polygonum baldschuanicum*)
LEAF MINERS 144

S

Skimmia
HORSE CHESTNUT SCALE **40**, 139
IRON DEFICIENCY AND LIME-INDUCED CHLOROSIS **24**, 141
MAGNESIUM DEFICIENCY **22**, 147
PHYTOPHTHORA 161
VINE WEEVIL **29**, 188
WATERLOGGING **34**, 190

Solanum
FLEA BEETLES **29**, 128
GREY MOULD **34**, 135
LEAF SPOT (FUNGAL) **12**, **13**, 144
VERTICILLIUM WILTS **29**, 187

Stranvaesia
FIREBLIGHT **35**, 128

HONEY FUNGUS **34**, 138

Sumach (*Rhus*)
FROST DAMAGE **28**, **44**, 130
LEAF SPOT (FUNGAL) **12**, **13**, 144
VERTICILLIUM WILTS **29**, 187

Sun rose (*Cistus*)
APHIDS 99

Sweet bay (*Laurus*) *see* Laurel

T

Trumpet vine (*Campsis*)
BROWN SCALE **36**, 112
GLASSHOUSE RED SPIDER MITE **22**, **23**, 133

V

Viburnum
BLACKFLY 108
CAPSID BUGS **29**, **46**, 115
CARNATION TORTRIX MOTH **20**, 115
GLASSHOUSE THRIPS 134
GREY MOULD **34**, 135
HONEY FUNGUS **34**, 138
LEAF SPOT (FUNGAL) **12**, **13**, 144
VIBURNUM BEETLE 187
VIBURNUM SCALE 187
VIBURNUM WHITEFLY **15**, 187

W

Weigela
BROWN SCALE **36**, 112
CAPSID BUGS **29**, **46**, 115
HONEY FUNGUS **34**, 138
LEAF AND BUD EELWORM **25**, 143
LEAF SPOT (FUNGAL) **12**, **13**, 144

Wisteria
BROWN SCALE **36**, 112
FROST DAMAGE **28**, **44**, 130
GLASSHOUSE RED SPIDER MITE **22**, **23**, 133
HONEY FUNGUS **34**, 138
LEAF SPOT (FUNGAL) **12**, **13**, 144
MAGNESIUM DEFICIENCY **22**, 147
POWDERY MILDEWS **25**, 167
WISTERIA SCALE 193

Y

Yucca
BLACKFLY 108
SNAILS **29**, 178
YUCCA LEAF SPOT **18**, 194

HERBACEOUS PERENNIALS

PERENNIALS DO NOT suffer from many serious problems. The annual dying-back of foliage and stems gives an opportunity for the removal and disposal of plant tissue harbouring many diseases and pests. Regular division of established clumps also helps to keep plants healthy. A number of common problems may affect a wide range of unrelated plants (*see right*). Some, however, are more host-specific (*see below*). Resistant species or cultivars may be suggested in the entry for the problem in question in the *A–Z of Pests, Diseases and Disorders*.

COMMON PROBLEMS

The most common non-host-specific ailments affecting herbaceous perennials are:

APHIDS 99	LEAF SPOTS (FUNGAL) 144
CATERPILLARS 115	POWDERY MILDEWS 167
CROWN ROT 121	RABBITS 168
DEER 123	SLUGS 178
DOWNY MILDEW 125	SNAILS 178
DROUGHT 125	SWIFT MOTH 182
FROST DAMAGE 130	VIRUSES 189
GREY MOULD 135	WATERLOGGING 190

A

Acanthus
POWDERY MILDEWS **25**, 167

Agapanthus
VIRUSES **42**, 189

Alstroemeria
VIRUSES **42**, 189

Alyssum
CLUBROOT **55**, 119
DOWNY MILDEW 125
FLEA BEETLES **29**, 128
FOOT AND ROOT ROT **55**, 129
FUSARIUM WILT 131
POWDERY MILDEWS **25**, 167
SCLEROTINIA **36**, 175
WHITE BLISTER **14**, 191

Anchusa
SCLEROTINIA **36**, 175

Anemone
DOWNY MILDEW 125
GREY MOULD **45**, 135
PLUM RUST **22**, 163
POWDERY MILDEWS **25**, 167
SMUTS **35**, 178
VIRUSES **42**, 189

Astilbe
POWDERY MILDEWS **25**, 167

Aubrieta
APHIDS 99
DOWNY MILDEW 125
FLEA BEETLES **29**, 128
POWDERY MILDEWS **25**, 167
WHITE BLISTER **14**, 191

Auricula
(*Primula auricula*)
CROWN ROT **20**, 121
GLASSHOUSE LEAFHOPPER **20**, 133
ROOT APHIDS **55**, 171
VINE WEEVIL **29**, 188

C

Californian poppy
(*Romneya*)
CATERPILLARS 115
PHYTOPHTHORA 161
VERTICILLIUM WILTS **29**, 187

Campanula
DOWNY MILDEW 125
GREY MOULD **45**, 135
LEAF SPOT (FUNGAL) **12**, **13**, 144
PHYTOPHTHORA 161
POWDERY MILDEWS **25**, 167
RUSTS 174
VERTICILLIUM WILTS **29**, 187
VIRUSES **42**, 189

Candytuft (*Iberis*)
CLUBROOT **55**, 119
DOWNY MILDEW 125
LEAF SPOT (FUNGAL) **12**, **13**, 144
POWDERY MILDEWS **25**, 167

Canna
HONEY FUNGUS **34**, 138
VIRUSES **42**, 189

Carnation (*Dianthus*)
APHIDS 99
FUSARIUM WILT 131
GREY MOULD **45**, 135
LEAF SPOT (FUNGAL) **12**, **13**, 144
LEAFY GALL **39**, 145
POWDERY MILDEWS **25**, 167
RUSTS 174
SLUGS **29**, 178
SMUTS **35**, 178
SNAILS **29**, 178
VERTICILLIUM WILTS **29**, 187
VIRUSES **42**, 189

Centaurea
LETTUCE DOWNY MILDEW **22**, 146
POWDERY MILDEWS **25**, 167
RUSTS 174
SCLEROTINIA **36**, 175

Chrysanthemum
APHIDS 99, 148
CAPSID BUGS **29**, 115
CATERPILLARS 115
CHRYSANTHEMUM EELWORM 118
CHRYSANTHEMUM LEAF MINER **25**, 118
CHRYSANTHEMUM WHITE RUST **14**, 118
CROWN GALL **38**, 120
EARWIGS **46**, 126
FUSARIUM WILT 131
LEAF SPOT (FUNGAL) **12**, **13**, 144
MELON COTTON APHID **43**, 148
POWDERY MILDEWS **25**, 167
RUSTS 174

Columbine (*Aquilegia*)
APHIDS 99
AQUILEGIA SAWFLY 102
LEAF MINERS 144
LEAF SPOT (FUNGAL) **12**, **13**, 144
POWDERY MILDEWS **25**, 167

Convolvulus
FROST DAMAGE (FOLIAGE AND STEMS) **28**, 130
WATERLOGGING **34**, 190

Coral flower (*Heuchera*)
LEAF SPOT (FUNGAL) **12**, **13**, 144
LEAFY GALL **39**, 145
RUSTS 174
VINE WEEVIL **29**, 188

Cosmos
FROST DAMAGE **28**, **44**, 130
POWDERY MILDEWS **25**, 167

Cranesbill (*Geranium*)
CAPSID BUGS **29**, 115
DOWNY MILDEW 125
GERANIUM SAWFLY 132
LEAFY GALL **39**, 145
POWDERY MILDEWS **25**, 167
VINE WEEVIL **29**, 188

D

Daisy (*Bellis perennis*)
FASCIATION **43**, 127
POWDERY MILDEWS **25**, 167

Daylily (*Hemerocallis*)
HEMEROCALLIS GALL MIDGE **45**, 136
LEAF SPOT (FUNGAL) **12**, **13**, 144

Delphinium
DELPHINIUM LEAF MINER **18**, 123
DELPHINIUM MOTH 123
FASCIATION **43**, 127
FOOT AND ROOT ROTS **55**, 129
GREY MOULD **45**, 135
LEAF SPOT (BACTERIAL) **13**, **29**, 144
LEAF SPOT (FUNGAL) **12**, 144
POWDERY MILDEWS **25**, 167
SCLEROTINIA **36**, 175
SLUGS **29**, 178

E

Elephant's ears (*Bergenia*)
LEAF AND BUD EELWORM **25**, 143
LEAF SPOT (FUNGAL) **12**, **13**, 144
VINE WEEVIL **29**, 188

Epimedium
LEAF-CUTTING BEES **30**, 144
LEAF SPOT (FUNGAL) **12**, **13**, 144
MAGNESIUM DEFICIENCY **22**, 147
VINE WEEVIL **29**, 188

Euphorbia
FASCIATION **43**, 127
GREY MOULD **45**, 135

F

Foxglove (*Digitalis*)
CATERPILLARS 115

DOWNY MILDEW 125
GLASSHOUSE LEAFHOPPER **20**, 133
LEAF SPOT (FUNGAL) **12**, **13**, 144
POWDERY MILDEWS **25**, 167

G

Gentian
RUSTS 174

Geranium *see* **Cranesbill**

Geum
APHIDS 99
DOWNY MILDEW 125
GEUM SAWFLIES 132

Globe thistle (*Echinops*)
APHIDS 99
LEAF MINERS 144

Goatsbeard
(*Aruncus dioicus*)
ARUNCUS SAWFLY **32**, 102
POWDERY MILDEWS **25**, 167

Golden rod (*Solidago*)
POWDERY MILDEWS **25**, 167

H

Hellebore (*Helleborus*)
APHIDS 99
DOWNY MILDEW 125
HELLEBORE BLACK DEATH **20**, 136
HELLEBORE LEAF BLOTCH **20**, 136
SLUGS **29**, 178

Himalayan poppy
(*Meconopsis*)
DOWNY MILDEW 125
PHYTOPHTHORA 161
SLUGS **29**, 178

Hollyhock (*Alcea*)
CATERPILLARS 115
HOLLYHOCK RUST **17**, 137
SLUGS **29**, 178
SNAILS **29**, 178

Honesty (*Lunaria*)
Clubroot **55**, 119
Powdery Mildews **25**, 167
Viruses **42**, 189
White Blister **14**, 191

Hosta
Grey Mould **45**, 135
Leaf Spot (Fungal) **12**, **13**, 144
Slugs **29**, 178
Snails **29**, 178
Vine Weevil **29**, 188
Viruses **42**, 189

House leek (*Sempervivum*)
Rusts 174
Sempervivum Leaf Miner 176
Vine Weevil **29**, 188

I

Ice plant (*Sedum spectabile*)
Vine Weevil **29**, 188

J

Japanese anemone (*Anemone* × *hybrida*, *A. hupehensis*)
Grey Mould **45**, 135
Leaf and Bud Eelworm **25**, 143
Powdery Mildews **25**, 167
Viruses **42**, 189

L

Lily-of-the-valley (*Convallaria*)
Grey Mould **45**, 135

Leaf Spot (Fungal) **12**, **13**, 144

Lobelia
Damping Off **34**, 123
Foot and Root Rots **55**, 129
Glasshouse Red Spider Mite **22**, **23**, 1323
Slugs **29**, 178
Snails **29**, 178
Viruses **42**, 189

Lupin (*Lupinus*)
Foot and Root Rots **55**, 129
Glasshouse Red Spider Mite **22**, **23**, 133
Leaf Spot (Fungal) **12**, **13**, 144
Lupin Aphid **41**, 147
Powdery Mildews **25**, 167
Sclerotinia **36**, 175
Slugs **29**, 178
Snails **29**, 178
Viruses **42**, 189

Lychnis
Grey Mould **45**, 135

M

Michaelmas daisy (*Aster novi-belgii*)
Fusarium Wilt 131
Grey Mould **45**, 135
Leaf Spot (Fungal) **12**, **13**, 144
Michaelmas Daisy Mite **45**, 149
Powdery Mildews **25**, 167

P

Peony (*Paeonia*)
Caterpillars 115

Honey Fungus **34**, 138
Leaf Spot (Fungal) **12**, **13**, 144
Peony Wilt **43**, 160
Viruses **42**, 189

Penstemon
Frost Damage (Foliage and Stems) **28**, 130
Grey Mould **45**, 135
Leaf and Bud Eelworm **25**, 143
Leaf Spot (Fungal) **12**, **13**, 144

Periwinkle (*Vinca*)
Periwinkle Rust **15**, 160

Phlox
Crown Gall **38**, 120
Leaf Spot (Fungal) **12**, **13**, 144
Leafy Gall **39**, 145
Phlox Eelworm **33**, 160
Powdery Mildews **25**, 167
Verticillium Wilts **29**, 187

Phygelius
Capsid Bugs **29**, 115
Figwort Weevils 128

Pinks *see* **Carnation**

Polyanthus (*Primula*)
Aphids 99
Brown Core of Primulas 111
Bryobia Mites 112
Caterpillars 115
Glasshouse Leafhopper **20**, 133
Glasshouse Red Spider Mite **22**, **23**, 133
Grey Mould **45**, 135
Leaf Spot (Fungal) **12**, **13**, 144
Slugs **29**, 178
Sparrows **46**, 179
Vine Weevil **29**, 188
Viruses **42**, 189

Primrose *see* **Polyanthus**

Pulmonaria
Powdery Mildews **25**, 167

Pyrethrum (*Tanacetum coccineum*)
Aphids 99
Chrysanthemum Eelworm 118
Chrysanthemum Leaf Miner **25**, 118
Grey Mould **45**, 135
Viruses **42**, 189

R

Red-hot poker (*Kniphofia*)
Aphids 99
Snails **29**, 178

Rock cress (*Arabis*)
Flea Beetles **29**, 128
Viruses **42**, 189
White Blister **14**, 191

S

Sage (*Salvia*)
Capsid Bugs **29**, 115
Foot and Root Rots **55**, 129
Leafhoppers 144
Powdery Mildews **25**, 167
Slugs **29**, 178
Verticillium Wilts **29**, 187

Saxifrage (*Saxifraga*)
Rusts 174
Vine Weevil **29**, 188

Snapdragon (*Antirrhinum*)
Antirrhinum Rust **15**, 99
Damping Off **34**, 123
Downy Mildew 125

Fusarium Wilt 131
Leaf Spot (Bacterial) **13**, 29, 144
Leaf Spot (Fungal) **12**, **13**, 144
Powdery Mildews **25**, 167
Sclerotinia **36**, 175
Shot-hole **18**, **30**, 176
Verticillium Wilts **29**, 187
Wilts 193

Stachys
Powdery Mildews **25**, 167
Stachys Case-Bearing Caterpillar 180

V

Verbascum
Figwort Weevils 128
Mullein Moth 151
Powdery Mildews **25**, 167
Waterlogging **34**, 190

Violet (*Viola*)
Aphids 99
Glasshouse Red Spider Mite **22**, **23**, 133
Leaf Spot (Fungal) **12**, **13**, 144
Powdery Mildews **25**, 167
Violet Gall Midge **27**, 188

W

Water lily (*Nymphaea*)
China Mark Moth **30**, 117
Chironomid Midges 117
Water Lily Aphid 188
Water Lily Beetle **30**, 190
Water Lily Leaf Spot 190

LAWNS

Many gardeners happily tolerate a certain quantity of weeds in the lawn, and indeed some would say that a crop of daisies adds to a lawn's appeal. A possible exception is moss, which can be pernicious, particularly on damp ground; there are several effective moss-killers available, often combined with a lawn-feed mixture. Eradication of severe moss infestation may also have to be combined with re-seeding.

Pests and diseases, however, are always disfiguring and controls are often necessary. In most cases, chemical control must be combined with cultural improvements if the lawn's health is to be improved in the long term.

COMMON PROBLEMS

The most common ailments affecting lawns are:

Ants **59**, 99
Burrowing Bees **58**, 113
Chafer Grubs **58**, 116
Earthworms **58**, 125
Leatherjackets **58**, 145
Moles **59**, 150
Slime Mould **58**, 177
Turf Dollar Spot **58**, 185
Turf Fairy Rings **59**, 186
Turf Ophiobolus Patch 186
Turf Red Thread **59**, 186
Turf Snow Mould **59**, 186

Turf Thatch Fungal Mycelium **58**, 186
Turf Toadstools 186

BULBOUS PLANTS

THE PLANTING OF bulbs, corms and tubers in a suitable site is crucial to their health. Many will not tolerate soils which become heavy or wet in winter, soon succumbing to disease. In most cases, too, it is essential for the foliage to be left in place for several weeks after flowering in order to allow the bulbs to build and store up energy reserves for the following year.

Division of established clumps allows for the removal of less vigorous or unhealthy-looking portions before problems can spread. Many bulbs are prone to virus infections, and regular inspection and removal of those showing symptoms is advised. Other problems are more host-specific; the particular problems affecting various plants are listed below. Resistant species or cultivars may be suggested in the entry for the problem in question in the *A–Z of Pests, Diseases and Disorders*.

COMMON PROBLEMS

The most common non-host-specific ailments affecting bulbs are:

SLUGS 178   VIRUSES 189

A

Acidanthera
GLADIOLUS CORM ROT **55**, 131
GLADIOLUS DRY ROT 132
GREY MOULD **45**, 135
LEAF SPOT (FUNGAL) **12**, **13**, 144

Autumn crocus (*Colchicum*)
SLUGS **29**, **43**, 178
SMUTS **35**, 178
TULIP GREY BULB ROT 185

B

Bluebell (*Hyacinthoides*)
BLUEBELL RUST 109
NARCISSUS BULB FLY **54**, 151
VIRUSES **42**, 189

C

Colchicum *see* **Autumn crocus**

Crinum
NARCISSUS LEAF SCORCH **29**, 152

Crocosmia *see* **Montbretia**

Crocus
BLUE MOULD ON BULBS **54**, 109
GLADIOLUS CORE ROT 132
GLADIOLUS DRY ROT 133
GREY MOULD **45**, 135
MICE **51**, 149
NARCISSUS BASAL ROT 151
SPARROWS **46**, 179
SQUIRRELS **52**, 180
TULIP GREY BULB ROT 185
VIOLET ROOT ROT **57**, 188

Cyclamen, hardy
BACTERIAL SOFT ROT 104
FOOT AND ROOT ROTS **55**, 129
FUSARIUM WILT 131
GREY MOULD **45**, 135
LEAF SPOT (FUNGAL) **12**, **13**, 144
POWDERY MILDEWS **25**, 167
SLUGS **29**, **43**, 178

D

Daffodil (*Narcissus*)
BLINDNESS OF BULBS **41**, 108
BULB SCALE MITE 112
NARCISSUS BASAL ROT 151
NARCISSUS BULB FLY **54**, 151
NARCISSUS EELWORM **54**, 151
NARCISSUS LEAF SCORCH **29**, 152
NARCISSUS SMOULDER 152
NARCISSUS VIRUSES **23**, 152
POLLEN BEETLES **43**, 163
SLUGS **29**, **43**, 178
TULIP GREY BULB ROT 185
WEEDKILLER DAMAGE – CONTACT 191

Dahlia
BLACKFLY 108
CAPSID BUGS **29**, **46**, 115
CROWN GALL **38**, 120
DAHLIA SMUT 123
EARWIGS **46**, 126
FOOT AND ROOT ROTS **55**, 129
GLASSHOUSE RED SPIDER MITE **22**, **23**, 133
GREY MOULD **45**, 135
LEAF SPOT (FUNGAL) **12**, **13**, 144
LEAFY GALL **39**, 145
ONION THRIPS 155
PHYTOPHTHORA 161
POWDERY MILDEWS **25**, 167
SCLEROTINIA **36**, 175
VERTICILLIUM WILTS **29**, **33**, 187
VIRUSES **42**, 189

E

Erythronium
GREY MOULD **45**, 135
RUSTS 174

F

Freesia
APHIDS 99
BLUE MOULD ON BULBS **54**, 109
GLADIOLUS CORE ROT 132
GLADIOLUS DRY ROT 133
GLADIOLUS HARD ROT 133
GLASSHOUSE RED SPIDER MITE **22**, **23**, 133
GREY MOULD **45**, 135
LEAF SPOT (FUNGAL) **12**, **13**, 144
SLUGS **29**, **43**, 178
VIRUSES **42**, 189

Fritillary (*Fritillaria*)
LILY BEETLE 147
SLUGS **29**, **43**, 178
VIRUSES **42**, 189

G

Gladiolus
APHIDS 99
FUSARIUM BULB ROT 131
GLADIOLUS CORE ROT 132
GLADIOLUS CORM ROT **55**, 132
GLADIOLUS DRY ROT 133
GLADIOLUS SCAB AND NECK ROT 133
GLADIOLUS THRIPS **45**, 133
LEAF SPOT (FUNGAL) **12**, **13**, 144
SLUGS **29**, **43**, 178
SNAILS **29**, **43**, 178
VIRUSES **42**, 189

Grape hyacinth (*Muscari*)
BLUEBELL RUST 109
SMUTS **35**, 178

H

Hyacinth (*Hyacinthus*)
FUSARIUM BULB ROT 131
GREY MOULD **45**, 135
SLUGS **29**, **43**, 178
TULIP GREY BULB ROT 185
VIRUSES **42**, 189

I

Iris
APHIDS 99
BLUE MOULD ON BULBS **54**, 109
IRIS INK DISEASE 140
IRIS RHIZOME ROT **55**, 140
IRIS SAWFLY **29**, 141
IRIS SCORCH 141
LEAF MINERS 144
LEAF SPOT (FUNGAL) **12**, **13**, 144
NARCISSUS BASAL ROT 151
RUSTS 174
SLUGS **29**, **43**, 178
SNAILS **29**, **43**, 178
TULIP GREY BULB ROT 185
VIRUSES **42**, 189

L

Lily (*Lilium*)
APHIDS 99
LILY BEETLE 147
LILY DISEASE 147
SLUGS **29**, **43**, 178
VINE WEEVIL **29**, **53**, 188
VIRUSES **42**, 189

M

Montbretia (*Crocosmia*)
GLADIOLUS DRY ROT 133

GLASSHOUSE RED SPIDER MITE **22**, **23**, 133
IRIS INK DISEASE 140
LEAF SPOT (FUNGAL) **12**, **13**, 144

O

Ornithogalum
SMUTS **35**, 178

S

Snowdrop (*Galanthus*)
GLADIOLUS DRY ROT 133
NARCISSUS BULB FLY **54**, 151
NARCISSUS EELWORM **54**, 151
NARCISSUS LEAF SCORCH **29**, 152
SNOWDROP GREY MOULD 179

Solomon's seal (*Polygonatum*)
SLUGS **29**, **43**, 178
SOLOMON'S SEAL SAWFLY **31**, 179

T

Tulip (*Tulipa*)
APHIDS 99
BLUE MOULD ON BULBS **54**, 109
DEER **41**, 123
MICE **51**, 149
SQUIRRELS **52**, 180
TULIP FIRE **22**, **55**, 185
TULIP GREY BULB ROT 185
VIRUSES **42**, 189

W

Winter aconite (*Eranthis*)
SMUTS **35**, 178

ANNUALS AND BIENNIALS

MANY PROBLEMS affecting annuals and biennials can be avoided by careful initial choice of plants, good planting, and regular feeding and watering. To avoid the foot and root rots to which many are prone, do not grow the same, or closely related, plants on any site for more than one year. Some problems affect a wide range of unrelated plants (*see right*); others are more host-specific (*below*). Resistant cultivars may be suggested in the entry for the problem in question in the *A–Z of Pests, Diseases and Disorders*.

COMMON PROBLEMS

The most common non-host-specific ailments affecting annual and biennial plants are:

APHIDS 99
CUTWORMS 122
FOOT AND ROOT ROTS 129
FROST DAMAGE 130
LEAF SPOT (FUNGAL) 144
POTASSIUM DEFICIENCY 164
POWDERY MILDEWS 167

RABBITS 168
SLUGS 178
SNAILS 178

A

Anchusa *see* HERBACEOUS PERENNIALS 200–1

African marigold (*Tagetes erecta*)
CUTWORMS **53**, 122
FOOT AND ROOT ROTS **55**, 129
GLASSHOUSE LEAFHOPPER **20**, 133
GLASSHOUSE RED SPIDER MITE **22**, **23**, 133
SLUGS **29**, **43**, 178
SNAILS **29**, **43**, 178

Ageratum
POWDERY MILDEWS **25**, 167

B

Busy Lizzie (*Impatiens*)
APHIDS 99
DOWNY MILDEW 125
FOOT AND ROOT ROTS **55**, 129
GLASSHOUSE LEAFHOPPER **20**, 133
GLASSHOUSE RED SPIDER MITE **22**, **23**, 133
GLASSHOUSE WHITEFLY 134
GREY MOULD **45**, 135
LEAF SCORCH 144
LEAF SPOT (FUNGAL) **12**, **13**, 144
TARSONEMID MITES **27**, 183
VINE WEEVIL **29**, **53**, 188
WATERLOGGING **34**, 190
WESTERN FLOWER THRIPS **42**, 191

C

Cabbage, Ornamental *see* VEGETABLES AND HERBS 207–8

Campanula *see* HERBACEOUS PERENNIALS 200–1

Canna *see* HERBACEOUS PERENNIALS 200–1

Castor-oil plant (*Ricinus communis*)
HONEY FUNGUS **34**, 138

Centaurea *see* HERBACEOUS PERENNIALS 200–1

China aster (*Callistephus*)
APHIDS 99
CUTWORMS **53**, 122
FUSARIUM WILT 131
GREY MOULD **45**, 135
POWDERY MILDEWS **25**, 167
SCLEROTINIA **36**, 175
VIRUSES **23**, **25**, 189

Convolvulus *see* HERBACEOUS PERENNIALS 200–1

E

Evening primrose (*Oenothera*)
LEAF SPOT (FUNGAL) **12**, **13**, 144

F

French marigold (*Tagetes patula*) *As for* **African marigold**

Forget-me-not (*Myosotis*)
APHIDS 99
DOWNY MILDEW 125
POWDERY MILDEWS **25**, 167

L

Lobelia
DAMPING OFF **34**, 123
FOOT AND ROOT ROTS **55**, 129
GLASSHOUSE RED SPIDER MITE **22**, **23**, 133
SLUGS **29**, **43**, 178

M

Marigold (*Calendula*)
CROWN GALL **38**, 120
GREY MOULD **45**, 135
LEAF SPOT (FUNGAL) **12**, **13**, 144
POWDERY MILDEWS **25**, 167
SMUTS **35**, 178

N

Nasturtium (*Tropaeolum*)
BLACKFLY 108
CABBAGE WHITE BUTTERFLY 113
DOWNY MILDEW 125
FLEA BEETLES **29**, 128
LEAFY GALL **39**, 145
VIRUSES **23**, **25**, 189

Nemesia
PHYTOPHTHORA 161

Nicotiana
GLASSHOUSE LEAFHOPPER **20**, 133
POWDERY MILDEWS **25**, 167
VIRUSES **23**, **25**, 189

P

Pansy (*Viola*)
APHIDS 99
DOWNY MILDEW 125
FOOT AND ROOT ROTS **55**, 129
GLASSHOUSE RED SPIDER MITE **22**, **23**, 133
LEAF SPOT (FUNGAL) **12**, **13**, 144
PANSY SICKNESS 156
POWDERY MILDEWS **25**, 167
RUSTS 174
SLUGS **29**, **43**, 178
SMUTS **35**, 178

Petunia
APHIDS 99
DAMPING OFF **34**, 123
PETUNIA FOOT ROT 160
POWDERY MILDEWS **25**, 167
SLUGS **29**, **43**, 178
VIRUSES **23**, **25**, 189

Poppy (*Papaver*)
DOWNY MILDEW 125
PEDICEL NECROSIS 159
POWDERY MILDEWS **25**, 167
VERTICILLIUM WILTS **29**, **33**, 187

S

Salvia
APHIDS 99
CAPSID BUGS **29**, **46**, 115
SLUGS **29**, **43**, 178

Statice
GREY MOULD **45**, 135
POWDERY MILDEWS **25**, 167

Stocks (*Matthiola*)
CLUBROOT **55**, 119
DOWNY MILDEW 125
FLEA BEETLES **29**, 128
GREY MOULD **45**, 135
VIRUSES **23**, **25**, 189
WHITE BLISTER **14**, 191

Sweet pea (*Lathyrus*)
APHIDS 99
CROWN GALL **38**, 120
CUCUMBER MOSAIC VIRUS **21**, **25**, 121
DOWNY MILDEW 125
FOOT AND ROOT ROTS **55**, 129
FUSARIUM WILT 131
GREY MOULD **45**, 135
LEAF SPOT (FUNGAL) **12**, **13**, 144
LEAFY GALL **39**, 145
POLLEN BEETLES **43**, 163
POWDERY MILDEWS **25**, 167
SLUGS **29**, **43**, 178
SNAILS **29**, **43**, 178
VIRUSES **23**, **25**, 189

Sweet William (*Dianthus barbatus*)
LEAF SPOT (FUNGAL) **12**, **13**, 144
RUSTS 174

T

Tobacco plant *see* **Nicotiana**

V

Verbena
APHIDS 99
FOOT AND ROOT ROTS **55**, 129
GLASSHOUSE LEAFHOPPER **20**, 133
GLASSHOUSE RED SPIDER MITE **22**, **23**, 133
WESTERN FLOWER THRIPS **42**, 191

Viola *see* **Pansy**

W

Wallflower (*Cheiranthus*)
CLUBROOT **55**, 119
CROWN GALL **38**, 120
DOWNY MILDEW 125
FLEA BEETLES **29**, 128
GREY MOULD **45**, 135
LEAFY GALL **39**, 145
RABBITS 168
VIRUSES **23**, **25**, 189
WHITE BLISTER **14**, 191

Z

Zinnia
FOOT AND ROOT ROTS **55**, 129
GREY MOULD **45**, 135
LEAF SPOT (FUNGAL) **12**, **13**, 144

GREENHOUSE AND HOUSE PLANTS

A SHELTERED environment not only protects plants, but also pests and diseases. They may therefore build up very rapidly, and exotic species may flourish that would have little chance to cause serious problems out of doors. However, an enclosed environment also lends itself to effective and thorough treatment, and to the use of biological controls (*see pp.90–91*) to eradicate pests.

Some common problems may affect a wide range of unrelated plants (*see right*). Others are more host-specific (*see below*). Resistant species or cultivars may be suggested in the entry for the problem in question in the *A–Z of Pests, Diseases and Disorders*.

COMMON PROBLEMS

The most common non-host-specific ailments affecting plants grown in the greenhouse or conservatory, or as house plants, are:

APHIDS 99	POWDERY MILDEWS 167
CARNATION TORTRIX MOTH 115	ROOT MEALYBUGS 171
FOOT AND ROOT ROTS 129	SCALE INSECTS 175
FUNGUS GNATS 131	SLUGS 178
GLASSHOUSE LEAFHOPPER 133	SPRINGTAILS 180
GLASSHOUSE RED SPIDER MITE 133	VINE WEEVIL 188
GLASSHOUSE WHITEFLY 134	VIRUSES 189
LEAF SCORCH 144	WATERLOGGING 190
MEALYBUGS 148	WESTERN FLOWER THRIPS 191

A

Abutilon
APHIDS 99
GLASSHOUSE WHITEFLY 134
SOFT SCALE 179
VIRUSES **23**, 189

African violet (*Saintpaulia*)
APHIDS 99
CROWN ROT 121
GREY MOULD **34**, 135
LEAF SCORCH 144
MEALYBUGS **32**, **37**, 148
PHYTOTOXICITY 161
POWDERY MILDEWS **49**, 167
ROOT MEALYBUGS **55**, 171
TARSONEMID MITES **27**, 183
VINE WEEVIL **53**, 188
WESTERN FLOWER THRIPS **42**, 191

Amaryllis (*Hippeastrum*)
BULB SCALE MITE 112
NARCISSUS BULB FLY **54**, 151
NARCISSUS LEAF SCORCH **29**, 152
VIRUSES **23**, 189

Angels' trumpets (*Brugmansia/Datura*)
GLASSHOUSE RED SPIDER MITE **22**, **23**, 133
GLASSHOUSE WHITEFLY 134
MEALYBUGS **32**, **37**, 148
POTATO BLIGHT **28**, **56**, 164
TOMATO BLIGHT **51**, 183
VIRUSES **23**, 189

Arum lily (*Zantedeschia*)
APHIDS 99
BACTERIAL SOFT ROT 104
LEAF SPOT (FUNGAL) **12**, **13**, 144
PHYTOPHTHORA 161
VIRUSES **23**, 189

Asparagus fern
CARNATION TORTRIX MOTH **20**, 115
MEALYBUGS **32**, **37**, 148
OLEANDER SCALE 154

Aspidistra
GLASSHOUSE RED SPIDER MITE **22**, **23**, 133
LEAF SPOT (FUNGAL) **12**, **13**, 144

B

Banana (*Musa*)
GLASSHOUSE RED SPIDER MITE **22**, **23**, 133
LEAF SCORCH 144
LEAF SPOT (FUNGAL) **12**, **13**, 144
MEALYBUGS **32**, **37**, 148

Begonia
APHIDS 99
FOOT AND ROOT ROTS **55**, 129
GREY MOULD **34**, 135
LEAF AND BUD EELWORM **25**, 143
LEAF SCORCH 144
LEAF SPOT (FUNGAL) **12**, **13**, 144
OEDEMA **19**, 154
PHYTOPHTHORA 161
POWDERY MILDEWS **49**, 167
TARSONEMID MITES **27**, 183
VINE WEEVIL **53**, 188
VIRUSES **23**, 189

Bougainvillea
APHIDS 99
MEALYBUGS **32**, **37**, 148

C

Cacti
GLASSHOUSE RED SPIDER MITE **22**, **23**, 133
MEALYBUGS **32**, **37**, 148
PHYTOPHTHORA 161
ROOT MEALYBUGS **55**, 171
WATERLOGGING **34**, 190

Calceolaria
APHIDS 99
FOOT AND ROOT ROTS **55**, 129
PHYTOPHTHORA 161
VIRUSES **23**, 189

Canna *see* HERBACEOUS PERENNIALS 200–1

Carnation (*Dianthus*)
APHIDS 99
CARNATION TORTRIX MOTH **20**, 115
FUSARIUM WILT 131
GLASSHOUSE RED SPIDER MITE **22**, **23**, 133
LEAFY GALL **39**, 145
POWDERY MILDEWS **49**, 167

Christmas cactus (*Schlumbergera*)
MEALYBUGS **32**, **37**, 148

Chrysanthemum
CHRYSANTHEMUM EELWORM 118
CHRYSANTHEMUM LEAF MINER **25**, 118
CHRYSANTHEMUM PETAL BLIGHT **45**, 118
CHRYSANTHEMUM RAY BLIGHT 118
CHRYSANTHEMUM WHITE RUST **14**, 118
CROWN GALL **38**, 120
FUSARIUM WILT 131
GLASSHOUSE LEAFHOPPER **20**, 133
GLASSHOUSE RED SPIDER MITE **22**, **23**, 133
GLASSHOUSE WHITEFLY 134
GREY MOULD **34**, 135
LEAF SPOT (FUNGAL) **12**, **13**, 144
LEAFY GALL **39**, 145

Melon Cotton Aphid 43, 148
MELON COTTON APHID **43**, 148
PHYTOPHTHORA 161
RUSTS 174
SCLEROTINIA **36**, 175
VERTICILLIUM WILTS **29**, **33**, 187
VIRUSES **23**, 189
WESTERN FLOWER THRIPS **42**, 191

Cineraria (*Senecio* × *hybridus*)
APHIDS 99
CHRYSANTHEMUM LEAF MINER **25**, 118
CROWN GALL **38**, 120
DOWNY MILDEW 125
FOOT AND ROOT ROTS **55**, 129
FUSARIUM WILT 131
GREY MOULD **34**, 135
LEAF SPOT (FUNGAL) **12**, **13**, 144
LEAFY GALL **39**, 145
LETTUCE DOWNY MILDEW **22**, 146
PHYTOPHTHORA 161
RUSTS 174
SCLEROTINIA **36**, 175
VERTICILLIUM WILTS **29**, **33**, 187
VIRUSES **23**, 189

Citrus fruits (*Citrus, Fortunella*)
APHIDS 99
FLUTED SCALE 129
GLASSHOUSE RED SPIDER MITE **22**, **23**, 133
GLASSHOUSE THRIPS 134
MEALYBUGS **32**, **37**, 148
PHYTOPHTHORA 161
SOFT SCALE 179

Clivia
LEAF SPOT (FUNGAL) **12**, **13**, 144
MEALYBUGS **32**, **37**, 148
VERTICILLIUM WILTS **29**, **33**, 187

Coleus (*Solenostemon*)
APHIDS 99
GREY MOULD **34**, 135
VERTICILLIUM WILTS **29**, **33**, 187

Croton (*Codiaeum*)
FROST DAMAGE **28**, **44**, 130
GLASSHOUSE RED SPIDER MITE **22**, **23**, 133
MEALYBUGS **32**, **37**, 148
WATERLOGGING **34**, 190

Cyclamen
APHIDS 99
BACTERIAL SOFT ROT 104
CARNATION TORTRIX MOTH **20**, 115
FOOT AND ROOT ROT **55**, 129
FUSARIUM WILT 131
GLASSHOUSE RED SPIDER MITE **22**, **23**, 133
GREY MOULD **34**, 135
LEAF SPOT (FUNGAL) **12**, **13**, 144
MEALYBUGS **32**, **37**, 148
POWDERY MILDEWS **49**, 167
TARSONEMID MITES **27**, 183
VINE WEEVIL **53**, 188
WATERLOGGING **34**, 190
WESTERN FLOWER THRIPS **42**, 191

D

Dracaena
LEAF SPOT (FUNGAL) **12**, **13**, 144

E

Echeveria
MEALYBUGS **32**, **37**, 148
RUSTS 174
VINE WEEVIL **53**, 188

Epiphyllum
APHIDS 99
MEALYBUGS **32**, **37**, 148

F

Ferns
APHIDS 99
FOOT AND ROOT ROTS **55**, 129
HEMISPHERICAL SCALE **16**, 137
LEAF AND BUD EELWORM **25**, 143
LEAF SCORCH 144
MEALYBUGS **32**, **37**, 148
RUSTS 174
SMUTS **35**, 178
SOFT SCALE 179

Frangipani (*Plumeria*)
GLASSHOUSE RED SPIDER MITE **22**, **23**, 133
MEALYBUGS **32**, **37**, 148

Fuchsia
APHIDS 99
FOOT AND ROOT ROTS **55**, 129
FUCHSIA GALL MITE 130
FUCHSIA RUST **18**, 131
GLASSHOUSE LEAFHOPPER **20**, 133
GLASSHOUSE RED SPIDER MITE **22**, **23**, 133
GLASSHOUSE WHITEFLY 134
LEAF SPOT (FUNGAL) **12**, **13**, 144
ROOT MEALYBUGS **55**, 171
VINE WEEVIL **53**, 188

G

Gardenia
APHIDS 99
BUD, DRY 112
GLASSHOUSE RED SPIDER MITE **22**, **23**, 133
GREY MOULD **34**, 135
IRON DEFICIENCY AND LIME-INDUCED CHLOROSIS **24**, 141
MAGNESIUM DEFICIENCY **22**, 147
MEALYBUGS **32**, **37**, 148
SOFT SCALE 179

Gazania
APHIDS 99
CHRYSANTHEMUM LEAF MINER **25**, 118
GLASSHOUSE RED SPIDER MITE **22**, **23**, 133
GLASSHOUSE WHITEFLY 134

Gerbera
APHIDS 99
CHRYSANTHEMUM LEAF MINER **25**, 118
FOOT AND ROOT ROTS **55**, 129
GLASSHOUSE WHITEFLY 134
LEAF SPOT (FUNGAL) **12**, **13**, 144
VERTICILLIUM WILTS **29**, **33**, 187

Gloxinia (*Sinningia*)
FOOT AND ROOT ROTS **55**, 129
GLASSHOUSE LEAFHOPPER **20**, 133
POWDERY MILDEWS **49**, 167
VINE WEEVIL **53**, 188
WESTERN FLOWER THRIPS **42**, 191

H

Hibiscus
APHIDS 99
GLASSHOUSE RED SPIDER MITE **22**, **23**, 133
VIRUSES **23**, 189

Hot water plant (*Achimenes*)
APHIDS 99
GLASSHOUSE LEAFHOPPER **20**, 133
WESTERN FLOWER THRIPS **42**, 191

K

Kalanchoe
APHIDS 99
FOOT AND ROOT ROT **55**, 129
POWDERY MILDEWS **49**, 167
VINE WEEVIL **53**, 188

Kangaroo vine (*Rhoicissus*)
MEALYBUGS **32**, **37**, 148
POWDERY MILDEWS **49**, 167
VINE WEEVIL **53**, 188

L

Lantana
GLASSHOUSE WHITEFLY 134

Lapageria
GLASSHOUSE RED SPIDER MITE **22**, **23**, 133
MEALYBUGS **32**, **37**, 148

M

Marguerite (*Argyranthemum frutescens*)
APHIDS 99
CHRYSANTHEMUM LEAF MINER **25**, 118
FUSARIUM WILT 131
LEAF SPOT (FUNGAL) **12**, **13**, 144
POWDERY MILDEW **49**, 167

Mimosa (*Acacia*)
CARNATION TORTRIX MOTH **20**, 115

GLASSHOUSE RED SPIDER MITE **22**, **23**, 133
MEALYBUGS **32**, **37**, 148

Money tree (*Crassula*)
CARNATION TORTRIX MOTH **20**, 115
GREY MOULD **34**, 135
MEALYBUGS **32**, **37**, 148
VINE WEEVIL **53**, 188

O

Oleander (*Nerium*)
APHIDS 99
CARNATION TORTRIX MOTH **20**, 115
GLASSHOUSE RED SPIDER MITE **22**, **23**, 133
GREY MOULD **34**, 135
HEMISPHERICAL SCALE **16**, 136
MEALYBUGS **32**, **37**, 148
OLEANDER SCALE 154

Orchids
APHIDS 99
FOOT AND ROOT ROTS **55**, 129
GLASSHOUSE RED SPIDER MITE **22**, **23**, 133
LEAF SPOT (FUNGAL) **12**, **13**, 144
SCALE INSECTS 175
SLUGS **29**, **43**, 178
VIRUSES **23**, 189

P

Palms
BANDED PALM THRIPS **22**, 104
FOOT AND ROOT ROTS **55**, 129
GLASSHOUSE RED SPIDER MITE **22**, **23**, 133
GLASSHOUSE THRIPS 134
HEMISPHERICAL SCALE **16**, 137
LEAF SPOT (FUNGAL) **12**, **13**, 144
MEALYBUGS **32**, **37**, 148
OLEANDER SCALE 154
SOFT SCALE 179
WATERLOGGING **34**, 190

Papyrus (*Cyperus*)
MEALYBUGS **32**, **37**, 148

Passion flower (*Passiflora*)
CUCUMBER MOSAIC VIRUS 121
GLASSHOUSE LEAFHOPPER **20**, 133
GLASSHOUSE RED SPIDER MITE **22**, **23**, 133

MAGNESIUM DEFICIENCY **22**, 147
MEALYBUGS **32**, **37**, 148
VIRUSES **23**, 189

Pelargonium
APHIDS 99
BLACK LEG OF CUTTINGS 108
CARNATION TORTRIX MOTH **20**, 115
CROWN GALL **38**, 120
GLASSHOUSE LEAFHOPPER **20**, 133
GLASSHOUSE WHITEFLY 134
LEAF SPOT (FUNGAL) **12**, **13**, 144
LEAFY GALL **39**, 145
MEALYBUGS **32**, **37**, 148
OEDEMA **19**, 154
ROOT MEALYBUGS **55**, 171
RUSTS 174
TARSONEMID MITES **27**, 183
VERTICILLIUM WILTS **29**, **33**, 187
VINE WEEVIL **53**, 188
WESTERN FLOWER THRIPS **42**, 191

Pepper, ornamental (*Capsicum*)
APHIDS 99
FOOT AND ROOT ROTS **55**, 129
GLASSHOUSE RED SPIDER MITE **22**, **23**, 133
GLASSHOUSE WHITEFLY 134
VIRUSES **23**, 189

Plumbago
CARNATION TORTRIX MOTH **20**, 115
GLASSHOUSE WHITEFLY 134
MEALYBUGS **32**, **37**, 148

Poinsettia (*Euphorbia pulcherrima*)
GLASSHOUSE WHITEFLY 134
LEAFY GALL **39**, 145
SOFT SCALE 179

R

Rubber plant (*Ficus elastica*)
BANDED PALM THRIPS **22**, 104
GLASSHOUSE RED SPIDER MITE **22**, **23**, 133
IRREGULAR WATERING **36**, 141
LEAF SCORCH 144
MEALYBUGS **32**, **37**, 148
SOFT SCALE 179
WATERLOGGING **34**, 190

S

Schefflera
BANDED PALM THRIPS **22**, 104

GLASSHOUSE RED SPIDER MITE **22**, **23**, 133
IRREGULAR WATERING **36**, 141
MEALYBUGS **32**, **37**, 148
SOFT SCALE 179
WATERLOGGING **34**, 190

Spider plant (*Chlorophytum*)
HEMISPHERICAL SCALE **16**, 137
SOFT SCALE 179

Stephanotis
GLASSHOUSE RED SPIDER MITE **22**, **23**, 133
HEMISPHERICAL SCALE **16**, 137
MEALYBUGS **32**, **37**, 144
SOFT SCALE 179
VIRUSES **23**, 189

Streptocarpus
APHIDS 99
GLASSHOUSE LEAFHOPPER **20**, 133
GREY MOULD **34**, 135
MEALYBUGS **32**, **37**, 144
POWDERY MILDEWS **49**, 167
ROOT MEALYBUGS **55**, 171
TARSONEMID MITES **27**, 183
VINE WEEVIL **53**, 188
WATERLOGGING **34**, 190
WESTERN FLOWER THRIPS **42**, 191

W

Wandering Jew (*Tradescantia*)
GLASSHOUSE LEAFHOPPER **20**, 133
VINE WEEVIL **53**, 188

Wax plant (*Hoya*)
GLASSHOUSE RED SPIDER MITE **22**, **23**, 133
MEALYBUGS **32**, **37**, 148
VIRUSES **23**, 189

Weeping fig (*Ficus benjamina*)
BANDED PALM THRIPS **22**, 104
GLASSHOUSE RED SPIDER MITE **22**, **23**, 133
IRREGULAR WATERING **36**, 141
LEAF SCORCH 144
MEALYBUGS **32**, **37**, 148
SOFT SCALE 179
WATERLOGGING **34**, 190

Winter cherry (*Solanum capsicastrum*)
APHIDS 99
GLASSHOUSE RED SPIDER MITE **22**, **23**, 133
GLASSHOUSE WHITEFLY 134
MAGNESIUM DEFICIENCY **22**, 147

FRUITS AND NUTS

MANY GARDENERS are happy to use chemicals on ornamentals, but when it comes to cropping plants, most prefer to avoid them wherever possible. Good cultivation, preventive measures taken at the correct time, and the use of resistant cultivars will all help. Some problems affect many fruits (*see right*). Others are more host-specific; plants so affected are listed below. More resistant species or cultivars may be suggested in the entry for the problem in question in the *A–Z of Pests, Diseases and Disorders*.

COMMON PROBLEMS

The most common problems affecting plants and trees bearing edible fruits and nuts are:

APHIDS 99
BIRD DAMAGE ON FRUITS 107
CORAL SPOT 120
GREY MOULD 135
HONEY FUNGUS 138
MAGNESIUM DEFICIENCY 147
POWDERY MILDEWS 167

RABBITS 168
SQUIRRELS 180
WASPS 190

A

Apple
APPLE BITTER PIT **48**, 100
APPLE BLOSSOM WEEVIL **46**, 100
APPLE CANKER **36**, 100
APPLE CAPSID **47**, 100
APPLE LEAF MINER **19**, 101
APPLE SAWFLY **47**, **48**, 101
APPLE SCAB **16**, 101
APPLE SOOTY BLOTCH 102
APPLE WATER CORE 102
BLOSSOM WILT **41**, 109
BROWN ROT **47**, **49**, 111
CODLING MOTH **47**, 119
CORAL SPOT **41**, 120
CROWN GALL **38**, 120
FIREBLIGHT **35**, 128
FRUIT TREE RED SPIDER MITE **20**, **40**, 130
HIGH TEMPERATURE INJURY 137
HONEY FUNGUS **34**, 138
IRON DEFICIENCY AND LIME-INDUCED CHLOROSIS **24**, 141
LEOPARD MOTH 146
MAGNESIUM DEFICIENCY **22**, 147
MUSSEL SCALE **36**, 151
NITROGEN DEFICIENCY 152
PAPERY BARK **40**, 156
PHYTOPHTHORA 161
POTASSIUM DEFICIENCY **24**, 164
POWDERY MILDEWS **49**, 167
PROLIFERATION **44**, 167
ROSY APPLE APHID **48**, 174
SILVER LEAF 177
SPLITTING 180
VIRUSES **23**, 189
WINTER MOTH **31**, 193
WOOLLY APHID **35**, 194

Apricot *see* **Peach**

B

Blackberry
BLACKBERRY CANE SPOT 108

CROWN GALL **38**, 120
FROST DAMAGE (BUDS AND FLOWERS) **44**, 130
FROST DAMAGE (FOLIAGE AND STEMS) **28**, 130
GREY MOULD **50**, 135
LEAFHOPPERS 144
LEAF SPOT (FUNGAL) **12**, **13**, 144
POWDERY MILDEWS **49**, 167
RASPBERRY BEETLE **50**, 168
RED BERRY MITE **49**, 169
RUSTS 174
VIRUSES **23**, 189

C

Cherry
APPLE LEAF MINER **19**, 101
BACTERIAL CANKER **35**, 104
BLOSSOM WILT **41**, 109
BROWN ROT **47**, 112
CHERRY BLACKFLY **24**, 117
CROWN GALL **38**, 120
FROST DAMAGE (BUDS AND FLOWERS) **44**, 130
FROST DAMAGE (FOLIAGE AND STEMS) **28**, 130
HONEY FUNGUS **34**, 138
IRON DEFICIENCY AND LIME-INDUCED CHLOROSIS **24**, 141
LEAF SPOT (FUNGAL) **12**, **13**, 144
MAGNESIUM DEFICIENCY **22**, 147
NITROGEN DEFICIENCY 153
PEAR AND CHERRY SLUGWORM **19**, 158
PHOSPHORUS DEFICIENCY 161
POTASSIUM DEFICIENCY 164
POWDERY MILDEWS **49**, 167
SHOT-HOLE **18**, **30**, 176
SHOT-HOLE BORERS **33**, 177
SILVER LEAF 177
SPLITTING 180
VIRUSES **23**, 189
WINTER MOTH **31**, 193
WITCHES' BROOMS **41**, 193

Citrus *see* GREENHOUSE & HOUSE PLANTS 204–5

Currant, black
AMERICAN GOOSEBERRY MILDEW **25**, **49**, 98
BIG BUD MITES **41**, 107
BLACKCURRANT GALL MIDGE **26**, 108
CAPSID BUGS **29**, **46**, 115
CORAL SPOT **41**, 120
CURRANT BLISTER APHID 121
CURRANT CLEARWING MOTH **33**, 122
GLASSHOUSE RED SPIDER MITE **22**, **23**, 133
GOOSEBERRY DIEBACK 134
GREY MOULD **50**, 135
HONEY FUNGUS **34**, 138
LEAF SPOT (BACTERIAL) **13**, **29**, 144
LEAF SPOT (FUNGAL) **12**, **13**, 144
VIRUSES **23**, 189
WOOLLY VINE SCALE **39**, 194

Currant, red/white
BROWN SCALE **36**, 111
CAPSID BUGS **29**, **46**, 115
CORAL SPOT **41**, 120
CURRANT AND GOOSEBERRY LEAF SPOT 121
CURRANT BLISTER APHID **27**, 121
CURRANT CLEARWING MOTH **33**, 122
DIEBACK 124
GOOSEBERRY SAWFLY **31**, 135
GREY MOULD **50**, 135
HONEY FUNGUS **34**, 138
LEAF SPOT (FUNGAL) **12**, **13**, 144
VIRUSES **23**, 189
WOOLLY VINE SCALE **39**, 194

F

Fig
CORAL SPOT **41**, 120
DIEBACK 124
FROST DAMAGE **28**, **44**, 130
GLASSHOUSE RED SPIDER MITE **22**, **23**, 133
GREY MOULD **50**, 135
LEAF SPOT (FUNGAL) **12**, **13**, 144

MEALYBUGS **32**, **37**, 148
SOFT SCALE 179
VIRUSES **23**, 189

G

Gooseberry
AMERICAN GOOSEBERRY MILDEW **25**, **49**, 98
APHIDS 99
BROWN SCALE **36**, 112
BULLFINCHES **46**, 113
CROWN GALL **38**, 120
CURRANT AND GOOSEBERRY LEAF SPOT 121
DIEBACK 124
FROST DAMAGE (BUDS AND FLOWERS) **44**, 130
FROST DAMAGE (FOLIAGE AND STEMS) **28**, 130
GOOSEBERRY SAWFLY **31**, 135
GREY MOULD **50**, 135
HONEY FUNGUS **34**, 138
IRON DEFICIENCY AND LIME-INDUCED CHLOROSIS **24**, 141
RUSTS 174
VIRUSES **23**, 189

Grape *see* **Vine**

H

Hazel nut
APHIDS 99
BIG BUD MITES **41**, 107
LEAF SPOT (FUNGAL) **12**, **13**, 144
NUT WEEVIL **52**, 153
POWDERY MILDEWS **49**, 167
SQUIRRELS **52**, 180
VIRUSES **23**, 189
WINTER MOTH **31**, 193

M

Mulberry
MULBERRY BACTERIAL BLIGHT 150
MULBERRY CANKER **38**, 150

P

Peach, nectarine, apricot
BACTERIAL CANKER **35**, 104
BLOSSOM WILT **41**, 109
BROWN SCALE **36**, 111
CROWN GALL **38**, 120
FROST DAMAGE (BUDS AND FLOWERS) **44**, 130
FROST DAMAGE (FOLIAGE AND STEMS) **28**, 130
GLASSHOUSE RED SPIDER MITE **22**, **23**, 133
HONEY FUNGUS **34**, 138
IRON DEFICIENCY AND LIME-INDUCED CHLOROSIS **24**, 141
NITROGEN DEFICIENCY 153
PEACH APHIDS 157
PEACH LEAF CURL **26**, 157
PHOSPHORUS DEFICIENCY 161
PHYTOPHTHORA 161
PLUM RUST **22**, 163
POTASSIUM DEFICIENCY 164
POWDERY MILDEWS **49**, 167
SHOT-HOLE **18**, **30**, 176
SILVER LEAF 177

Pear
APPLE AND PEAR CANKER **36**, 100
APPLE POWDERY MILDEW **25**, 101
APPLE SCAB **16**, 101
BLOSSOM WILT **41**, 109
BORON DEFICIENCY 109
BROWN ROT **47**, **49**, 112
BULLFINCHES **46**, 113
CODLING MOTH **47**, 119
CORAL SPOT **41**, 120
FIREBLIGHT **35**, 128
FROST DAMAGE (BUDS AND FLOWERS) **44**, 130
HONEY FUNGUS **34**, 138
MAGNESIUM DEFICIENCY **22**, 147
NITROGEN DEFICIENCY 153
PEAR AND CHERRY SLUGWORM **19**, 158
PEAR BEDSTRAW APHID 158
PEAR LEAF BLISTER MITE **16**, 158
PEAR MIDGE **48**, 158

PEAR RUST 159
PEAR STONY PIT VIRUS **48**, 159
PEAR SUCKER 159
PHOSPHORUS DEFICIENCY 161
POTASSIUM DEFICIENCY 164
SOOTY MOULD 179
VIRUSES **23**, 189
WINTER MOTH **31**, 193

Plum, damson, gage
BACTERIAL CANKER **35**, 104
BLOSSOM WILT **41**, 109
BROWN SCALE **36**, 112
BULLFINCHES **46**, 113
CROWN GALL **38**, 120
FROST DAMAGE **28**, **44**, 130
FRUIT TREE RED SPIDER MITE **20**, **40**, 130
HONEY FUNGUS **34**, 138
IRON DEFICIENCY AND LIME-INDUCED CHLOROSIS **24**, 141
MEALY PLUM APHID 148
NITROGEN DEFICIENCY 153
PHOSPHORUS DEFICIENCY 161
PHYTOPHTHORA 161
PLUM LEAF-CURLING APHID **27**, 162
PLUM LEAF GALL MITE **16**, 163
PLUM MOTH **48**, 162
PLUM RUST **22**, 163

PLUM SAWFLY 163
POTASSIUM DEFICIENCY 164
POWDERY MILDEWS **49**, 167
SHOT-HOLE **18**, **30**, 176
SHOT-HOLE BORERS **33**, 177
SILVER LEAF 177
WINTER MOTH **31**, 193

Q

Quince
BROWN ROT **47**, **49**, 112
CROWN GALL **38**, 120
FIREBLIGHT **35**, 128
HONEY FUNGUS **34**, 138
POWDERY MILDEWS **49**, 167
QUINCE LEAF BLIGHT **48**, 168
SILVER LEAF 177
WINTER MOTH **31**, 193

R

Raspberry, loganberry, other hybrid berries
APHIDS 99
BROWN SCALE **36**, 112
CROWN GALL **38**, 120

FROST DAMAGE **28**, **44**, 130
GLASSHOUSE RED SPIDER MITE **22**, **23**, 133
GREY MOULD **50**, 135
HONEY FUNGUS **34**, 138
PHYTOPHTHORA 161
POOR POLLINATION 164
RASPBERRY BEETLE **50**, 168
RASPBERRY CANE BLIGHT 168
RASPBERRY CANE SPOT **34**, 169
RASPBERRY LEAF AND BUD MITE **18**, 169
RASPBERRY SPUR BLIGHT 169
RUSTS 174
SOUTHERN GREEN SHIELDBUG 179
VIRUSES **23**, 189
WEEDKILLER DAMAGE – BROADLEAVED **27**, 190
WEEDKILLER DAMAGE – GLYPHOSATE 191

S

Strawberry
APHIDS 99

GLASSHOUSE RED SPIDER MITE **22**, **23**, 133
GREY MOULD **34**, 135
LEAF SPOT (FUNGAL) **12**, **13**, 144
MILLIPEDES 149
POOR POLLINATION 164
SLUGS **29**, 178
STRAWBERRY BLACK EYE 181
STRAWBERRY GREEN PETAL 181
STRAWBERRY GREY MOULD **50**, 181
STRAWBERRY LEAF SPOT **12**, **17**, 181
STRAWBERRY MITE 181
STRAWBERRY RED CORE 181
STRAWBERRY SEED BEETLES 181
STRAWBERRY VIRUSES 182
VINE WEEVIL **29**, **53**, 188

V

Vine (grape)
BROWN SCALE **36**, 112
CROWN GALL **38**, 120
DOWNY MILDEW 125
GLASSHOUSE RED SPIDER MITE **22**, **23**, 133

GREY MOULD **50**, 135
HIGH TEMPERATURE INJURY 137
MEALYBUGS **32**, **37**, 148
OEDEMA **19**, 154
POWDERY MILDEWS **49**, 167
SHANKING OF GRAPES 176
VINE ERINOSE MITE **15**, 188
WOOLLY VINE SCALE **39**, 194

W

Walnut
CORAL SPOT **41**, 120
FROST DAMAGE (FOLIAGE AND STEMS) **28**, 130
SLIME FLUX AND WETWOOD 177
SQUIRRELS **52**, 180
VIRUSES **23**, 189
WALNUT BLISTER MITE 189
WALNUT BLOTCH 189
WALNUT SOFT SHELL 189

VEGETABLES AND HERBS

SOME PROBLEMS affect a wide range of vegetables (*right*); others are more host-specific (*below*). Even more so than with fruit, the choice of pest or disease-resistant cultivars can help ensure healthy crops and minimize the need for chemicals. Resistant cultivars may be suggested in the entry for the problem in question in the *A–Z of Pests, Diseases and Disorders*.

COMMON PROBLEMS

The most common non-host-specific ailments affecting vegetable crops are:

APHIDS 99
GREY MOULD 135
LEAF SPOT (FUNGAL) 144
POWDERY MILDEWS 167
SLUGS 178
VIRUSES 189

A

Artichoke, globe
BLACKFLY 108
GREY MOULD **50**, 135
LEAF SPOT (FUNGAL) **12**, **13**, 144
LETTUCE DOWNY MILDEW **22**, 146

Artichoke, Jerusalem
ROOT APHIDS **55**, 171
SCLEROTINIA **36**, 175
SLUGS **36**, **57**, 178

Aubergine
APHIDS 99
FOOT AND ROOT ROTS **55**, 129
GLASSHOUSE RED SPIDER MITE **22**, **23**, 133
GLASSHOUSE WHITEFLY 134
VIRUSES **23**, 189

B

Basil
FOOT AND ROOT ROTS **55**, 129
GREY MOULD **50**, 135

Bay see **Sweet bay** see SHRUBS AND CLIMBERS 198–9

Bean, broad
BEAN SEED BEETLES **51**, 105
BLACKFLY 108
CHOCOLATE SPOT 118
DOWNY MILDEW 125
FOOT AND ROOT ROTS **55**, 129
GREY MOULD **50**, 135
IRON DEFICIENCY AND LIME-INDUCED CHLOROSIS **24**, 141
MICE **51**, 149
PEA AND BEAN WEEVIL **31**, 156
RUSTS 174
SCLEROTINIA **36**, 175

VIRUSES **23**, 189
WILTS 193

Bean, dwarf/French
BEAN ANTHRACNOSE 105
BEAN SEED FLY **51**, 105
BLACKFLY 108
FOOT AND ROOT ROTS **55**, 129
GLASSHOUSE RED SPIDER MITE **22**, **23**, 133
GREY MOULD **50**, 135
HALO BLIGHT 136
ROOT APHIDS **55**, 171
RUSTS 174
VIRUSES **23**, 189
WILTS 193

Bean, runner
BEAN ANTHRACNOSE 105
BEAN SEED FLY **51**, 105
BLACKFLY 108
FOOT AND ROOT ROTS **55**, 129
FROST DAMAGE (FOLIAGE AND STEMS) **28**, 130

GLASSHOUSE RED SPIDER MITE **22**, **23**, 133
GREY MOULD **50**, 135
HALO BLIGHT 136
IRON DEFICIENCY AND LIME-INDUCED CHLOROSIS **24**, 141
NECTAR ROBBING **46**, 152
ROOT APHIDS **43**, 171
RUSTS 174
SCLEROTINIA **36**, 175
SLUGS **36**, **57**, 178
SOUTHERN GREEN SHIELDBUG 179
VIRUSES **23**, 189
WILTS 193

Beetroot
BEET LEAF MINER **17**, 106
BORON DEFICIENCY 109
DOWNY MILDEW 125
FOOT AND ROOT ROTS **55**, 129
FUSARIUM ROT 131
IRON DEFICIENCY AND LIME-INDUCED CHLOROSIS **24**, 141

LEAF SPOT (FUNGAL) **12**, **13**, 144
RUSTS 174
VIOLET ROOT ROT **57**, 188
VIRUSES **23**, 189

Brassicas
BOLTING **44**, 109
BORON DEFICIENCY 109
BRASSICA DARK LEAF SPOT **13**, 111
BRASSICA LIGHT LEAF SPOT 111
CABBAGE MOTH **30**, 113
CABBAGE ROOT FLY 113
CABBAGE WHITE BUTTERFLY 113
CABBAGE WHITEFLY **32**, 114
CLUBROOT **55**, 119
DOWNY MILDEW 125
FLEA BEETLES **29**, 128
GREY MOULD **50**, 135
IRON DEFICIENCY AND LIME-INDUCED CHLOROSIS **24**, 141
MAGNESIUM DEFICIENCY **22**, 147

INDIVIDUAL PLANT PROBLEMS

MEALY CABBAGE APHID **32**, 148
NITROGEN DEFICIENCY 152
OEDEMA **19**, 154
PIGEONS **30**, 161
POWDERY MILDEWS **49**, 167
SOFT ROT 179
TURNIP GALL WEEVIL 187
VIRUSES **23**, 189
WEEDKILLER DAMAGE –
    BROADLEAVED **39**, 190
WHIPTAIL OF BRASSICAS 191
WHITE BLISTER **14**, 191

C

Capsicum (peppers)
APHIDS 99
BLOSSOM END ROT 108
FOOT AND ROOT ROTS **55**,
    129
GLASSHOUSE RED SPIDER
    MITE **22**, **23**, 133
GLASSHOUSE WHITEFLY 134
OEDEMA **19**, 154
SCLEROTINIA **36**, 175
VIRUSES **23**, 189

Carrot
CARROT CAVITY SPOT 115
CARROT FLY **51**, 115
CARROT MOTLEY DWARF
    VIRUS **25**, 115
DOWNY MILDEW 125
POWDERY MILDEWS **49**, 167
ROOT APHIDS **55**, 171
SCLEROTINIA **36**, 175
SPLITTING 180
VIOLET ROOT ROT **57**, 188
VIRUSES **23**, 189

Celery, celeriac
BOLTING **44**, 109
BORON DEFICIENCY 109
CARROT FLY **51**, 115
CELERY LEAF MINER **19**, 116
CELERY LEAF SPOT **12**, 116
DOWNY MILDEW 125
FOOT AND ROOT ROT **55**, 129
GREY MOULD **50**, 135
SCLEROTINIA **36**, 175
SLUGS **36**, **57**, 178
VIOLET ROOT ROT **57**, 188
VIRUSES **23**, 189

Courgette see **Marrow**

Cucumber, melon
APHIDS 99
CUCUMBER GUMMOSIS 121
CUCUMBER MOSAIC VIRUS
    21, **25**, 121
DOWNY MILDEW 125
FOOT AND ROOT ROT **55**, 129
FUSARIUM WILT 131
GLASSHOUSE RED SPIDER
    MITE **22**, **23**, 133
GLASSHOUSE WHITEFLY 134
GREY MOULD **50**, 135
LEAF SPOT (BACTERIAL) **13**,
    29, 144

LEAF SPOT (FUNGAL) **12**, **13**,
    144
POWDERY MILDEWS **49**, 167
SCLEROTINIA **36**, 175
VERTICILLIUM WILTS **33**, 187
VIRUSES **23**, 189
WEEDKILLER DAMAGE –
    BROADLEAVED **39**, 190

L

Leek
ALLIUM LEAF MINER 98
FOOT AND ROOT ROTS **55**, 129
LEAF SPOT (FUNGAL) **12**, **13**,
    144
LEEK MOTH **23**, 145
LEEK RUST **17**, 145
ONION FLY 154
ONION THRIPS 155
ONION WHITE ROT **52**, 155
SMUTS **35**, 178

Lettuce
APHIDS 99
BACTERIAL SOFT ROT 104
BOLTING **44**, 109
CUTWORMS **53**, 122
FOOT AND ROOT ROT **55**, 129
GREY MOULD **50**, 135
LEAF SPOT (FUNGAL) **12**, **13**,
    144
LETTUCE DOWNY MILDEW
    22, 146
LETTUCE ROOT APHID 146
SLUGS **36**, **57**, 178
VIRUSES **23**, 189

M

**Marrow, courgette,
pumpkin**
APHIDS 99
CUCUMBER MOSAIC VIRUS
    21, **25**, 121
FOOT AND ROOT ROT **55**, 129
GREY MOULD **50**, 135
POWDERY MILDEWS **49**, 167
SLUGS **36**, **57**, 178
VIRUSES **23**, 189

Melon see **Cucumber**

Mint
CATERPILLARS 115
LEAFHOPPERS 144
MINT BEETLE **31**, 149
MINT RUST 149

O

Onion, shallot, garlic
ALLIUM LEAF MINER 98
BACTERIAL SOFT ROT 104
BOLTING 109
CABBAGE MOTH **30**, 113
LEEK MOTH **23**, 145
ONION EELWORM **52**, 154

ONION FLY 154
ONION NECK ROT **52**, 155
ONION THRIPS 155
ONION WHITE ROT **52**, 155
RUSTS 174
SMUTS **35**, 178
VIRUSES **23**, 189

P

Parsley
CARROT FLY **51**, 115
CARROT MOTLEY DWARF
    VIRUS **25**, 115
CELERY LEAF MINER **19**, 116
DOWNY MILDEW 125
LEAF SPOT (FUNGAL) **12**, **13**,
    144
MAGNESIUM DEFICIENCY **22**, 147
POWDERY MILDEWS **49**, 167
VIRUSES **23**, 189

Parsnip
CARROT FLY **51**, 115
CELERY LEAF MINER **19**, 116
DOWNY MILDEW 125
LEAF SPOT (FUNGAL) **12**, **13**,
    144
PARSNIP CANKER **57**, 156
POWDERY MILDEWS **49**, 167
ROOT APHIDS **55**, 171
SCLEROTINIA **36**, 175
SPLITTING 180
VIOLET ROOT ROT **57**, 188
VIRUSES **23**, 189

Pea
APHIDS 99
DOWNY MILDEW 125
FOOT AND ROOT ROT **55**,
    129
FUSARIUM WILT 131
GREY MOULD **50**, 135
MICE **51**, 149
PEA AND BEAN WEEVIL **31**,
    156
PEA LEAF AND POD SPOT **14**,
    51, 156
PEA MOTH **51**, 157
PEA THRIPS 157
POWDERY MILDEWS **49**, 167
SLUGS **36**, **57**, 178
VIRUSES **23**, 189

Potato
APHIDS 99
COLORADO BEETLE 119
FROST DAMAGE (FOLIAGE
    AND STEMS) **28**, 130
LEAF SPOT (FUNGAL) **12**, **13**,
    144
MAGNESIUM DEFICIENCY **22**,
    147
MILLIPEDES 149
POTATO BLACK LEG 164
POTATO BLIGHT **28**, **56**,
    164
POTATO COMMON SCAB **56**, 165
POTATO CYST EELWORMS **55**,
    165
POTATO DRY ROT **57**, 165

POTATO EARLY BLIGHT 165
POTATO GANGRENE **57**, 166
POTATO HOLLOW HEART 166
POTATO INTERNAL RUST SPOT
    166
POTATO POWDERY SCAB **56**,
    166
POTATO SILVER SCURF 166
POTATO SPRAING **56**, 166
POTATO VIRUSES 166
POTATO WART DISEASE 166
SCLEROTINIA **36**, 175
SLUGS **36**, **57**, 178
SPLITTING 180
VERTICILLIUM WILTS **33**, 187
VIOLET ROOT ROT **57**, 188
WIREWORMS **54**, **56**, 193

Pumpkin see **Marrow**

R

Radish
CABBAGE ROOT FLY 113
DAMPING OFF **34**, 123
DOWNY MILDEW 125
FLEA BEETLES **29**, 128
LEAF SPOT (FUNGAL) **12**, **13**,
    144
SLUGS **36**, **57**, 178
VIRUSES **23**, 189
WHITE BLISTER **14**, 191

Rhubarb
BLACKFLY 108
CROWN ROT 121
CUCUMBER GUMMOSIS 121
DOWNY MILDEW 125
HONEY FUNGUS **34**, 138
LEAF SPOT (FUNGAL) **12**, **13**,
    144
RHUBARB GUMMING 171
SLUGS **36**, **57**, 178
SNAILS **29**, 178
VIRUSES **23**, 189

Rocket, salad
FLEA BEETLES **29**, 128

Rosemary
FROST DAMAGE **28**, 130
HONEY FUNGUS **34**, 138
LEAFHOPPERS 144
ROESMARY BEETLE 174

S

Spinach, perpetual
BEET LEAF MINER 106
BOLTING **44**, 109
CATERPILLARS 115
DOWNY MILDEW 125
IRON DEFICIENCY AND LIME-
    INDUCED CHLOROSIS **24**, 141
LEAF SPOT (FUNGAL) **12**, **13**,
    144
MAGNESIUM DEFICIENCY **22**, 147
POWDERY MILDEWS **49**, 167
RUSTS 174
SLUGS **36**, **57**, 178
VIRUSES **23**, 189

Swede
BORON DEFICIENCY 109
BRASSICA DARK LEAF SPOT
    13, 111
CABBAGE ROOT FLY 113
CLUBROOT **55**, 119
DOWNY MILDEW 125
FLEA BEETLES **29**, 128
LEAF SPOT (FUNGAL) **12**, **13**,
    144
POWDERY MILDEWS **49**,
    167
VIOLET ROOT ROT **57**, 188
VIRUSES **23**, 189
WHITE BLISTER **14**, 191

Sweetcorn
BORON DEFICIENCY 109
MICE **51**, 149
SWEETCORN SMUT **51**, 182
VIRUSES **23**, 189

T

Tomato
APHIDS 99
BLOSSOM END ROT **50**, 108
FUSARIUM WILT 131
GLASSHOUSE LEAFHOPPER **20**,
    133
GLASSHOUSE RED SPIDER
    MITE **22**, **23**, 133
GLASSHOUSE WHITEFLY 134
GREY MOULD **50**, 135
OEDEMA **19**, 154
SLUGS **36**, **57**, 178
SNAILS **29**, 178
SOUTHERN GREEN SHIELDBUG
    179
SPLITTING 180
TOMATO BLIGHT **51**, 183
TOMATO BLOTCHY RIPENING
    51, 183
TOMATO GREENBACK **50**, 184
TOMATO LEAF MOULD 184
TOMATO MOTH 184
TOMATO PITH NECROSIS
    185
VERTICILLIUM WILTS **33**,
    187
VIRUSES **23**, 189
WEEDKILLER DAMAGE –
    BROADLEAVED **27**, 190

Turnip
BORON DEFICIENCY 109
BRASSICA DARK LEAF SPOT
    13, 111
CABBAGE ROOT FLY 113
CLUBROOT **55**, 119
DOWNY MILDEW 125
FLEA BEETLES **29**, 128
LEAF SPOT (FUNGAL) **12**, **13**,
    144
POWDERY MILDEWS **49**,
    167
TURNIP GALL WEEVIL 186
VIOLET ROOT ROT **57**, 188
VIRUSES **23**, 189
WHITE BLISTER **14**, 191

REFERENCE SECTION

In ADDITION TO a comprehensive index, conversion charts and glossary of terms, this section contains current information on the types of insecticides and fungicides formulated for use in the garden.

REFERENCE SECTION

CONTROLS CURRENTLY APPROVED

ALWAYS READ THE LABEL – USE GARDEN CHEMICALS SAFELY

It is important to identify each pest or disease correctly in order to apply appropriate control measures – some can be kept down by good cultivation techniques such as crop rotation, destruction of crop residues, or hand removal of pests. Encouraging natural enemies or supplementing them with biological controls also reduces the need for spraying. Use chemicals only when they are really necessary.

Always select a recommended chemical, and read the instructions on the label carefully. Do not purchase large amounts of pesticides that will take many years to use up. If only a few plants require treatment, a ready-to-use spray bottle is the best answer. When spraying edible plants, check that the pesticide is suitable and note the period of time that must be left between treatment and harvesting.

ORGANIC INSECTICIDES

Fatty acids/insecticidal soaps
A modern version of the traditional soft soap spray for use against small insects, including aphids, whiteflies, red spider mite, thrips, scale insects, and mealybugs.

Ferric phosphate Pelleted bait for controlling slugs and snails.

Plant oils Extracted from plants, including sunflower and oil seed rape. Available as sprays for use against a similar range of pests as fatty acids.

Plant oil winter wash Spray for dormant fruit trees and bushes to control overwintering eggs of aphids, apple sucker, mussel scale, and other pests.

Pyrethrum Extracted from the flowers of *Pyrethrum cinerariifolium*. Available as sprays and dusts for controlling aphids, whiteflies, thrips, leafhoppers, ants, small caterpillars, and other small insects.

SYNTHETIC INSECTICIDES

Acetamiprid A systemic and contact insecticide for spraying on ornamental plants for controlling aphids, whitefly, scale insects, mealybugs and thrips. A ready-diluted formulation controls red spider mite, small caterpillars and lily beetle. A compost drench can be used to control vine weevil grubs and sap-feeding insects on container-grown ornamental plants. Also available in fertilizer sticks for pot plants.

Aluminium sulphate Applied to soil as crystals for the control of slugs and snails.

Bendiocarb A dust for controlling ants, woodlice, and wasp nests.

Bifenthrin A synthetic compound related to pyrethrum. Available as sprays for controlling aphids, whiteflies, thrips, leafhoppers, small caterpillars, beetles, and red spider mite.

Imidacloprid A systemic insecticide absorbed into plants through the roots and foliage. Controls leatherjackets and chafer grubs in lawns.

Metaldehyde Pelleted baits or a liquid drench for controlling slugs and snails.

Thiacloprid A systemic insecticide available as a spray concentrate and ready-diluted spray for controlling a wide range of pests, including aphids, whitefly, scale insects, mealybugs, leafhoppers, capsid bugs, thrips, small caterpillars, sawfly larvae and leaf beetles. A compost drench formulation can be used on container-grown ornamentals to control vine weevil grubs and sap-feeding insects. The spray formulations can be used on edible plants listed by the manufacturer; the recommendations for use on edible fruits refers to the ready-to-use product.

Thiamethoxam A systemic insecticide for use on container-grown ornamentals to control aphids, scale insects, whitefly, thrips, mealybugs and leaf beetles. Available as a spray concentrate or a ready-diluted spray. The latter also contains **abermectin**, which gives additional control of red spider mite. Fertilizer sticks containing thiamethoxam are available for pot plants.

ORGANIC FUNGICIDES

Bordeaux mixture A mixture of copper sulphate and calcium hydroxide approved under some organic regimes. Labelled in UK for control of potato and tomato blight, celery leaf spot, apple and pear canker, bacterial canker of *Prunus*, peach leaf curl, and some soft fruit rusts.

Sulphur Available in UK as dust, or as a liquid suspension with fatty acids. The dust may be used for control of a wide range of powdery mildews, and storage rots of bulbs, corms, etc. The liquid formulation is used for powdery mildew and pest control on roses and other ornamentals. Sulphur as chips or powder (not the fungicide form) can also be used in gardens to acidify soil that is excessively alkaline.

SYNTHETIC FUNGICIDES

Cheshunt compound A copper-ammonium complex for control of damping-off problems in seedlings.

Copper oxychloride Protects against similar diseases to Bordeaux mixture.

Trifloxystrobin A recently introduced fungicide for the control of lawn diseases.

Mancozeb A protectant fungicide labelled in UK for control of lettuce downy mildew, potato and tomato blight, apple and pear scab, peach leaf curl, blackcurrant leaf spot, and some rusts.

Myclobutanil A systemic fungicide available in UK in several formulations for control of rose black spot, rusts, and powdery mildew on ornamentals, apple and pear scab, and some soft fruit powdery mildews.

Penconazole A systemic fungicide for control of powdery mildew on ornamentals and rose black spot; also provides some control of rusts on ornamentals. Some plants, for example fuchsias, may be sensitive to this product.

GLOSSARY

A

ABDOMEN The hind part of an insect's body.

ACARICIDE A pesticide effective against mites.

ACID (of soil) With a pH value of less than 7; see also *Alkaline* and *Neutral*.

ADVENTITIOUS (of roots) Arising above ground from a stem or leaf.

AERATE (of soil) 1. Loosen by mechanical means in order to allow air to enter.
2. Alter soil texture and structure by creating more air spaces.

ALKALINE (of soil) With a pH value of more than 7; see also *Acid* and *Neutral*.

ALTERNATE HOST One of two host plants needed for some pathogens or pests to complete their life cycles. The alternate host is often unrelated to the other host; most pathogens and pests are able to complete their life cycles using a single host plant.

ANAEROBIC (of an organism) Able to live and grow without oxygen.

ANNUAL A plant or fungus that completes its life cycle, from germination through to death, in one year.

ANTENNA (*pl.* antennae) A jointed sensory organ on the head of insects and some other arthropods.

ANTHER The part of a stamen that produces pollen; it is usually borne on a filament.

ANTHRACNOSE A term used to describe several unrelated fungal diseases which are seen as dark spots or lesions on foliage, stems, pods or fruits.

ARTHROPODS Invertebrate animals which have jointed legs, e.g. insects, mites, woodlice, centipedes and millipedes.

ASEXUAL REPRODUCTION A form of reproduction not involving fertilization.

AXIL The upper angle between a leaf and stem, between a main stem and a lateral branch, or between a stem and a bract; where an axillary bud develops.

B

BACTERIUM (*pl.* bacteria) A microscopic single-celled organism. Some are beneficial to plants, others may be pathogenic or cause secondary rotting.

BACTERICIDE A chemical capable of controlling bacteria.

BARK The surface layer of the trunk and branches of woody plants, protecting the tissues within; usually composed of dead corky cells.

BARK-RINGING The controlled removal of a partial ring of bark from the trunk or branches of certain fruit trees, to reduce vigorous growth and encourage fruit cropping. Complete removal or girdling of a trunk, as in damage by rabbits, squirrels, voles and vine weevil grubs, can lead to the death of woody plants. See also *Girdling*.

BASAL PLATE The flattened base of a bulb, from which the roots grow.

BIENNIAL A plant that produces leafy growth in the first year, and then flowers, sets seed and dies in the next year.

BIOLOGICAL CONTROL The control of pests, diseases and weeds by the use of natural enemies such as predators, parasites and pathogens.

BLEEDING The oozing of sap through a pruning cut or wound.

BLIGHT A term loosely used to describe a fungal, or occasionally an insect, attack.

BLIND SHOOT A shoot that does not form a terminal flower bud, or one where the growing point has been destroyed.

BOLT To produce flowers and seed prematurely, especially in vegetables.

BORDEAUX MIXTURE A fungicide mixture containing copper sulphate and lime.

BRACT A modified leaf at the base of a flower or flower cluster, sometimes resembling a flower petal or leaves. Bracts may also be reduced to scale-like structures.

BROADLEAVED Of trees or shrubs that bear broad, flat leaves, rather than needle-like foliage as in conifers.

BROAD-SPECTRUM Used to describe a chemical which has an effect on a wide range of often unrelated organisms.

BUD A condensed shoot containing an embryonic leaf, leaf cluster or flower.

BULB A storage organ consisting mainly of fleshy scales and swollen, modified leaf-bases on a much reduced stem. Bulbs usually, but not always, grow from underground.

C

CALLUS The corky wound tissue formed by a plant in response to wounding. Callus frequently grows in from the edges of a wound created when a tree limb is removed, or at the base of a cutting.

CATERPILLAR The larval stage of moths and butterflies.

CERTIFIED STOCK Plants sold guaranteed to be of good quality and free from pests and diseases, especially virus infection, at the time of sale.

CHEMICAL In a garden situation, a fungicide, pesticide, weedkiller, repellent or fertilizer used to control pathogens, pests or weeds, or to increase the soil fertility.

CHESHUNT COMPOUND A fungicide mixture of ammonium carbonate and copper sulphate.

CHLOROPHYLL The green pigment in plants that absorbs light, providing the energy for photosynthesis.

CHLOROSIS Yellowing of plant tissue due to a deficiency or loss of chlorophyll (the green pigment). It may be brought about by a range of physiological disorders, nutrient deficiencies, pests and diseases.

CHLOROTIC Showing the symptoms of chlorosis.

CHRYSALIS The resting stage between a fully-fed caterpillar and the adult moth or butterfly. See also *Pupa*.

COCOON A protective outer covering, usually of silk, produced by caterpillars and some other larvae before they pupate.

COLOUR-BREAKING A term used to describe a patterned change in flower colour, where the original colour is "broken" into elaborate feathered patterns. It is usually caused by virus infections.

COMPANION PLANTING Positioning plants together that are reputed to have a beneficial effect on neighbouring plants by discouraging pests and diseases or improving growth.

COMPLETE METAMORPHOSIS See *Metamorphosis*.

COMPOST 1. A potting medium comprising a mixture of loam, sand, peat, leaf mould, coir, bark or other ingredients. 2. An organic material, rich in humus, formed by decomposed plant remains and other organic matter, including material from the kitchen and garden, used as a soil improver or mulch. 3. A growing medium available in different forms.

COMPOUND Made up of several or many parts, e.g. a leaf that is divided into two or more leaflets.

CONTACT Used to describe a chemical that remains on the outside of a plant and is absorbed by a pest or pathogen when it is sprayed or comes into contact with a treated surface.

CORM A bulb-like underground storage organ, consisting mainly of a swollen stem base and often surrounded by a papery tunic.

COTYLEDON A seed leaf; the first leaf or leaves to emerge from some seeds after germination, often markedly different from mature leaves. Flowering plants (angiosperms) are classified into monocotyledons (one) and dicotyledons (two) depending on how many cotyledons are contained in the mature seed. In gymnosperms (conifers) they are often produced in whorls.

CRUCIFEROUS Of the family Cruciferae, to which brassicas belong.

CULTIVAR A type of plant which has developed from the original species in cultivation (rather than in the wild), either by accident or by deliberate breeding; a contraction of "cultivated variety". See also *Variety*.

D

DECIDUOUS Plants that shed their leaves at the end of the growing season (in autumn and winter) and renew them in spring.

DEFOLIATION Loss of leaves.

DERRIS An insecticide derived from the roots of *Derris* and *Lonchocarpus* spp. Its active ingredient is rotenone. No longer used in the UK.

DIAPAUSE A resting period in the life cycle of invertebrate animals, the onset and duration of which is controlled by environmental factors such as day length and temperature. It enables them to survive adverse conditions, such as winter, and has similarities to hibernation in vertebrate animals.

DIEBACK The death of shoots, often spreading down from the tip of the stem, caused by damage or disease.

DORMANCY 1. The state of temporary cessation of growth in plants, and slowing down of other activities, usually during winter.
2. Seed dormancy: non-germination of seed when placed in conditions suitable for germination, due to physical, chemical, or other factors inherent in the seed.

DRUPES, DRUPELETS See *Stone fruits*.

DRENCH Pest or disease control applied as a liquid to the roots of a plant to control soil-borne pests or pathogens.

DUST Pest or disease control applied as a fine dust, usually to the plant, but occasionally to the soil round it.

E

ECDYSIS The process in invertebrate animals of shedding their outer skin as they grow and increase in size.

ECTOPARASITE A parasite that feeds on the outside of a plant or animal's body.

ELYTRON (*pl.* elytra) A beetle's wing case; when not in flight these are folded back and form a covering for the top of the abdomen.

ENDOPARASITE A parasite that feeds inside a plant or animal's body.

EYE On a stem or, more commonly, on a tuber, e.g. a potato, a dormant or latent growth bud that is visible. Of a fruit, e.g. apple, the end opposite the stalk.

F

f. [Lat.] *forma*; a variant within a species.

FAMILY A category in animal and plant classification, grouping together related genera. All these genera have characteristics which are all constant, or which are clearly different from those of other families.

FERTILIZATION The fusion of a pollen grain nucleus (male) with an ovule (female) to form a fertile seed.

FERTILIZER A plant food (usually in concentrated form), naturally or synthetically produced, applied to the soil or to plant foliage. A balanced fertilizer contains a balance of nitrogen, phosphate and potash. A complete fertilizer contains both these and other nutrients.

FLEECE, HORTICULTURAL Synthetic material composed of very fine fibres woven or compressed together into a fabric which can be draped over plants to protect them from weather extremes and certain pests.

FOLIAR FEED Liquid fertilizer formulated for application direct to healthy plant foliage. Nutrients applied in this manner are very rapidly taken up by the plant.

FRASS Insect excrement pellets, especially of caterpillars.

FREE-LIVING An organism such as a nematode which is associated with but not attached to its host.

FROST HARDY A plant able to withstand temperatures down to −5°C (23°F).

FROST TENDER A plant vulnerable to frost damage.

FRUITING BODIES (of fungi) Spore-bearing structures of larger fungi, commonly known as toadstools, mushrooms or bracket fungi. May be annual or perennial.

FULLY HARDY See *Hardy*.

FUNGICIDE A pesticide capable of controlling a fungus.

FUNGUS (*pl.* fungi) A very variable group of organisms, which obtain their food materials from living or dead organic matter (see also *Parasite* and *Saprophyte*). They vary greatly in size, from microscopic to clearly visible with the naked eye.

G

GALL An abnormal growth produced by a plant in response to chemicals secreted by an animal, fungus or bacterium that lives within the galled tissue.

GENUS (*pl.* genera) A category in animal and plant classification, grouping related species. Denoted by the first part of the scientific name. See also *Species*.

GERMINATION The physical and chemical changes that take place as a seed starts to grow and develop into a plant.

GIRDLING The removal of bark all round a trunk, stem or branch, by animals or physical injury. See also *Bark-ringing*.

H

HABIT 1. Characteristic, natural form of growth of a plant – upright, prostrate, weeping, etc.
2. The characteristic growth or general appearance of a plant.

HALF HARDY 1. Plants not able to tolerate frost, but generally able to withstand lower temperatures than

frost-tender plants.
2. Not tolerating frost, but withstanding temperatures down to 0°C (32°F).

HARDY Plants tolerating year-round climatic conditions in temperate regions, including frost, without protection.

HAULM Used to describe the top-growth of plants such as potatoes and legumes.

HEAVY (of soil) Having a high proportion of clay, and often poorly drained.

HERBACEOUS 1. A non-woody plant in which the above-ground part dies down to the rootstock at the end of the growing season. 2. Dying down at the end of the growing season.

HERBICIDE A chemical used to control or kill weeds.

HERBIVORE A plant-feeding animal.

HERMAPHRODITE Animals with both female and male sexual organs, e.g. slugs, snails and earthworms.

HONEYDEW A sugary liquid excreted by sap-feeding insects including many, but not all, aphids, whiteflies, suckers, mealybugs and scale insects.

HOST The plant on which a pest or pathogen develops or feeds; also the insect or other animal in which parasitic animals develop.

HOST RANGE The range of host plants that a given pest or pathogen will use. The host range is often restricted to closely-related plants.

HYPHA (*pl.* hyphae) The minute thread-like growths formed by fungi.

I

IMMUNE A plant with characteristics that prevent a given pest or pathogen from attacking or colonizing it. A plant immune to one disease or pest may, however, be susceptible to others.

INCOMPLETE METAMORPHOSIS See *Metamorphosis*.

INORGANIC 1. Used generally, not of plant or animal origin, i.e. a mineral or synthetically-produced material. 2. Of a chemical compound, one that does not contain carbon. Many man-made pesticides contain carbon-based chemicals and so are organic compounds as defined by a chemist, but not in the sense in which gardeners use the word organic. Inorganic fertilizers are refined from naturally-occurring minerals or produced artificially. See also *Organic*.

INSECT Invertebrate animal whose body is divided into a head, thorax and abdomen. Adult insects have three pairs of jointed legs on the thorax and usually two pairs of wings, e.g. moths, butterflies, beetles, ants, bees, wasps, sawflies, some aphids, whitefly and earwigs.

INSECTICIDAL SOAP A pesticide containing fatty acids, used against some pests, diseases and weeds.

INSECTICIDE A pesticide capable of controlling insects.

INSTAR The stages between ecdysis in the development of larvae and nymphs of insects and mites.

INTEGRATED CONTROL A combination of chemical and biological controls with good cultivation techniques, aimed at keeping pests and diseases below the level at which damage occurs.

INVERTEBRATE Animals without backbones, e.g. insects, mites, woodlice, molluscs, nematodes, earthworms and millipedes.

L

LARVA (*pl.* larvae) The immature feeding-stage in the life cycle of an insect undergoing complete metamorphosis, e.g. moth caterpillars, beetle grubs, fly maggots. See also *Nymph*.

LATENT INFECTION Infection by a pathogen which does not result in any visible symptoms on the plant.

LEACHING The loss of nutrients when they are washed down through the soil to areas out of reach of plant roots, often caused by excessive watering or rain.

LEGUME A plant of the family Leguminosae, to which peas and beans belong, or of one of its three sub-families, Caesalpiniaceae, Mimosaceae and Papilionaceae.

LENTICEL A breathing pore found in the stems of woody plants. Often enlarged and clearly visible to the naked eye.

LESION Point or area of damage caused by pests, pathogens or physical injury.

LIME Compounds of calcium; the amount of lime in soil determines whether it is alkaline, neutral or acid.

M

MANDIBLES The biting mouthparts of insects, especially leaf-eating types.

METAMORPHOSIS The changes undergone during an insect or mite's development. Incomplete metamorphosis, as shown by aphids, capsid bugs, earwigs and mites, involves little more than an increase in size and the gradual development of sexual organs and sometimes wings. Complete metamorphosis involves a dramatic change of form with the larva bearing no resemblance to the adult, e.g. in moths, flies, beetles and sawflies.

MOISTURE SHELLS Layers of humid air which accumulate or build up round a plant or its leaves as they transpire and increase humidity locally.

MOLLUSCICIDE A pesticide capable of controlling slugs and snails.

MULCH 1. A material applied in a layer to the soil surface to suppress weeds, conserve moisture, and maintain a cool, even root temperature. In addition to organic materials such as manure, bark and garden compost, which also enrich the soil, polythene, foil and gravel may also be used. 2. A layer of organic matter applied to the soil over or around a plant to conserve moisture, protect the roots from frost, reduce the growth of weeds, and enrich the soil.

MULTI-STEMMED Tree or shrub with several main stems arising either directly from the ground or from a short main stem.

MYCELIUM A compacted mass of fungal hyphae, often developed into a distinct "sheet" of fungal growth.

MYCOPLASMA Microscopic organism closely related and with similar characteristics to a virus.

MYCORRHIZA (*pl.* mycorrhizae) Soil fungi that live in mutually beneficial association with plant roots, helping them to absorb nutrients from the soil.

N

NECROSIS The deterioration and death of plant tissues.

NECROTIC Dead, frequently brown or black, areas of plant tissue.

NECTAR A sweet, sugary liquid secreted by nectary glandular tissue, usually found in the flower, but sometimes on the leaves or stem.

NEMATICIDE A pesticide capable of controlling nematodes or eelworms.

NEMATODES Microscopic, unsegmented, worm-like animals, also known as eelworms. Some are used as pest controls, e.g. of slugs and vine weevil.

NEUTRAL (of soil) With a pH value of 7,

REFERENCE SECTION

the point at which soil is neither acid nor alkaline. See also *Acid* and *Alkaline*.

NITRATE A salt of nitric acid, having a high nitrogen content that is available to plants. Either produced by the activity of bacteria in the soil or manufactured.

NITRITE A salt of nitric acid in which the nitrogen is not readily available to plants.

NODE The point on a stem where one or more buds, leaves, shoots or flowers are attached.

NODULE A small swelling.

NOSE (of a bulb or corm) The pointed upper end.

NOTIFIABLE A pest or disease is described as notifiable when it is obligatory to notify the Department for Environment, Food and Rural Affairs (DEFRA) as soon as it is discovered.

NUTRIENTS Minerals (mineral ions) used to develop proteins and other compounds required for plant growth.

NYMPH The immature stage of an insect or mite that undergoes incomplete metamorphosis. See also *Larva*.

O

ORGANIC 1. Of garden chemicals, referring to compounds containing carbon derived from plant or animal organisms.
2. Loosely, mulches, composts, or similar materials derived from plant or animal-derived materials.
3. A system of crop production and gardening without using synthetic or non-organic materials.

OVICIDE A pesticide capable of controlling the egg stage of pests.

OVIPOSITOR The egg-laying organ of an insect.

P

PARASITE An organism that gets all or part of its food from another plant or animal over a period of time, often without killing its host. Parasitic insects that eventually kill the insect in which they have developed are more correctly termed parasitoids.

PARTHENOGENESIS The ability of some female pests to reproduce without fertilization. Many aphids have this ability, as have vine weevils, some sawflies, scale insects and mealybugs. Males are rare or non-existent in these species.

PATHOGEN An organism which is parasitic and causes disease in another organism. Pathogens include certain bacteria, fungi, viruses and mycoplasmas.

PATHOVAR A distinct strain or type of a named pathogen.

PEDICEL The stalk of an individual flower.

PERENNIAL Living for at least three seasons. In this book the term when used as a noun, and unless qualified, denotes a herbaceous perennial. A woody-based perennial dies down only partially, leaving a woody stem at the base.

PERSISTENCE (of chemicals) The length of time a chemical persists in an active form either on the plant or in the soil or compost after it has been applied.

PESTICIDE A chemical substance; a term usually applied to chemicals that control pests, but in its broader definition also includes fungicides, herbicides, animal repellents and wood preservatives.

PETIOLE The stalk of a leaf.

pH The scale by which the acidity or alkalinity of soil is measured. See also *Acid, Alkaline, Neutral*.

PHEROMONE Volatile chemicals produced by insects as a means of communicating with others of the same species. Males often locate virgin females by tracking down the source of the pheromone or scent released by the female.

PHLOEM Conducting elements within plant tissue, largely associated with the transport of food materials in solution, produced in the leaves and distributed to the rest of the plant.

PHOTOSYNTHESIS The synthesis of carbohydrates in green plants from carbon dioxide and water, using light energy from sunshine or artificial light absorbed by the green pigment chlorophyll.

PHYTOTOXIC Harmful to plants. Usually used to describe a chemical (usually an insecticide or fungicide) which can damage a plant onto which it has been sprayed.

POLLEN The male cells of a plant, formed in the anthers.

POLLINATION The transfer of pollen from the anthers to the stigma of the same or different flowers, often carried out by insects, especially bees, or by air movement, resulting in fertilization of the embryonic seeds in the ovary.

POLYPHAGOUS Pests that feed on a wide range of plants which may be botanically unrelated.

POME FRUIT A firm, fleshy fruit, with seeds enclosed in a central core, as in apples and pears.

PROBOSCIS Insect mouthparts, used for sucking up liquids, especially nectar.

PROLEGS The clasping legs on the abdomen of a caterpillar or sawfly larva.

PUPA (*pl.* pupae) The resting, non-feeding stage in the life cycle of an insect undergoing complete metamorphosis. Also known as a chrysalis.

PUPATION The period of time during which larval tissues are dissolved and reconstructed into the form of the adult insect.

PYCNIDIUM (*pl.* pycnidia) Minute raised, rounded fungal fruiting body, produced on the surface of infected plant tissue.

R

RACE (of fungi) A specific strain of a named fungus which may vary in its physical or pathogenic characteristics from other races of that fungus.

REPELLENTS Substances that deter pests without causing them harm.

RESISTANT A plant is described as resistant to a pest or disease when it is attacked by it, but shows no ill effects. A fungus may also develop resistance to a fungicide, or to a closely-related group of fungicides. Fungicide resistance occurs most commonly when a particular fungicide or closely-related ones have been used repeatedly for many years. Similar resistance problems also occur with some pests and weeds that are no longer controlled by insecticides and herbicides that were formerly effective.

RHIZOME A specialized, usually horizontally creeping, swollen or slender underground stem that acts as a storage organ and produces aerial shoots at its apex and along its length.

RHIZOMORPH A cord-like fungal structure, which is very resilient and withstands extremes of temperature and moisture. Honey fungus rhizomorphs are dark and tough, and grow from host to host.

ROOT The part of a plant, normally underground, that anchors it and through which water and nutrients are absorbed.

ROTATION Planning where plants are grown, to ensure that different types of plant are grown on a given piece of land from season to season. By rotating plants some pest and disease outbreaks can be avoided. Rotation is a particularly effective method of avoiding problems with soil-borne pests and pathogens.

RUSSETING A discoloration, generally brown, of a plant surface (usually the skin of fruits) which is often roughened.

S

SAPROPHYTE An organism (most commonly a fungus) which lives on dead or decaying organic material but does not attack living material.

SCLEROTIUM (*pl.* sclerotia) The resilient fungal resting bodies formed by some fungi to allow them to overwinter successfully and withstand extremes of weather.

SHOOT The aerial part of a plant which bears leaves. A side-shoot arises from a bud along the length of a main shoot.

SPECIES A category in plant and animal classification, comprising individuals in the same genus that are alike and naturally breed with each other. Denoted by the second part of the scientific name. See also *Genus* and *Family*.

SPORE Microscopic reproductive structure of fungi and bacteria. Certain non-flowering plants, such as ferns, also produce spores.

SPUR A short fruiting branch on fruit bushes and trees, particularly apples and pears.

SSP. Subspecies; a higher category in plant classification than *forma* or variety.

STAMEN The male reproductive organ in a plant, comprising the pollen-producing anther and usually its supporting filament or stalk.

STEM The main axis of a plant, usually above ground and supporting leaves, flowers and fruits.

STIGMA The female part of a flower that receives pollen during the process of pollination.

STOLON A horizontally spreading or arching stem, usually above ground, which roots at its tip to produce a new plant, e.g. strawberry.

STONE FRUITS Fruits, also known as "drupes", with one or more seeds ("stones") surrounded by fleshy, usually edible, tissue. They are common in the genus *Prunus* (e.g. apricots, plums, cherries).

STOOL A group of shoots emerging from the base of a single plant. In propagation the term refers to the roots and stem bases of a plant such as a chrysanthemum after it has finished flowering.

STRAIN (of bacteria, etc.) A distinct form of a named pathogen.

STYLETS The mouthparts of sap-feeding insects and mites, used for piercing plant tissues.

SYMBIOSIS A relationship between two organisms, where neither harms the other and both derive benefit from the relationship, e.g. nitrogen-fixing bacteria in root nodules.

SYSTEMIC 1. Insecticides and fungicides that are absorbed into a plant's sap and translocated within the plant. Systemic fungicides are, however, rarely able to move great distances but may move from one leaf surface through the leaf to the lower surface. 2. Of diseases, some pathogens, in particular the viruses, that are systemic within the host plant, moving with the plant sap and being found in all or a large part of the host.

T

TENDER Vulnerable to low temperatures. Tender plants may be categorized as: cool-growing, down to a minimum temperature of 10°C (50°F); intermediate, minimum 13°C (55°F); or warm-growing, minimum 18°C (64°F). See also *Frost tender*.

THORAX The mid-part of an insect's body, between the head and abdomen.

TOLERANT Having the ability to tolerate an attack by a pest or a pathogen. A pest or pathogen may be described as tolerant if able to withstand the effects of a pesticide.

TRACE ELEMENTS Chemical elements needed by plants, but only in very small amounts, e.g. boron, copper, manganese, iron, molybdenum and zinc.

TRANSPIRATION Loss of water by evaporation from the leaves and stems of plants.

TUBER A swollen, usually underground, organ derived from a stem or root, used for food storage.

V

VARIEGATED Irregularly marked with various colours; particularly of leaves patterned with markings in white, yellow or other colours.

VARIETY A plant which has developed with a slight difference from a wild plant species. Plant varieties are often brought into cultivation once developed.

VECTOR See *Virus vector*.

VIRUS A sub-microscopic particle often responsible for causing plant virus diseases.

VIRUS VECTOR An insect, mite or nematode that transmits plant viruses, usually on its mouthparts or in its saliva. Some fungal and bacterial diseases can also be spread by plant pests.

VIVIPAROUS Giving birth to live young. Most aphid reproduction is by this method.

W

WETTING AGENT A material used to lower the surface tension of liquids on a plant surface, so allowing a spray to form a film over the surface rather than forming into droplets. Many garden pesticide formulations contain wetting agents (also known as wetters or spreaders).

WIND SNAP Wind causing a tree to be broken off at ground level; this can be made more likely by decay round the base of the trunk.

WIND THROW Wind causing a tree to be uprooted. Decay in the tree roots can make the tree unstable and therefore susceptible to wind damage.

WORM Various animals with elongate, cylindrical bodies. Eelworms or nematodes have simple, unsegmented bodies; earthworms are more advanced animals with bodies divided into many segments.

X

XYLEM Vascular tissue under bark taking water and nutrients up a stem.

Z

ZOOSPORE A motile stage, capable of self-propulsion, in the life cycle of certain fungi.

REFERENCE SECTION

INDEX

References to illustrations and tables are in **bold**. Sub-headings are arranged with symptoms first.

A

Abies alba see fir, silver
abutilon 204
acacia, false (*Robinia*) **39**, 196
Acer see maple, sycamore
Acer pseudoplatanus see sycamore
acer gall mite **17**, 98
acer tar spot **13**, 183
Achimenes see hot water plant
acidanthera 202
acid-loving plants **24**, 141
acorn gall wasp **52**, 98
adelgids 75, 98
Aesculus hippocastanum see horse chestnut
African marigold (*Tagetes*) 203
African violet (*Saintpaulia*) 204
agapanthus 200
Alcea see hollyhock
alder (*Alnus*) 98, 194, 196
  leaf symptoms **32**
alder sucker **32**, 98
algae 98
allium ornamental 145
allium leaf miner 98
almond (*Prunus dulcis*) 157, 177, 196
  symptoms **26**, **33**
Alnus see alder
alstroemeria 200
alternaria leaf spot 98
alyssum 200
amaryllis (*Hippeastrum*) 112, 152, 204
  symptoms **29**, **54**
Ambleyseius spp. **91**, 150
amelanchier 198
American gooseberry mildew **25**, **49**, 98
anchusa 200
anemone 99, 163, 200
  leaf symptoms **22**
anemone rust *see* plum rust
anemone smut 99
Anemone hybrids *see* Japanese anemone

angels' trumpets (*Datura*) 204
animals
  beneficial **60–3**, 79, 90, **90**, 143
  pesticide injury 83
  plant relationships 66–7, **67**, 68–9
anthocorid bugs **60**
anthracnose 99
Antirrhinum see snapdragon
antirrhinum rust **15**, 99
ants **32**, 59, 83, 87, 99
Aphidius spp. **91**
Aphidoletes aphidomyza **91**, 131
aphids 73, 74, 75, 99
  biological control **60–1**, 90–1, **91**
  chemical control 81
  life cycle **74**, 77, **77**
apple 100–2, 119, 130, 134, 146, 158, 174–5, 206
  flower symptoms **43**, **44**, **46**
  fruit symptoms **47**, **48**
  leaf symptoms **16**, **19**, **20**, **25**, **31**
  stem symptoms **35**, **36**, **40**, **41**
  fruit protection 88
  replanting 96, 174
  resistant cultivars 100, 101, 158
apple and pear canker **36**, 100, 158
apple bitter pit **48**, 100
apple blossom weevil **46**, 75, 100
apple capsid **47**, 100
apple frost damage 100
apple fruit split **48**, 101
apple leaf miner **19**, 101
apple powdery mildew **25**, 101
apple sawfly **47**, **48**, 75, 101
apple scab **16**, 81, 101
apple scald **47**, 101
apple sooty blotch 101
apple sucker **43**, 102
apple water core 102
apricot *see* peach
Aquilegia see columbine
aquilegia sawfly 102
Arabis see rock cress
arabis mosaic virus 102
Araucaria araucana see monkey puzzle
arbutus leaf spot **14**, 102

Arbutus unedo see strawberry tree
artichoke, globe 107, 207
artichoke, Jerusalem 207
arum lily (*Zantedeschia*) 204
Aruncus dioicus see goatsbeard
aruncus sawfly **30**, 77, 102
ash (*Fraxinus*) 102, 196
ash catkin gall mite 102
ash sucker 102
asparagus beetle **40**, 103
asparagus fern 204
aspidistra 204
Aster see Michaelmas daisy
aster wilt 103
astilbe 200
aubergine 121, 124, 207
  stem symptoms **33**
  resistant cultivars 121
aubrieta 200
auricula (*Primula*) 200
auricula (*Primula*) 200
autumn crocus (*Colchicum*) 202
azalea *see* rhododendron
azalea gall 103
azalea whitefly 103

B

bacteria 67, 72–3, 103
  nitrogen-fixing **68**, 69
  as pest control 91, **91**
bacterial canker **35**, 104
bacterial leaf spot **13**, **29**, 144
bacterial soft rot **73**, 104
bait 82, 83
  chemicals **210**
banana (*Musa*) 204
banded palm thrips **22**, 104
barberry (*Berberis*) 107, 198
bark beetles **33**, 75, 104
barriers **78**, 86, 88–9, **88**, **89**
basal rots *see* foot rots
basil 207
bay, sweet bay (*Laurus*) 104–5, 196, 199, 207
  leaf symptoms **26**
bay sucker **26**, 104
bean, broad 69, 105, 107, 149, 152, 207
  flower symptoms **46**

leaf symptoms **12**, **31**, **32**
  seed symptoms **51**
bean, dwarf *see* bean, French
bean, French 105, 149, 207
  seed symptoms **51**
  nitrogen fixation 69
  resistant cultivars 105
bean, runner 105, 149, 152, 207
  symptoms **46**, **51**
  nitrogen fixation 69
  resistant cultivars 105
bean anthracnose 105
bean chocolate spot **12**, 105
bean halo blight 105
bean rust 105
bean seed beetle **51**, 105
bean seed fly **51**, 105
beech (*Fagus*) 106, 196
  bark symptoms **35**, **39**
  leaf symptoms **24**, **32**
beech bark disease **35**, 106
beech bark scale **35**, 106
beech gall midge 106
beech leaf-mining weevil **24**, 106
beech woolly aphid **32**, 106
beefsteak fungus 128
bees
  beneficial **62**
  burrowing 58, 113
  leaf-cutting **30**, 75, 144
  nectar robbing **46**, 152
beet leaf miner **17**, 106
beetles 75, 106
beetroot 106, 109, 207
  symptoms **17**, **44**
begonia 23, 107, 148, 204
begonia powdery mildew **23**, 107
Bellis perennis see daisy
beneficial creatures **60–3**, 79, 90, **90**, 143
Berberis see barberry
berberis sawfly 107
Bergenia see elephant's ears
Betula see birch
biennials 203
big bud mites **41**, 107
biological control 79, 90–1, **91**
birch (*Betula*) 107, 196

birch witches' brooms 107
birds 46, 47, 75, 107
  beneficial **63**, **90**
  chemical injury 83, **83**
  seed dispersal 67
bitter rot *see* gleosporium rot
black bean aphid **32**, 107
black leg of cuttings 108
blackberry 108, 169, 206
  symptoms **34**, **49**
blackberry cane spot **34**, 108
blackcurrant *see* currant, black
blackcurrant gall midge **26**, 108
blackfly 108
bleeding canker 108
blind shoots **40**, 108
blindness of bulbs **41**, 108
blossom end rot **50**, 108
blossom wilt **41**, 109
blotch *see* leaf spot
blue mould on bulbs **54**, 109
bluebell (*Hyacinthoides*) 109, 202
bluebell rust 109
blushing bracket *see* *Daedaleopsis confragosa*
bolting **44**, 109
boron deficiency 109
Botrytis see grey mould
box (*Buxus*) **26**, 110, 198
box blight
  cylindrocladium sp. 110
  volutella 110
box red spider mite 110
box sucker **26**, 110
bracket fungus **37**, **38**, 110
brassica bacterial soft rot 111
brassica dark leaf spot **13**, 111
brassica downy mildew 111
brassica light leaf spot 111
brassica ring spot 111
brassica wire stem 111
brassicas 98, 109, 110–11, 113–14, 129, 148, 161–2, 191, 207–9

leaf symptoms **13**, **14**, **29**, **30**, **32**
root symptoms **55**
stem symptoms **44**
plant protection 89, **89**
resistant cultivars **92**, 119
broom (*Cytisus*) 111, 198
stem symptoms **39**
broom gall mite **39**, 111
brown core of primula 111
brown rot (*Sclerotinia*) **47**, **49**, 112
brown scale **36**, 112
brown-tail moth 112
Brugmansia see Datura
bryobia mite 112
bud, dry 112
bud drop 112
buddleia (*Buddleja*) 198
buff-tip moth 112
bulb scale mite 112–13
bulbs 92, 108, 109, 125, 202
symptoms **41**, **54–5**
hot water treatment 89
bullfinch **46**, 74, 75, 113
busy Lizzie (*Impatiens*) 126–7, 203
leaf symptoms **30**
Buxus see box

C

cabbage *see* brassicas
cabbage moth **30**, 113
cabbage root fly 75, 113
plant protection 89, **89**
cabbage white butterfly 74, **76**, 113
cabbage whitefly **32**, 114
cacti **71**, 114, 148, 204
plant symptoms **22**
cactus corky scab **22**, 114
calceolaria 204
calcium 69
deficiency 114
Calendula see marigold
Californian poppy 200
Callistephus see China aster
resistant cultivars 103
Calluna see heather
camellia 114, 122, 198
leaf symptoms **15**, **22**, **26**
camellia gall **26**, 114
camellia petal blight 114
camellia yellow mottle **22**, 114
campanula 200
Campsis see trumpet vine

candytuft (*Iberis*) 200
canker **73**, 114
canna 200
Capsicum (ornamental pepper) 205
Capsicum (edible pepper) **50**, 108, 208
capsid bugs 115
symptoms **29**, **46**, 75
carabid beetle **63**
carnation (*Dianthus*) 123, 200, 204
carnation tortrix moth **20**, 86, 115
carpenteria 198
Carpinus see hornbeam
carrot 98, 109, 115, 208
symptoms **25**, **51**
carrot cavity spot 115
carrot fly **51**, 75, 89, 115
carrot motley dwarf virus **25**, 115
caryopteris 198
Castanea sativa see sweet chestnut
castor-oil plant (*Ricinus communis*) 203
catalpa 196
caterpillars 74, 75, 115–16
cats 89, 116
cauliflower 109
cavity decay 116
ceanothus 198
cedar (*Cedrus*) 116, 194
leaf symptoms **32**
cedar aphid **32**, 116
celeriac **12**, 208
celery 109, 116, 208
symptoms **12**, **19**, **36**
celery leaf miner **19**, 116
celery leaf spot **12**, 116
centaurea 200
centipedes **61**, 116
Cercidiphyllum japonicum 196
Cercis siliquastrum see Judas tree
chaenomeles 199
chafer grubs **53**, **58**, **91**, 75, 116
Cheiranthus see wallflower
chemicals 78–9, 80-3, **210**
damage 161, 190–1
trials **81**
see also pesticides
cherry 101, 103–4, 117, 158, 177, 206
bark symptoms **33**, **41**
leaf symptoms **19**, **24**
resistant cultivars 104
cherry, flowering *see* prunus, ornamental
cherry and pear slugworm 158

cherry blackfly **24**, 117
cherry laurel *see* laurel, cherry
cherry leaf scorch 117
cherry leaf spot 117
chicken of the woods *see Laetiporus sulphureus*
China aster (*Callistephus*) 203
resistant cultivars 103
China aster fusarium wilt 103
china mark moth **30**, 117
chironomid midges 117–18
chives **17**, 145
Chlorophytum see spider plant
chlorosis 117
Choisya see Mexican orange
Christmas cactus (*Schlumbergera*) 204
chrysanthemum (*Dendranthema*) 118, 126, 148, 200, 204
flower symptoms **43**, **45**
leaf symptoms **14**, **25**
hot water treatment 89, 118
chrysanthemum eelworm 89, 118
Chrysanthemum frutescens see marguerite
chrysanthemum leaf miner **25**, 118
chrysanthemum petal blight **45**, 118
chrysanthemum ray blight 118
chrysanthemum white rust **14**, 118
cineraria (*Senecio*) 98, 148, 204
Cistus see sun rose
citrus 204
clematis 118–19, 126, 198
flower symptoms **46**
clematis green petal 118
clematis wilt 119
click beetle 119
climbing plants 78, **93**, 198–9
clivia 204
cloches 88, **89**
clubroot **55**, 119
Codiaeum see croton
codling moth **47**, 75, 86, 119
Colchicum see autumn crocus
coleus (*Solenostemon*) 204

collar rot *see* phytophthora
Colletotrichum 119
Colorado beetle 119–20
columbine (*Aquilegia*) 200
companion planting **78**, **86**
composting **78**, 79, 96, **96**
conifer aphid 120, 122
conifer needle rust 152–3
conifer red spider mite 120
conifer woolly aphid *see* adelgids
conifers 98, 120, 122, 129, 152–3, 176
symptoms **37**
contact pesticides 82, **82**, **83**
Convallaria see lily-of-the-valley
convolvulus 200
copper deficiency 120
coral flower (*Heuchera*) 200
coral spot **41**, 120
corms 125, 202
corm symptoms **55**
Cornus see dogwood
Corylus see hazel
cosmos 200
cotinus 198
cotoneaster 120, 194, 198
stem symptoms **35**, **37**
cotoneaster webber moths **37**, 120
courgette *see* marrow
covers, protective 88–9, **88**, **89**
crab apple (*Malus*) 101, 196
leaf symptoms **16**
crane flies (leatherjackets) 120, 145
cranesbill (*Geranium*) 132, 200
leaf symptoms **25**
Crassula see money tree
Crataegus see hawthorn
crinum **29**, 152, 202
Crocosmia see montbretia
crocus 149, 179, 202
symptoms **46**, **55**
crop rotation **78**, 89, 96
croton (*Codiaeum*) 204
crown gall **38**, 120
crown rot 121
Cryptolaemus montrouzieri **91**, 143
cuckoo spit **39**, **43**, 75, 121

cucumber 121, 129, 148, 208
fruit symptoms **55**
leaf symptoms **21**, **25**, **43**
resistant cultivars 121
cucumber foot & root rot **55**, 129
cucumber gummosis 121
cucumber mosaic virus **21**, **25**, **72**, **73**, 121
Cupressus see cypress
currant, black 98–9, 108, 121–2, 206
bud symptoms **41**
leaf symptoms **25**, **26**
stem symptoms **33**
resistant cultivars 108
see also currants
currant blister aphid **27**, 121
currant clearwing moth **33**, 75, 122
currant, flowering (*Ribes*) 198
currant, red **31**, 206
see also currants
currant reversion 122
currant, white **31**, 206
currant & gooseberry leaf spot 121–2
see also currants
currants 120, 121–2, 134, 194
symptoms **27**, **39**
plant protection 88, **88**
see also currant, black; currant, red/white
cushion scale **15**, 122
cutworms **53**, 75, 122
cyclamen 204
cyclamen, hardy 202
Cyperus see papyrus
cypress (*Cupressus*) **35**, 122, 196
susceptible sp. 176
cypress aphid 120, 122
Cytisus see broom

D

Daedaleopsis confragosa **37**, 122
daffodil (*Narcissus*) 43, 112–13, 151–2, 202
symptoms **23**, **29**, **43**, **54**
hot water treatment 89
daffodil bulb scale mite 112–13
dahlia 107, 123, 126, 148, 202
symptoms **32**, **43**, **46**
dahlia smut 123
daisy (*Bellis perennis*) 200

damping off **34**, 123
damson *see* plum
Datura see angels'
   trumpets
daylily (*Hemerocallis*)
   136, 200
   flower symptoms **45**
deer **41**, **70**, 75, 88–9,
   123
delphinium 123–4, 200
   leaf symptoms **12**, **18**
delphinium black blotch
   12, 123
delphinium leaf miner
   18, 123
delphinium moth 123
delphinium viruses 124
Dendranthema see
   chrysanthemum
devil's coach horse **62**
Dianthus see carnation
dianthus anther smut
   124
Dianthus barbatus see
   sweet William
dianthus fusarium wilt
   124
didymella stem rot **33**,
   124
dieback 124
Digitalis see foxglove
diseases 72–3, **72**, **73**,
   74
dog lichens 58, **59**
dogs 124
dogwood (*Cornus*) 124,
   196, 198
dogwood anthracnose
   124
Douglas fir
   (*Pseudotsuga*
   menziesii) 125, 196
Douglas fir adelgid 124
downy mildew **73**, 125
dracaena 204
drought **28**, **44**, 70–1,
   70, 125
   bud disorders 112
dry rot of bulbs &
   corms 125
Dutch elm disease
   125–6

E

earthworms **58**, **61**, **66**,
   67, 125
earwigs **46**, 74, 75, 126
   traps 87, 88, **88**
echeveria 205
Echinops see globe
   thistle
ecology in the garden
   66–7
eelworm 73, 75, 96, 126
   biological control 91,
   91

hot water treatment 89
elaeagnus 120, 126, 198
elaeagnus sucker 126
elder (*Sambucus*) 198
elephant hawk moth
   30, 126
elephant's ears
   (*Bergenia*) 200
elm bark beetle **33**,
   104
elm gall mite **16**, 126
elm (*Ulmus*) 104, 126,
   127, 196
   symptoms **16**, **33**
Encarsia formosa **91**
enchytraeid worms 127
epimedium 200
Epiphyllum 114, 205
   plant symptoms **22**
Eranthis see winter
   aconite
Erica see heather
erythronium 202
escallonia 198
eucalyptus 127, 196
eucalyptus gall wasp
   127
eucalyptus sucker 127
euonymus 127, 198
   leaf symptoms **21**
euonymus scale **21**, 127
euphorbia 200
Euphorbia pulcherrima
   205
evening primrose
   (*Oenothera*) 203

F

Fagus see beech
fairy rings 185–6
false acacia (*Robinia*)
   39, 196
fasciation **43**, 127
fatsia 198
felt gall mites **19**,
   127–8
felted beech coccus *see*
   beech bark scale
ferns 205
fertilizers 68, 93
Ficus benjamina see
   weeping fig
Ficus elastica see rubber
   plant
fig (*Ficus*) 206
   weeping (*Ficus*
   benjamina) 205
figwort weevil 128
fir, Douglas 125, 196
fir, silver (*Abies alba*)
   98, 177, 196
fireblight **35**, 128
firethorn (*Pyracantha*)
   167–8, 194, 198
   symptoms **35**, **49**
   resistant cultivars 167

fish 83, **83**
Fistulina hepatica 128
flatworms 128
flea beetles **29**, 74, 75,
   89, 128
   sticky traps 87
flies 129
flower damage **42–6**,
   73, 159, 167
   frost **44**, 130
   pests 74-5
fluted scale 129
Fomes pomaceous **39**,
   129
fomes root & butt rot
   37, 129
food chains 67–8, **83**
foot and root rots **55**,
   129–30
forget-me-not (*Myosotis*)
   203
forsythia 130, 198
   stem symptoms **40**
forsythia gall **40**, 130
Fortunella see citrus
foxes 130
foxglove (*Digitalis*) 200
frangipani (*Plumeria*)
   205
Fraxinus see ash
freesia 202
French marigold *see*
   African marigold
fritillary (*Fritillaria*)
   202
froghopper *see* cuckoo
   spit
frogs **79**, 90
frost damage **28**, **44**,
   70, 71, 130
   apple 100
fruit cages 88, **88**
fruit damage **47–51**,
   111–12, 180
   pests 75
fruit plants 206–7
fruit tree red spider mite
   20, **40**, 130
fuchsia 114, 130–1, 205
   symptoms **18**, **30**,
   46
   fuchsia, hardy **30**, 114,
   126, 198
fuchsia gall mite 130
fuchsia rust **18**, 131
fumigation **80**
fungi 67, 72–3, **72**, **73**
   mycorrhizal
   associations 69
fungicides **210**
   phytoxicity 73, 81, 161
   see also pesticides
fungus gnat **53**, 87, 131
   biological control **91**
fusarium bulb rot *see*
   gladiolus corm rot;
   narcissus basal rot
fusarium wilt 131

G

gage *see* plum
Galanthus see
   snowdrop
gall midges 75, 131
gall mites 75, 131
gall wasps 75, 77, 132
gall-forming pests 74,
   75, 77, 131–2
ganoderma 132
garden, gardening 66–7
   hygiene 79, 92, **94**,
   96, **96**
   techniques 78-9, 86-9,
   92-3
   tool disinfection 73,
   73
   see also pruning
gardenia 205
garlic 145, 208
gazania 205
gentian 200
Geranium see cranesbill
geranium downy
   mildew **25**, 132
geranium sawfly 132
gerbera 205
geum 200
geum sawfly 132
ghost spot *see* grey
   mould
gladiolus 132–3, 202
   symptoms **45**, **55**
gladiolus core rot 132
gladiolus corm rot **55**,
   133
gladiolus dry rot 133
gladiolus hard rot 133
gladiolus scab & neck
   rot 133
gladiolus thrips **45**, 133
glasshouse leafhopper
   20, 133
glasshouse plants *see*
   greenhouse plants
glasshouse red spider
   mite **22**, **23**, **43**, 77,
   81, 133
   biological control **91**
glasshouse thrips 134
glasshouse whitefly 77,
   81, 87, **91**, 134
gleditsia gall midge **27**,
   134
Gleditsia see honey
   locust
gleosporium rot 134
globe artichoke 107
globe thistle (*Echinops*)
   200
gloxinia (*Sinningia*) 205
glyphosate damage 191
goatsbeard (*Aruncus*)
   102, 200
   leaf symptoms **30**
golden rod (*Solidago*)
   200

gooseberry 98–9, 121,
   134, 194, 206
   symptoms **25**, **31**, **49**
   resistant cultivars 99
gooseberry dieback 134
gooseberry mildew *see*
   American gooseberry
   mildew
gooseberry rust 134
gooseberry sawfly **31**,
   135
grape *see* vine (grape)
grape hyacinth
   (*Muscari*) 109, 202
grapefruit *see* citrus
grease bands 87
green spruce aphid
   135
green tiger beetle **63**
greenfly *see* aphids
greenhouse plants 77,
   133–4, 136–7, 148,
   183, 204–5
   flower symptoms **42**,
   43
   leaf symptoms **16**, **20**,
   22, **23**, **27**, **32**
   high temperature
   injury 135–6
   pest control **79**, 81,
   85, 87
grey mould (*Botrytis*)
   135
   symptoms **34**, **42**, **45**,
   50
   chemical control 81
Grifola frondosa 135
ground beetles 136
gummosis 121
gypsy moth 136

H

hail damage 71
halo blight 105
hardenbergia 205
hawthorn (*Crataegus*)
   100, 119, 136,
   196
   symptoms **19**, **37**
hawthorn gall midge
   136
hazel (*Corylus*) 107,
   151, 178, 206
   symptoms **41**, **52**
heather (*Erica*) 198
hebe 136, 198
hebe downy mildew
   136
Hedera see ivy
hedgehog **63**, **67**, 90
Helianthemum see rock
   rose
hellebore (*Helleborus*)
   135, 200
hellebore "Black Death"
   20, 136

hellebore leaf blotch **20**, 136
Helleborus see hellebore leaf symptoms **20** resistant spp. 136
Hemerocallis see daylily
hemerocallis gall midge **45**, 75, 136
hemispherical scale **16**, 137
herbaceous plants 114, 143, 200–1 symptoms **20**, **39**
Heterobasidion annosum see fomes root & butt rot
Heterorhabditis megidis 91, **91**
Heuchera see coral flower
hibiscus 204
high temperature injury 71, 137
Himalayan poppy (*Meconopsis*) 201
Hippeastrum see amaryllis symptoms **29**, **54**
holly (*Ilex*) 121, 137, 196 leaf symptoms **15**, **17**
holly leaf blight 137
holly leaf miner **17**, 137
hollyhock (*Alcea*) 137, 201 leaf symptoms **17** resistant spp. 137
hollyhock rust **72**, 137
holm oak leaf miner 137
honesty (*Lunaria*) 201
honey fungus **34**, **73**, 138 resistant spp. 138
honey locust 133, 196–7 leaf symptoms **27**
honeydew **32**, 138
honeysuckle (*Lonicera*) 138, 199
honeysuckle aphid 138
hornbeam (*Carpinus*) 138, 197
hornbeam witches' broom 138
horse chestnut (*Aesculus*) 138, 197 symptoms **20**, **40**
horse chestnut bleeding canker **23**, **38**, 138
horse chestnut leaf blotch **20**, 139
horse chestnut leaf-mining moth 139
horse chestnut marginal leaf scorch 139
horse chestnut scale **40**, 139

horticultural fleece **88**, 89
hosta 201
hot water plant (*Achimenes*) 205
house leek (*Sempervivum*) 201
house plants 104, 131, 133–4, 136–7, 148, 183, 204–5 symptoms **22**, **37**, **53**
hoverflies **61**, **90**, 139
Hoya see wax plant
hyacinth (*Hyacinthus*) 202
Hyacinthoides see bluebell
Hyacinthus see hyacinth
hydrangea 139, 199 leaf symptoms **15**
hydrangea leaf spot 139
hydrangea powdery mildew 140
hydrangea scale **15**, 140
hypericum 199
Hypoaspis miles **91**

I

Iberis see candytuft
ice plant (*Sedum*) 201
Ilex see holly
Impatiens see busy Lizzie
injury, physical **40**, **70**
Inonotus hispidus 140
insecticides **210** organic 83, 90 phytoxicity 73, 81, 161 *see also* pesticides
insects 74–75, **74**, **75**, **77** beneficial **60–3**, **60**, 143 life cycles 76–7, **76**
integrated control **78–9** complementary practices 86–9, **86**, **87**, **88**, **89**
Ipomoea see morning glory
iris 140–1, 202 symptoms **15**, **29**, **55**
iris ink disease 140
iris leaf spot 140
iris rhizome rot **55**, 140
iris rust **15**, 141
iris sawfly **29**, 141
iris scorch 141
iron 69 deficiency **24**, 141
ivy (*Hedera*) 141–2, 199
ivy leaf spot 141
ivy on tree trunks 142

J

jade plant (*Crassula*) 205
Japanese anemone (*Anemone*) 201
Japanese quince (*Chaenomeles*) 199
jasmine (*Jasminum*) 199
Jerusalem sage (*Phlomis*) 199
Judas tree (*Cercis siliquastrum*) 197
Juglans see walnut
juniper (*Juniperus*) 142, 197 symptoms **32**, **37**
juniper aphid 120, 142
juniper scale **37**, 142
juniper webber moth **32**, 142
Juniperus see juniper

K

kabatina shoot blight 142
kalanchoe 205
kangaroo vine 205
keithia thujina needle blight **28**, 142
Kniphofia see red-hot poker
knopper galls *see* acorn gall wasp
kumquat *see* citrus

L

laburnum 142, 197 leaf symptoms **18**
laburnum leaf miners **18**, 142
lacebug **21**, 161, 170
lacewing **63**, 143
lackey moth **30**, 143
ladybird **60**, **77**, **91**, 143
Laetiporus sulphureus **38**, 143
lantana 205
lapageria 205
larch (*Larix*) 143, 197 needle symptoms **32**
larch adelgid **32**, 98, 143
Larix see larch
Lathyrus see sweet pea
laurel (*Laurus*) 199
laurel, cherry (*Prunus laurocerasus*) 197
Laurus see bay, sweet bay
lavatera 142, 199
lavatera leaf & stem rot 143

lavender (*Lavandula*) 143, 199
lavender shab 143
lawns **78**, 116, 126, 185–6, 201 susceptible spp. 185 symptoms **58**, **59**
leaf & bud eelworm **25**, 143
leaf curl **26**, 157–8
leaf damage 12–32, 75, 143–5, 162 frost **28**, **44**, **70**, 130 weedkillers **34**, 190–1
leaf miners 144
leaf scorch 144
leaf spot (bacterial) **13**, **29**, 144
leaf spot (fungal) **73**, 144 symptoms **12**, **13**
leaf weevils 145
leaf-cutting bees **30**, 75, 144
leafhoppers 74, 75, 144
leafy gall **39**, 145
leatherjackets **53**, **58**, 75, **91**, 145 *see also* crane flies
leek 145–6, 208 symptoms **17**, **23**, **52** resistant cultivars 146
leek moth **23**, 145
leek rust **17**, 145
leek white tip 146
lemon *see* citrus
leopard moth 75, 146
lettuce 109, 146, 208 symptoms **22**, **28**, **44**, **55** resistant cultivars 146
lettuce downy mildew **22**, 146
lettuce grey mould **28**, 146
lettuce root aphid 146
lettuce viruses 146
lightning damage **70**
lilac (*Syringa*) 146–7, 167, 199 symptoms **19**, **22**, **33**
lilac blight **33**, 146
lilac leaf miner **19**, 147
Lilium see lily
lily (*Lilium*) 147, 202 resistant spp. 147
lily beetle 147
lily disease 147
lily viruses 147
lily-of-the-valley (*Convallaria*) 201
lime (*Tilia*) 147, 197 leaf symptoms **14**
lime nail gall mite **14**, 147
lime-induced chlorosis **24**, 68, 141
liquidambar 197

Liriodendron 197
lobelia 201, 203
loganberry *see* raspberry
Lonicera see honeysuckle
Lunaria see honesty
lupin (*Lupinus*) 147, 201 stem symptoms **41**
lupin aphid **41**, 147
Lupinus see lupin
lychnis 201

M

magnesium 69 deficiency **22**, 147
magnolia 120, 197
mahonia 148, 199 leaf symptoms **12**, **24**
mahonia rust **12**, **24**, 148
Malus spp. 101, 196 leaf symptoms **16**
manganese 69 deficiency **24**, 141
maple (*Acer*) 98, 120, 182, 197 leaf symptoms **13**, **17**
marguerite 205
marigold (*Calendula*) 203
marrow 208 resistant cultivars 121
Matthiola see stock
mealy cabbage aphid **32**, 148
mealy plum aphid 148
mealybugs **32**, **37**, 74, 75, 148 biological control **91**
Meconopsis see Himalayan poppy
melon 148, 208 plant symptoms **43**
melon cotton aphid **43**, 148
Meripilus giganteus **39**, 148
metamorphosis 76–7, **76**, **77**
Mexican orange (*Choisya*) 199
mice **51**, **79**, 89, 149 chemical control **208–9**
Michaelmas daisy (*Aster*) 149, 201 flower symptoms **45** susceptible cultivars 149
Michaelmas daisy mite **45**, 74, 75, 149
Michaelmas daisy wilt 149

REFERENCE SECTION

midges *see* chironomid midges; gall midges
millipedes **54**, 149
mimosa 205
mint (*Mentha*) **31**, 149, 208
mint beetle **31**, 149
mint rust 149
mistletoe **67**
mites 149
mites, bryobia 112
mock orange (*Philadelphus*) 107, 199
moles **59**, 89, 150
molybdenum deficiency 111
money tree (*Crassula*) 205
monkey puzzle (*Araucaria araucana*) 197
montbretia (*Crocosmia*) 125, 202
morning glory (*Ipomoea*) 199
moths 150
mountain ash (*Sorbus aucuparia*) 150, 197
mountain ash blister mite 150
mulberry **38**, 150, 206
mulberry bacterial blight 150
mulberry canker **38**, 150
mullein moth 151
Musa see banana
Muscari see grape hyacinth
mussel scale **36**, 151
mycoplasmas 72, 151
mycorrhiza 69
Myosotis see forget-me-not

N

Narcissus see daffodil
narcissus basal rot 151
narcissus bulb fly **54**, 89, 151
narcissus eelworm **54**, 89, 151
narcissus leaf scorch **29**, 152
narcissus smoulder 152
narcissus viruses **23**, 152
nasturtium (*Tropaeolum*) 107, 203
  leaf symptoms **32**
nectar robbing **46**, 152
nectarine *see* peach
nectria canker 152
needle blights of pines 152

needle rusts of conifers 152
nematodes *see* eelworm
nemesia 203
nerine 152
Nerium see oleander
New Zealand flax (*Phormium*) 161, 199
Nicotiana see tobacco plant
nitrogen 93
  deficiency 152
  fixation 68, **68**, 69
nut weevil **52**, 75, 153
Nymphaea see water lily

O

oak (*Quercus*) 153–4, 197
  symptoms **17**, **24**, **39**, **52**
oak gall wasp **17**, 153
oak phylloxera 153
oak powdery mildew **24**, 153
oak processionary moth 153
oedema **19**, 154
Oenothera see evening primrose
oleander (*Nerium*) 154, 205
oleander scale 154
onion 145, 154-5, 208
  leaf symptoms **17**, **23**, **52**
  resistant cultivars 154
onion bolting 154
onion downy mildew 154
onion eelworm **52**, 154
onion fly 75, 154–5
onion neck rot **52**, 155
onion thrips 155
onion white rot **52**, 155
Opuntia **22**, 114
orange *see* citrus
orchid **12**, 155, 205
orchid viruses **12**, 155
organic gardening 79
oribatid mites 155
ornamental allium 145
ornithogalum 202
overwintering of pests 77

P

Paeonia see peony
palms 205
pansy (*Viola*) 155–6, 203
  leaf symptoms **13**
pansy downy mildew 155

pansy leaf spot **13**, 155
pansy sickness 156
Papaver see poppy
papery bark **40**, 156
papyrus (*Cyperus*) 205
parsley 115, 208
  symptoms **19**, **25**, **51**
parsnip 115, 156, 208
  symptoms **19**, **51**, **57**
  resistant cultivars 156
parsnip canker **57**, 156
parsnip viruses 156
parthenogenesis 76–7
Passiflora see passion flower
passion flower (*Passiflora*) 205
pea 105, 149, 156–7, 208
  leaf symptoms **14**, **28**, **30**, **31**
  pod symptoms **51**
  nitrogen fixation 69
  resistant cultivars 156, 157
pea & bean weevil **31**, 156
pea downy mildew 156
pea leaf & pod spot **14**, **51**, 156
pea marsh spot 156
pea moth **51**, 75, 157
pea powdery mildew 157
pea thrips 157
pea viruses 157
pea wilt **28**, 157
peach 157-8, 206
  leaf symptoms **26**
peach aphids 157
peach leaf curl **26**, 157
pear 109, 119, 128, 158–9, 206–7
  flower symptoms **41**, **44**
  fruit symptoms **47**, **48**
  leaf symptoms **16**, **19**, **31**
  stem symptoms **35**, **36**
  fruit protection 88
  ornamental (*Pyrus salicifolia*) 177, 197
  resistant cultivars 158–9
pear & apple canker **36**, 100, 158
pear & cherry slugworm **19**, 158
pear bedstraw aphid 158
pear boron deficiency 158
pear leaf blister mite **16**, 158
pear midge **48**, 158
pear rust 159
pear scab **48**, 159
pear stony pit virus **48**, 159

pear sucker 159
pedicel necrosis 159
pelargonium 154, 159–60, 205
  leaf symptoms **15**, **19**, **21**
pelargonium rust **15**, 159
pelargonium viruses **21**, 159
penstemon 201
peony (*Paeonia*) 160, 201
  flower symptoms **43**
peony wilt **43**, 160
pepper, edible (*Capsicum*) **50**, 108, 208
pepper, ornamental (*Capsicum*) 205
perennials, herbaceous 200–1
periwinkle (*Vinca*) **15**, 160, 201
periwinkle rust **15**, 160
pestalotiopsis 160
pesticides 78–9, 80–3, **210**
  application **80**, **82**, 83–5, **84**, **85**
  pathogen, resistance to 80–1
  selective 83, 90
  see also chemicals
pests 74–5, **74**, **75**
  chemical control 78–9, 80–5
  complementary control 86–9
  life cycles 76–7, **76**
petunia 160, 203
petunia foot rot 160
petunia viruses 160
Phasmarhabditis hermaphrodita 91, **91**
Phellinus ignarius 160
Phellinus pomaceous see Fomes pomaceous
pheromone traps 86–7, **87**, 119
Philadelphus see mock orange
Phlomis see Jerusalem sage
phlox **33**, 89, 160, 201
phlox eelworm **33**, 89, 160
phlox powdery mildew 160
phomopsis canker 160
Phormium (New Zealand flax) 161, 199
phormium mealybug 161

phosphorus 69, 93
  deficiency 161
photosynthesis 66–7
phygelius 201
phytophthora 73, **73**, 161
Phytoseiulus persimilis **91**, 150
phytotoxicity 73, 81, 161
Picea see spruce
pieris lacebug 161
pigeons 30, 89, 161–2
pine (*Pinus*) 98, 120, 152, 162, 197
  symptoms **39**
pine, needle blight 152
pine adelgid **39**, 98, 161
pine aphid 98, 120
pineapple gall adelgid **38**, 162
pink *see* carnation
Pinus see pine
plane (*Platanus*) 162, 197
plane anthracnose 162
plant associations 67, **67**
plant metabolism 66–7
plant nutrition 67, 68–9, 93
plant quality 73, 92, **92**
plant raising 92–3, **93**, 123
plant requirements 67, 70
Platanus see plane
plum 148, 162–3, 177, 207
  bark symptoms **33**, **40**, **41**
  fruit symptoms **48**
  leaf symptoms **16**, **19**, **20**, **22**, **27**
  resistant cultivars 104, 177
  susceptible cultivars 162
plum blossom wilt 162
plum leaf gall mite **16**, 162
plum leaf-curling aphid **27**, 162
plum moth **48**, 75, 86, 162
plum pox (sharka) 163
plum rust **22**, 163
plum sawfly 75, 163
Plumeria see frangipani
pocket plum 163
poinsettia (*Euphorbia pulcherrima*) 205
pollen beetle **43**, 74, 75, 163–4
pollination 60, 67
  nectar robbing bees **46**, 152
  poor 71, 164
polyanthus (*Primula*) 144–5, 201

symptoms **13**, **46**
Polygonatum see
Solomon's seal
Polygonum
baldschuanicum see
Russian vine
pome fruits, fireblight
35, 128
pond plants 117–18
see also water lily
ponds **83**, 90
poplar (*Populus*) 100,
164, 197
poplar canker 164
poplar yellow blister
164
poppy (*Papaver*) 107,
203
Populus see poplar
pot plants **54**, **55**,
204–5
potash deficiency 24,
164
potassium 69
deficiency **24**, 164
potato 98, 164–7, 178,
208
leaf symptoms **28**
tuber symptoms **51**,
54, **55**, **56**, **57**
resistant cultivars 165,
166, 178
susceptible cultivars
164, 166
potato black leg 164
potato blight **28**, **51**,
56, 164–5
potato common scab
56, 165
potato cyst eelworm **55**,
165
potato dry rot **57**, 165
potato early blight
165–6
potato gangrene **57**, 165
potato hollow heart 166
potato internal rust spot
166
potato powdery scab
56, 166
potato silver scurf **56**,
166
potato spraing **56**, 166
potato viruses 166
potato wart disease 166
powdery mildew **25**,
49, **73**, 167
primrose (*Primula*)
144–5, 201
Primula see primrose,
polyanthus
symptoms **13**, **46**
Primula auricula 200
privet (*Ligustrum*) 167,
199
leaf symptoms **22**
privet thrips **22**, 167
proliferation **44**, 167

pruning 94–5, **94**, **95**
poor **35**, **40**, **71**
Prunus dulcis see
almond
Prunus laurocerasus see
laurel, cherry
prunus, ornamental 177,
197
leaf symptoms **18**, **30**
Pseudotsuga menziesii
see Douglas fir
pulmonaria 201
pumpkin *see* marrow
Pyracantha see firethorn
pyracantha leaf miner
167
pyracantha scab **49**,
167
pyrethrum (*Tanacetum*)
201
Pyrus salicifolia see
ornamental pear
pythium root rot 168

Q

Quercus see oak
quince 168, 207
fruit symptoms **48**
quince leaf blight **48**,
168
quince powdery mildew
168

R

rabbits 70, 75, 88–9, 168
radish 109, 208
raspberry 164, 168–9,
207
cane symptoms **34**, **40**
fruit symptoms **50**
leaf symptoms **18**
resistant cultivars 169
raspberry beetle **50**, 75,
168
raspberry cane blight
168
raspberry cane spot &
spur blight **34**, 169
raspberry leaf & bud
mite **18**, 169
raspberry rust **40**, 169
raspberry spur blight
169
raspberry viruses **18**,
169
rats 83, 169
red berry mite **49**, 169
red spider mite 75, 120,
130, 133–4, 150
symptoms **20**, **22**, **23**,
40, **43**
chemical control 81
red thread *see* turf red
thread

redcurrant *see* currant,
red
red-hot poker
(*Kniphofia*) 201
repellents 89
replant problems *see*
rose sickness
replanting 96, 174
resistant cultivars 92
see also individual
pathogens and pests
rhododendron 112,
170–1, 199
flower symptoms **41**,
42
leaf symptoms **14**, **15**,
21
susceptible spp. 171
rhododendron bud blast
41, 170
rhododendron lacebug
21, 170
rhododendron leaf spot
14, 170
rhododendron
leafhopper 170
rhododendron petal
blight **42**, 170
rhododendron powdery
mildew **15**, 171
rhododendron rust 171
Rhoicissus see kangaroo
vine
rhubarb 171, 208
rhubarb gumming 171
Rhus see sumach
Ribes see flowering
currant
Ricinus communis see
castor-oil plant
Robinia pseudoacacia
39, 196
Robin's pin cushion **39**,
171
rock cress (*Arabis*) 201
rock rose
(*Helianthemum*) 199
Romneya 200
root aphids **55**, 75, 171
root damage **53**–**7**,
129–30, 171
root feeding pests 74–5
root knot eelworm 171
root mealybug **55**, 75,
171
root rots **55**, 96, 129–30
rose (*Rosa*) **39**, 128,
172–4, 199
flower symptoms **43**,
44
leaf symptoms **13**, **16**,
18, **20**, **21**, **27**
stem symptoms **35**, **39**,
40
replanting 96, 174
resistant cultivars 172,
173
rose aphid **43**, 172

rose balling **44**, 172
rose black spot **13**, 81,
172
rose canker & die back
40, 172
rose grey mould *see*
grey mould; rose
canker
rose leafhopper **21**, 173
rose leaf-rolling sawfly
27, 172
rose powdery mildew
20, 81, 173
rose root aphid **35**,
173–4
rose rust **16**, 173
rose sickness/replant
problems 96, 173
rose slugworm **18**, 174
rose viruses **21**, 174
rosemary (*Rosmarinus*)
173, 199, 208
rosemary beetle 173
Rosmarinus see
rosemary
rosy apple aphid **48**,
174
rotation of crops **78**, 89,
96
rowan (*Sorbus*
aucuparia) 150, 197
rubber plant (*Ficus*
elastica) 205
Russian vine (*Solanum*)
199
rusts **73**, 174

S

sage (*Salvia*) 201
Saintpaulia see African
violet
salad rocket 208
Salix see willow
salt damage 175
Salvia see sage, salvia
salvia (*Salvia*) 203
Sambucus see elder
sap-feeding pests 74, 75
sawflies 74, 75, 175
saxifrage 201
scale insects 74, 75, 175
scefflera **22**, 104, 205
Schlumbergera 204
sciarid flies *see* fungus
gnat
sclerotinia **36**, 175
see also brown rot
scorch 175
Sedum see ice plant
seedlings 92–3, **93**, 123
symptoms **34**, **51**, **53**,
54
seeds 67
damage **51**, **52**
seiridium canker **35**,
176

Sempervivum see
houseleek
sempervivum leaf miner
176
Senecio see cineraria
shallot 145, 154, 208
symptoms **17**, **52**
shanking of grapes 176
sharka *see* plum pox
shieldbugs 176
shot-hole **18**, **30**, 176
shot-hole borers **33**, 177
shrew **62**, 90
shrubs 198–9
silver fir *see* fir, silver
silver fir adelgid 177
silver leaf 177
Sinningia see gloxinia
slime flux & wetwood
177
slime mould **58**, 177
slugs **57**, 75, 77, 178
symptoms **29**, **36**, **43**
biological control 91,
91
trap **82**, **83**, 86, 88
small bulb fly *see*
narcissus bulb fly
small ermine moths **38**,
178
smut **35**, **44**, 178
snails **74**, 75, 77, **90**,
178
symptoms **29**, **43**
trap 86, 88
snapdragon
(*Antirrhinum*) 99,
201
leaf symptoms **15**
resistant cultivars 99
snow 71
snowdrop (*Galanthus*)
152, 179, 202
snowdrop grey mould
179
soft scale 179
soils 68–9, **68**, **69**, 93
problems **53**–**7**
Solanum capsicastrum
see winter cherry
Solenostemon see coleus
Solidago see golden rod
Solomon's seal
(*Polygonatum*) 179,
202
leaf symptoms **31**
Solomon's seal sawfly
31, 179
sooty mould 179
Sorbus aucuparia see
mountain ash, rowan
southern green
shieldbug 179
sparrow **46**, 74, 75, 179
spider plant
(*Chlorophytum*) 205
spinach beet 106
leaf symptoms **17**

REFERENCE SECTION

spinach, perpetual 208
splitting 180
sprekelia 152
springtails **54**, 180
spruce (*Picea*) 98, 120,
135, 162, 197
  stem symptoms **38**
squirrels **52**, 75, 180
stachys 180, 201
stachys case-bearing
  caterpillar 180
statice 203
Steinernema feltiae 91,
  91
Steinernema kraussei
  91, **91**
stem & bulb eelworm
  180
stem damage 33–41,
  130, 180
stephanotis 205
stock (*Matthiola*) 203
storage rots 180
stranvaesia 199
strawberry 109, 164,
  181–2, 207
  fruit symptoms **50**
  leaf symptoms **12**, **17**
  plant protection 88,
    88
  susceptible cultivars
    182
strawberry black eye
  181
strawberry green petal
  181
strawberry grey mould
  50, 181
strawberry leaf spot **12**,
  17, 181
strawberry mite 181
strawberry red core 181
strawberry seed beetle
  181
strawberry tree
  (*Arbutus*) 102, 187,
  197
  leaf symptoms **15**
strawberry viruses 182
strawberry yellows 182
streptocarpus 205
succulents **22**, 114,
  148
sucker 75, 182
sumach (*Rhus*) 199
sun rose (*Cistus*) 199
swede 89, 109, 148, 208
  leaf symptoms **32**
  resistant cultivars 119
sweet chestnut
  (*Castanea sastiva*)
  197
sweet pea (*Lathyrus*)
  203
sweet pepper *see*
  Capsicum
sweet William
  (*Dianthus*) 203

sweetcorn 149, 182,
  208
  seed symptoms **51**
sweetcorn smut **51**, 182
swift moth **53**, 75, 182
sycamore (*Acer*) 98,
  182, 197
  symptoms **13**, **17**, **36**
sycamore sooty bark
  disease **36**, 182
systemic pesticides 82,
  82, **83**, **210**

T

tachinid fly **61**
Tagetes see African
  marigold
tar spot of acer **13**, 183
tarsonemid mites **27**,
  183
Taxus see yew
tayberry *see* raspberry
temperature extremes
  71
thrips 74, 75, 87, 183
thrushes **63**, **90**
thuja **28**, 142, 197
thunderworms 183
Tilia see lime
toads **63**, **79**, 90
toadstools 69, 186
tobacco plant
  (*Nicotiana*) 203
tomato 124, 183–5, 208
  symptoms **28**, **33**, **50**,
    51, **55**
  resistant cultivars 184,
    185
tomato blight **28**, **51**,
  183
tomato blotchy ripening
  51, 183
tomato cyst eelworm
  see potato cyst
  eelworm
tomato fruit splitting **50**,
  184
tomato ghost spot 184
tomato greenback **50**,
  184
tomato leaf mould 184
tomato moth 184
tomato pith necrosis
  185
tomato viruses 185
tortrix moth *see*
  carnation tortrix
  moth
Tradescantia see
  wandering Jew
traps 86–8, **86**, 88
  sticky **78**, **79**, 87, **87**
trees 196–7
  bark symptoms **33**,
    34, **37**, **38**, **39**, **40**,
    41

leaf symptoms **19**, **30**,
  31
Tropaeolum see
  nasturtium
trumpet vine (*Campsis*)
  199
tulip (*Tulipa*) 185, 202
  symptoms **22**, **55**
tulip fire **22**, **55**, 185
tulip grey bulb rot 185
tulip tree 197
turf *see* lawns
turf dollar spot **58**, 185
turf fairy ring **59**, 186
turf ophiobolus patch
  186
turf red thread **59**, 186
turf snow mould **59**,
  186
turf thatch fungus
  mycelium **58**, 186
turf toadstool 186
turnip 208
turnip gall weevil 186
two-spotted mite *see*
  glass-house red
  spider mite

U

Ulmus see elm

V

vapourer moth **31**, 187
vegetables 109, 114,
  207
  symptoms **50–2**, **53–4**,
    55–7
  crop rotation 78, 89,
    96
  see also individual
    plants
verticillium wilt **29**, **33**,
  187
viburnum 107, 187,
  199
  leaf symptoms **15**
viburnum beetle 187
viburnum scale 187
viburnum whitefly **15**,
  187
Vinca see periwinkle
vine (grape) 167, 176,
  188, 194, 207
  symptoms **15**, **29**, **47**,
    49, **53**
vine erinose mite **15**,
  188
vine weevil **29**, **53**, 75,
  77, 188
  control 87, **91**
violet (*Viola*) 188, 201
  symptoms **27**, **57**
violet gall midge **27**,
  188

violet root rot **57**, 188
viruses 72, **72**, 73, **73**,
  189
  flower symptoms **42**,
    73
  leaf symptoms **23**, **25**,
    27
  vectors 73, 188
voles **36**, 75, 189

W

wallflower
  (*Cheiranthus*) 203
walnut 180, 189, 207
  nut symptoms **52**
walnut blister mite 189
walnut blotch 189
walnut soft shell 189
wandering Jew
  (*Tradescantia*) 205
wasps **47**, **49**, **61**, 75,
  190
  beer trap **86**
  parasitic **60**, **62**
water core 102
water lily (*Nymphaea*)
  117, 190, 201
  leaf symptoms **30**
water lily aphid 190
water lily beetle **30**,
  190
water lily leaf spot 190
watering 93
  irregular **26**, **36**, **57**,
    71, 141
waterlogging **34**, 71, **71**,
  94, 190
wax plant (*Hoya*) 205
weather 70–1, **70**, **71**
  disease development
    73
  pest populations 77
weed control 94
weedkillers, damage **23**,
  34, **39**, 190
weeping fig (*Ficus*) 205
weevils 191
western flower thrips
  42, **91**, 191
whiptail of brassicas 191
white blister **14**, 191
whitecurrant *see* currant,
  white
whitefly 87, 191–2
  chemical control 81
wildlife, pest control 83
willow (*Salix*) 192–3,
  197
  symptoms **17**, **18**, **36**,
    41
  resistant spp. 192–3
willow anthracnose 192
willow bark aphid **36**,
  192
willow bean gall sawfly
  17, 192

willow black canker 192
willow leaf beetle **18**,
  192
willow scab **41**, 192
willow watermark
  disease 192–3
wilt **73**, 193
wind damage **45**, 71,
  193
winter aconite
  (*Eranthis*) 193, 202
winter aconite smut 193
winter cherry
  (*Solanum*) 205
winter moth **31**, 87, 193
wireworms **54**, **56**, 75,
  77, 193
  traps 88
wisteria 199
wisteria scale 193
witches' brooms **41**, 193
wolf spider **62**
woodlice **53**, 67, 194
woolly aphid **35**, 194
woolly vine scale **39**,
  194
worm casts *see*
  earthworms
worms
  cutworms **53**, 75, 122
  earthworms **58**, **61**,
    66, 67, 126
  eelworms 73, 75, 96,
    126
  enchytraeid worms
    127
  flatworms 128
  thunderworms 183
  see also wireworms
wound paints 95

Y

yellow-tail moth 194
yew (*Taxus*) 107, 197
yucca **18**, 194, 199
yucca leaf spot **18**, 194

Z

Zantedeschia see arum
  lily
Zelkova 125-6, 197
zinnia 194, 203
zucchini yellow mosaic
  virus 194

CONVERSION TABLES

METRIC MEASUREMENTS have been used throughout this book, so that the tiny sizes of some of the insects and mites mentioned can be given accurately. In addition, metric measurements are now given exclusively on most commercial packaging, and so must be used to calculate dosage and dilution rates for chemicals: however, the imperial conversions given here, although approximate, are suitable in most ordinary garden situations.

Five conversion tables are given here, which should cover all possible measurements needed when gardening.

WEIGHT

Metric	Imperial
5g	⅛oz
16g	½oz
25g	1oz
50g	2oz
85g	3oz
110g	4oz
140g	5oz
180g	6oz
200g	7oz
225g	8oz
250g	9oz
450g	16oz
1kg	2.2lb
5kg	11lb
10kg	22lb

POT SIZES

Metric	Imperial
5cm	2in
6cm	3in
9cm	3½in
10cm	4in
12cm	4½in
13cm	5in
14cm	5½in
15cm	6in
19cm	7in
21cm	8in
24cm	9in
25cm	10in
30cm	12in
38cm	15in
45cm	18in

LENGTH

Metric	Imperial
1mm	1/16in
2mm	1/16in
3mm	⅛in
4mm	⅛in
5mm/0.5cm	¼in
6mm	¼in
7mm	¼in
8mm	⅜in
9mm	⅜in
1cm	½in
2cm	¾in
2.5cm	1in
3cm	1¼in
3.5cm	1⅜in
4cm	1½in
4.5cm	1¾in
5cm	2in
6cm	2½in
7cm	3in
8cm	3in
9cm	3½in
10cm	4in
11cm	4½in
12cm	5in
13cm	5in
14cm	5½in
15cm	6in
16cm	6in
18cm	7in

LENGTH (CONTINUED)

Metric	Imperial
20cm	8in
22cm	9in
25cm	10in
30cm	12in
35cm	14in
40cm	16in
45cm	18in
50cm	20in
55cm	22in
60cm	24in
65cm	26in
70cm	28in
75cm	30in
80cm	32in
85cm	34in
90cm	1ft
1m	3ft
1.1m	3½ft
1.2m	4ft
1.35m	4½ft
1.5m	5ft
1.7m	5½ft
2m	6ft
2.2m	7ft
2.5m	8ft
3m	10ft
3.5m	11ft
4m	12ft
5m	15ft

VOLUME

Metric	Imperial
5ml	⅛fl oz
50ml	2fl oz
100ml	4fl oz
150ml	5fl oz
300ml	10fl oz
600ml	1 pint
1 litre	1.76 pints
2 litres	3.52 pints
3 litres	5.28 pints
4 litres	7.04 pints
5 litres	1.1 gallons
10 litres	2.2 gallons
50 litres	11 gallons
100 litres (1 hectolitre)	22 gallons

AREA

Metric	Imperial
1 sq cm	0.16 sq in
1 sq m	10 sq ft
2 sq m	20 sq ft
3 sq m	30 sq ft
4 sq m	40 sq ft
5 sq m	50 sq ft
6 sq m	60 sq ft
7 sq m	70 sq ft
8 sq m	80 sq ft
1 sq km	0.3 sq mile
1 hectare (1,000 sq km)	2.47 acres

REFERENCE SECTION

ACKNOWLEDGMENTS

AUTHORS' ACKNOWLEDGMENTS

Pippa Greenwood would like to thank all the colleagues past and present, and friends and family, who helped her to retrieve obscure but useful information and continued to put up with her whilst she was writing this book. In particular, Dr John Fletcher, her plant pathology "guru", who taught her so much and who really does share her fascination with diseased plants! She would also like to thank Justine for her word-processing help, and the numerous people at Dorling Kindersley who have been involved in this project and who have all been so tirelessly enthusiastic and such fun to work with.

Andrew Halstead is indebted to Dorothy Gibson for her speedy and accurate conversion of hand-written pages to a word-processed script. He would also like to thank the companies and colleagues who have given assistance and encouragement in writing this book.

PUBLISHER'S ACKNOWLEDGMENTS

PICTURE RESEARCH Melanie Simmonds, Louise Thomas and Victoria Walker

DESIGN ASSISTANCE Fay Singer and Heather Dunleavy

EDITORIAL ASSISTANCE James Nugent

ADDITIONAL ILLUSTRATIONS Karen Cochrane

CREEPY CRAWLY FONT Mark Bracey

INDEX Ella Skene

Dorling Kindersley would also like to thank Susanne Mitchell, Karen Wilson and Barbara Haynes at the Royal Horticultural Society and Ian Hodgson at *The Garden* magazine.

PHOTOGRAPHY CREDITS

The publisher would like to thank the following for their kind permission to reproduce photographs.

t=top, b=bottom, c=centre, l=left, r=right.

A–Z Botanical 39br. DW Bevan: 47c, 60tr. Martin Stankewitz: 67tr.

Alamy Images Brian Harris: 23tl.

Biofotos/Heather Angel 10 title page, 37bl, 38tc, 39cl, 43cb, 44cr, 59br, 61tl, 63tc, 63tr, 63cl, 72bl, 75tl, 96tl.

Bruce Coleman Jane Burton: 62bl. Kim Taylor: 63tl, 90tr.

Forestry Commission 36cr, 38br, 39tl, 40cr, 41tc, 70cr.

The Garden magazine Tim Sandall: 5bl, 12tc, 12cr, 12br, 13tl, 13cr, 13tr, 14bc, 14tr, 15tc, 15cr, 15bl, 16tl, 16tr, 16bl, 16bc, 16br, 17tr, 17tc, 17cl, 17cr, 17br, 18cr, 18br, 18tl, 18tc, 19bc, 19bl, 19tl, 20cr, 20c, 21tl, 21tc, 21bl, 22tr, 22cl, 22cr, 22bl, 22bc, 22tl, 23bl, 23bc, 23tr, 23cr, 23br, 24tc, 24tr, 24cl, 24br, 25tr, 25cl, 25c, 26cr, 26tl, 26bl, 27tl, 27tr, 27bc, 27cl, 27c, 27cr, 27bl, 28tr, 29cl, 29bc, 29c, 29cr, 29br, 30tc, 30tr, 30bl, 30cr, 31tr, 31cr, 31tl, 31br, 32tc, 32cr, 32tl, 33tr, 34tl, 34tr, 35c, 35cr, 35tl, 35bc, 36cl, 37tl, 37tr, 38cr, 38bl, 39tr, 39c, 40tr, 41bc, 43tr, 43br, 44bl, 45bl, 46bl, 47tc, 47c, 48tc, 48bl, 48c, 49tl, 50tr, 51tc, 51tr, 52cl, 53tr, 53tc, 53c, 53bl, 54tl, 54cr, 54br, 54tr, 55cl, 55cr, 56bc, 57br, 58bl, 58br, 75cr, 75cb. Derek St Romaine: 15tr, 27br, 42bl, 55tl.

Garden/Wildlife Matters 48cr, 48bc.

The Garden Picture Library John Glover: 20tr, 74bl.

Jerry Harpur 83bl.

Holt Studios John Adams: 28bl, 44tr. Gordon Roberts: 45tr. Inga Spence: 51bl. Nigel Cattlin: p6, 12tr, 12bl, 12bc, 13tc, 14tl, 14cr, 17bc, 21br, 23c, 24tl, 25tc, 25br, 25bl, 28tl, 28cr, 28bl, 28bc, 29tl, 29tc, 29tr, 33bl, 33bc, 35bl, 37br, 38tr, 40tc, 41cr, 42bc, 42tr, 43c, 43cl, 43bl, 44tl, 44tr, 47tr, 47br, 48tr, 49bc, 50tl, 50bl, 51cr, 52tl, 52br, 55tc, 55tr, 56br, 56tl, 56cr, 57cl, 58bc, 59tc, 59cl, 60bl, 60bc, 61cl, 62tl, 72bl, 80cl, 91tr, 96br.

Oxford Scientific Films 68cl. Marshall Black: 75tb. Scott Camazine: 72bl. Harry Fox: 61tc. Harold Taylor: 60br, 77bc, 77br. Terry Heathcote: 70bl. John McCammon: 81tr. Colin Milkins: 62br, 77tl. James Robinson: 75tr. Tony Tilford: 83tr. Thorsten Klapp/Okapia: 81bl, 81bc.

RHS Wisley 14br, 15br, 17c, 18bl, 19tr, 19cr, 20tl, 20bc, 21cl, 22tc, 26bc, 27bl, 30tl, 30bc, 30c, 30br, 31bl, 31bc, 32bl, 32br, 32cr, 32bc, 33br, 33tc, 33c, 33cr, 36bc, 36tl, 36tr, 38tl, 38bc, 39cr, 45tl, 45cl, 46tc, 46cl, 46tr, 46tl, 47bl, 48tl, 49cl, 51c, 51cl, 51bc, 52tc, 52bl, 53bc, 53cr, 54bc, 55br, 55bc, 58cr, 58c, 58tc, 61bl.

Photos Horticultural 15c, 18tr, 28br, 41cl, 43cr, 44bc, 44br, 45tc, 57bl, 62cr, p64, 71br, 75crb, 75br, 90bl. BT: 34br, 40c, 40cl, 47cr. RB: 49tr.

Harry Smith Collection 50cr.

Unwins Seeds Ltd Calabrese 'Trixie', 92c.

Jerry Young 63bl, 67br, 74tr.

Additional photography by **Peter Anderson**.

Every effort has been made to trace the copyright holders. Dorling Kindersley apologises for any unintentional omissions and would be pleased, in such cases, to add an acknowledgment in future editions.

USEFUL ADDRESSES

Alpine Garden Society, The AGS Centre, Avonbank, Pershore, Worcs WR10 3JP
www.alpinegardensociety.net

Arboricultural Association, Ampfield House, Ampfield, Romsey, Hants SO51 9PA
www.trees.org.uk

Crop Protection Association, 2 Swan Court, Cygnet Park, Hampton, Peterborough PE7 8GX
www.cropprotection.org.uk

DEFRA (Plant Health) CSL, Sand Hutton, York, YO4 1LZ   For local offices see:
www.defra.gov.uk/planth/senior.htm

Flora Locale, Denford Manor, Hungerford, Berks RG17 OUN   www.floralocale.org

Forest Research, Tree Health Division, Alice Holt Lodge, Farnham, Surrey GU10 4LW.
www.forestresearch.gov.uk

Garden Organic (HDRA), Ryton Organic Gardens, Coventry, Warwickshire CV8 3LG
www.gardenorganic.org.uk

Hardy Plant Society, c/o Little Orchard, Great Comberton, nr Pershore, Worcs WR10 3DP
www.hardy-plant.org.uk

National Gardens Scheme Charitable Trust, Hatchlands Park, East Clandon, Guildford, Surrey GU4 7RT   www.ngs.org.uk

National Society of Allotment and Leisure Gardeners, Odell House, Hunters Road, Corby, Northants NN17 5JE
www.nsalg.org.uk

National Vegetable Society, c/o 36 The Ridings, Ockbrook, Derbyshire DE22 3SF
www.nvsuk.org.uk

Plant Heritage (National Council for the Conservation of Plants and Gardens), 12 Home Farm, Loseley Park, Guildford, Surrey GU3 1HS
www.nccpg.com

Plantlife International, 14 Rollestone Street, Salisbury, Wilts SP1 1DX   www.plantlife.org.uk

Royal Caledonian Horticultural Society, c/o 6 Kirkliston Road, South Queensferry, Edinburgh EH30 9LT
www.rchs.co.uk

Royal Horticultural Society, 80 Vincent Square, London SW1P 2PE
email: gardeningadvice@rhs.org.uk
ww.rhs.org.uk

Royal Horticultural Society of Ireland, Cabinteely House, The Park, Cabinteely, Dublin 18, Eire   www.rhsi.ie

Royal National Rose Society, Chiswell Green, St Albans, Herts AL2 3NR
www.rosesociety.org

MAIL ORDER SUPPLIERS OF BIOLOGICAL CONTROLS AND/OR PHEROMONE TRAPS AND HORTICULTURAL FLEECES

Agralan Ltd, The Old Brickyard, Ashton Keynes, Swindon, Wilts SN6 6QR
www.agralan.co.uk

Biowise, Hoyle Depot, Graffham, Petworth, West Sussex GU28 0LR
www.biowise-biocontrol.co.uk

Defenders Ltd, Occupation Road, Wye, Ashford, Kent TN25 5EN
www.defenders.co.uk

Green Gardener, Brook Hill, Brundall Road, Blofield, Norfolk NR13 4LP
www.greengardener.co.uk

Just Green Ltd, Unit 14, Springfield Road Ind. Estate, Burnham-on-Crouch, Essex CM0 8UA
www.just-green.com

Organic Gardening Catalogue, Riverdene Business Park, Molesey Road, Hersham, Surrey KT12 4RG
www.organiccatalogue.com

Scarletts Plant Care, The Glasshouses, Fletchling Common, Newick, Lewes, East Sussex BN8 4JJ
www.ladybirdplantcare.co.uk